J. Haddad E. Saliba (Eds.)

Perinatal Asphyxia

With Contributions by

Ph. Arbeille · C. Boesch · B. Brunot · E:B. Cady · A. Calame
A. Constantinesco · A. Facello · G. Farmakides · C.L. Fawer
L. Fermont · A. Fignon · F. Gold · G. Greisen · J. Haddad
P. Lai · J. Laugier · J. Le Bidois · P. Leguyader · E. Martin
D. Maulik · J. Messer · F. Pierre · M.-F. Radvanyi-Bouvet
J.-P. Relier · E. Saliba · E. Scalais · E. Schneider
H. Schulman · S. Schulman

With 117 Figures, Partly in Colour

Springer-Verlag
Berlin Heidelberg New York
London Paris Tokyo
Hong Kong Barcelona
Budapest

JOSEPH HADDAD, M.D.
Service de Néonatologie
Hôpitale Universitaire de Strasbourg
F-67098 Strasbourg Cedex, France

ELIE SALIBA, M.D.
Unité 316 INSERM
University Hospital of Tours
49, Boulevard Béranger
F-37000 Tours, France

ISBN-13: 978-3-642-77898-8 e-ISBN-13: 978-3-642-77896-4
DOI: 10.1007/978-3-642-77896-4

Library of Congress Cataloging-in-Publication Data. Perinatal asphyxia / J. Haddad. E. Saliba [eds.] with contributions by Ph. Arbeille . . . [et al.]. p. cm. Includes bibliographical references and index: ISBN-13: 978-3-540-56135-4 1. Asphyxia neonetorum. 2. Fetal anoxia. I. Haddad, J. [Joseph] II. Saliba. E. Elie, 1950- . III. Arbeilla Ph. [DNLM: 1. Asphyxia Neonetorum. 2. Fetal Anoxia. H0 450 P445] RJ256.P47 1993 018.92:01 – dc20 DNLH/DLC 92-48709

Softcover reprint of the hardcover 1st edition 1993

Typesetting: Best-set Typesetter Ltd., Hong Kong

21/3130/SPS-5 4 3 2 1 0 – Printed on acid-free paper

Preface

It is difficult to determine the exact incidence of perinatal asphyxia, but it is becoming increasingly clear that ischemic or hypoxic insults, frequently unsuspected in the perinatal period, may lead to permanent functional loss with life-long neurologic and intellectual sequelae. In recent years efforts have been made to develop new and rational techniques to preserve the immature brain from the effects of asphyxia. These approaches to cerebral protection and the mechanisms by which they may provide such protection are discussed in this book by an outstanding cast of specialists representing obstetrics and pediatrics. Each of the authors has made noteworthy contributions relating to the understanding, detection, prevention and treatment or follow-up of perinatal asphyxia.

This book contains four parts: the pathophysiology of perinatal asphyxia; methods of investigation; follow-up; and management. Part I is a comprehensive review of neurochemical determinants of ischemic cell damage and provides the basic foundation necessary to understand much of the pathology that may occur and how the protective agents may act. Parts II and III incorporate methods of investigating the effects of asphyxia on the immature brain, both fetal and neonatal. The rapid advance in our understanding of cerebral pathology is related to the recent introductions of these methods. Parts IV and V involve the management and follow-up of perinatal asphyxia.

We hope that this book will provide all those involved in the care of the fetus and newborn with the information necessary to achieve a better understanding of the mechanisms of perinatal asphyxia. Perhaps a better understanding of these mechanisms will, in the future, enable physicians to offer more effective preventive measures and treatment of neurologic handicaps.

Tours, France
Strasbourg, France
October, 1992

Elie Saliba
Joseph Haddad

Contents

List of Contributors

ARBEILLE, PH., Unité INSERM 316, CHU Bretonneau, F-37044 Tours, France

BOESCH, C., Universitätsklinik Inselspital, Abteilung Magnetresonanz, Freiburger Straße 23, CH-3010 Bern, Switzerland

BRUNOT, B., Service de Médecine Nucléaire, University Hospital of Strasbourg, Hôpital de Hautepierre, Avenue Molière, F-67098 Strasbourg Cedex, France

CADY, E.B., Department of Medical Physics and Bio-Engineering, University College London, 1st Floor Shropshire House, 11–20 Capper Street, London WC1E 6AJ, UK

CALAME, A., Division de Néonatologie, Service de Pédiatrie, Centre Hospitalier Universitaire Vaudois, CH-1011 Lausanne, Switzerland

CONSTANTINESCO, A., Service de Médecine Nucléaire, University Hospital of Strasbourg, Hôpital de Hautepierre, Avenue Molière, F-67098 Strasbourg Cedex, France

FACELLO, A., Service de Médecine Nucléaire, University Hospital of Strasbourg, Hôpital de Hautepierre, Avenue Molière, F-67098 Strasbourg Cedex, France

FARMAKIDES, G., Department Obstetrics Gynecology, Winthrop-University Hospital, 259 First Street, Mineola, NY 11501, USA

FAWER, C.L., Division de Néonatologie, Service de Pédiatrie, Centre Hospitalier Universitaire Vaudois, CH-1011 Lausanne, Switzerland

FERMONT, L., Unité d'Explorations Cardiologiques, Institut de Puériculture de Paris, 26, Boulevard Brune, F-75014 Paris, France

FIGNON, A., Unité INSERM 316, CHU Bretonneau, F-37044 Tours, France

GOLD, F., Unité 316 INSERM, University Hospital of Tours, 49, Boulevard Béranger, F-37000 Tours, France

GREISEN, G., Department of Neonatology 5024, Rigshospitalet, Blegdamsvej 9, DK-2100 Copenhagen, Denmark

HADDAD, J., Service de Néonatologie, Hôpital Universitaire de Strasbourg, F-67098 Strasbourg Cedex, Francc

LAI, P., Department Obstetrics Gynecology, Winthrop-University Hospital, 259 First Street, Mineola, NY 11501, USA

LAUGIER, J., Unité 316 INSERM, University Hospital of Tours, 49, Boulevard Béranger, F-37000 Tours, France

LE BIDOIS, J., Unité d'Explorations Cardiologiques, Institut de Puériculture de Paris, 26, Boulevard Brune, F-75014 Paris, France

LEGUYADER, P., Department Obstetric and Gynecology, Hôpital de Lamentin, F-97100 Martinique, France

MARTIN, E., Magnetresonanz, Kinderspital Zürich, Steinwiesstrasse 75, CH-8032 Zürich, Switzerland

MAULIK, D., Truman Medical Center, 2301 Holmes Street, Kansas City, MO 64108, USA

MESSER, J., Service de Néonatologie, Hôpital Universitaire de Strasbourg, F-67098 Strasbourg Cedex, France

PIERRE, F., Départment de Gynécologie, Obstétrique et de la Reproduction, Clinique Universitaire du Beffroi, CHU de Tours, 23, Avenue de Roubaix, F-37100 Tours, France

RADVANYI-BOUVET, M.-F., U 29, Hôpital de Port-Royal, 123, Bd. Port Royal, F-75674 Paris, France

RELIER, J.-P., Service de Médecine Néonatale, Hospital de Cochin, 123 Boulevard Port Royal, F-75674 Paris Cedex, France

SALIBA, E., Unité 316 INSERM, University Hospital of Tours, 49, Boulevard Béranger, F-37000 Tours, France

SCALAIS, E., Clinique St. Vincent, Service de Pédiatrie, 207 Rue Francois Lefèvbre, B-4000 Rocourt, Liège, Belgique

SCHNEIDER, E., Department Obstetrics Gynecology, Winthrop-University Hospital, 259 First Street, Mineola, NY 11501, USA

SCHULMAN, H., Department Obstetrics Gynecology, Winthrop-University Hospital, 259 First Street, Mineola, NY 11501, USA

SCHULMAN, S., Department Obstetrics Gynecology, Winthrop-University Hospital, 259 First Street, Mineola, NY 11501, USA

I. Biochemical Mechanisms and Clinical Aspects of Perinatal Asphyxia

Biochemical Mechanisms of Hypoxic – Ischemic Brain Damage

E. Saliba and J. Laugier

Introduction

Cerebral hypoxia–ischemia remains a major cause of acute perinatal brain injury, leading to severe neurodevelopmental impairments. The incidence of hypoxic–ischemic encephalopathy at birth in full-term infants has been reported to be between two and four per 1000 deliveries in developed countries [3]. Up to 60% of small preterm neonates may suffer brain injury following perinatal asphyxia. In order to understand the therapeutic measures aimed at protecting the brain from hypoxia–ischemia, it is necessary to know the processes involved. In this chapter we will focus on those neurochemical aspects which occur during hypoxia–ischemia and recirculation.

Mechanisms of Cerebral Injury During Perinatal Asphyxia

Carbohydrate and Energy Metabolism

Because glucose and oxygen are the principal driving forces of energy production in the brain, the major biochemical effects of oxygen deprivation occur at the levels of carbohydrate and energy metabolism. Adenosine triphosphate (ATP) is the primary energy modulator of the cell. One of its major roles is to promote the energy required to preserve ionic gradients across plasma and intracellular membranes. The major ions that consume ATP during their movement are Na^+, K^+, and Ca^{2+}. Under normal conditions, ion pumping requires 50%–60% of total energy expenditure by the cell, a large proportion of which is used for Na^+/K^+ exchange. ATP is also utilized in the brain for synthetic processes involving neurotransmitters, structural and functional proteins, and membrane lipids. Under physiologic conditions, 1 mol glucose is catabolized in the presence of oxygen to yield 38 mol ATP. The first and most important process that generates ATP is the oxidative phosphorylation of nicotinamide adenine dinucleotide, reduced (NADH), which takes place within mitochondria. The glycolytic portion of the pathway that generates ATP occurs in the cytosol and generates only 2 mol of the 38 mol ATP. Glycolysis procedes under both aerobic and anaerobic conditions. When oxygen is not available, the cytochrome system of mitochondria becomes unsaturated, ATP production by oxidative phosphorylation is curtailed, and glycolysis becomes the only source of ATP production. However, as mentioned above, anaerobic glycolysis is an inefficient method of generating energy (2 mol ATP per 1 mol glucose). Thus, brain function cannot be maintained on this process alone.

Cerebral hypoxia–ischemia severe enough to produce irreversible tissue injury is always associated with major changes in carbohydrate and energy metabolism, including a decline in brain glucose level, a rapid increase in lactate concentration, and a fall in phosphocreatine, which is the principal storage form of high-energy phosphate in the brain. The latter is followed by a decrease in ATP. This initial fall in phosphocreatine relates primarily to the shift in creatine–phosphokinase reaction induced by the hydrogen ion and generated with lactate formation by anaerobic glycolysis. Later the creatine–phosphokinase reaction is driven by elevated concentrations of both adenosine diphosphate (ADP) and hydrogen ion. With the depletion in ATP, ADP and adenosine monophosphate (AMP) accumulate in proportion to the loss of ATP. Ultimately, the total adenine–nucleotid pool (ATP+ADP+AMP) also decreases as AMP is catabolized to adenosine. The depletion in cellular ATP during hypoxia–ischemia severely compromises the metabolic processes that require energy for their completion. Thus, ATP-dependent Na^+ extrusion through the plasma membrane in exchange for K^+ is curtailed, resulting in intracellular accumulation of Na^+ and Cl^-, as well as water. This leads to cytotoxic edema. Without the regeneration of energy, intracellular Na^+ and Cl^- ions and water will continue to accumulate, and electrochemical gradients cannot be reestablished. An important fact to know is that the depletion in ATP may persist well into the recovery period [36].

Does irreversible damage occur only from the complete exhaustion of substrate for energy supplies? At present it appears that this explanation is too simple and that other mechanisms triggered by the initial decrease in high-energy phosphates must be operating.

In the following pages we shall discuss four different kinds of potentially brain-damaging mechanisms, namely, the development of acidosis, the accumulation of cytotoxic amino acids, and the generation of oxygen-derived free radicals together with calcium ion intoxication.

Role of Acidosis

Hypoxia–ischemia of a degree sufficient to produce cellular injury is always associated with an accumulation of lactate and a reduction in intracellular pH. Tissue lactacidosis is the result of anaerobic glycolysis. In view of these findings, many authors have tried to correlate brain damage with the level of cerebral glucose supply. In a series of experiments, Myers and Yamaguchi [18, 19] showed that recovery from hypoxia–ischemia was impaired in juvenile monkeys that were given food or had a preischemic glucose infusion. These initial observations were confirmed in experimental studies examining clinical restitution, short-term metabolic recovery, postischemic blood reperfusion, and histopathologic findings [6, 10, 20, 22, 26]. Clinical studies in adult stroke patients also support the notion that hyperglycemia accentuates ischemic neuronal injury [23]. Recent studies have further characterized the effect of hyperglycemia on postischemic histopathology. Thus, hyperglycemia reduces the time lag before which histopathologic alterations are observed and also increases the severity of the brain injury with a specific type of lesion involving necrosis of the substantia nigra and pars reticulata.

On the other hand, glucose appears to have a protective role in immature animals. Research published in 1964 by Davies et al. and in 1975 by Vanucci demonstrated that the pretreatment of perinatal animals with glucose prolongs their survival when subjected to hypoxia or asphyxia and reduces permanent brain damage. Controlled studies from several other research laboratories have confirmed and extended the initial observation that hypoxic–ischemic glucose pretreatment is beneficial to immature animals and that hypoglycemia is harmful [34]. The mechanism that might explain the difference between glucose-treated adult animals and immature ones seems to be related to the differential effects of brain maturation. At least two factors related to the status of brain maturation may act in concert to limit lactate production in the perinatal brain with oxygen deprivation: less glucose uptake from the blood and less lactate production by glycolysis. Under normoxia, the calculated rate of cerebral glycolysis in the perinatal rat is approximately 10 μmol glucose/100 g per minute, one tenth of the rate calculated for adult rat cerebral cortex [32]. During total cerebral ischemia, glycolytic flux in the newborn rat brain increases fivefold compared with an eightfold increase in the adult rat brain [31].

Evidence from animal experiments suggests that lactic acid per se is not the primary factor causing tissue injury. Hypoxia – ischemia leads to cellular acidosis via sources of H^+ ions in addition to lactic acid.

It is not known what molecular mechanisms contribute to membrane dysfunction as a result of enhanced acidosis during ischemia. One can envisage a failure of energy, since a low intracellular pH inhibits mitochondrial phosphorylation [8]. It has also been suggested that glial cells may choose to regulate intracellular pH during acidosis, thereby sacrificing their volume regulation [27]. Thus, severe acidosis may exacerbate early damage with the failure of energy metabolism, progressive edema, and secondary compromise of microcirculation.

Free Radicals as Mediators of Reperfusion Damage

Free radicals are highly reactive compounds with a lone electron in an outer orbital. For aerobic cells, free radicals such as the superoxide anion ($O^{\cdot 2-}$) normally participate in a variety of cellular reactions: they are generated in the respiratory chain and take part in reactions catalyzed by cyclooxygenase and lipoxygenase. Free radicals are also formed in the xanthine oxidase reaction and they arise during the autooxidation of many substances, including catecholamines. These potentially destructive agents are extremely reactive and attack membrane structures (cell membrane and mitochondria) and also inflict direct injuries on the cellular deoxyribonucleic acid (DNA). During normal oxidative metabolism, several inactive systems effectively prevent the action of the free radicals. The superoxide dismutase prevents the accumulation of the superoxide radical in concentrations high enough to produce tissue damage and to activate the Haber-Weiss reaction that results in the formation of the most highly reactive radical, that is, the hydroxyl radical ($OH^{\cdot-}$). Other enzymes are present which protect cellular constituents from the deleterious effect of free radicals; these enzymes include endoperoxidase and catalase. Additional defenses against the potentially damaging effect of free radicals are provided by endogenous scavengers (i.e., cholesterol, vitamin E, and vitamin C). Thus, under normal conditions, cells are capable of rapidly destroying free radicals once they have formed via both enzymatic and nonenzymatic pathways.

During and after asphyxia (reoxygenation phase), an increase in the formation of free radicals can be anticipated. Free radicals can be produced within mitochondria when cytochrome oxidase is not fully saturated with oxygen, thereby liberating free radicals gradually. These oxygen free radicals cannot be consumed further and spurt into the cytoplasm. As mentioned previously, the energy-rich phosphated nucleotides (ATP, ADP, and AMP) are rapidly consumed and hypoxanthine is accumulated both in the blood stream and in tissues, including the brain [30]. Under aerobic conditions, hypoxanthine is converted to xanthine and to uric acid by the enzyme xanthine dehydrogenase. During reoxygenation after asphyxia or during reperfusion after ischemia, xanthine dehydrogenase is converted to xanthine oxidase. The transformation is induced by intracellular calcium, the concentration of which is increased by the lack of oxygen (see below). The xanthine oxidase system produces oxygen free radicals. The $O^{\cdot 2-}$ formed during this process may reduce Fe^{3+} which is then released from ferritin and transferrin. It is likely that a more injurious free radical species ($OH^{\cdot-}$) is formed when iron catalyzes the reaction between $O^{\cdot 2-}$ and H_2O_2 [7].

During hypoxia – ischemia there is an increase in arachidonic acid as a result of the gradual breakdown of membrane phospholipids. This is induced by an activation of tissue phospholipase via the elevation of intracellular free calcium ions levels [28]. Arachidonic acid stimulates two different oxidative enzymatic pathways. The lipooxygenase pathway metabolizes arachidonic acid to the leukotriene system with the generation of free oxygen radicals. The cyclooxygenase pathway, which catalyzes the formation of the prostaglandin intermediate PGG_2, also releases $O^{\cdot 2-}$.

Although free radical mechanisms provide a probable explanation for the development of brain damage, their role in the genesis of perinatal hypoxic – ischemic brain damage is not evident. At present, there is evidence supporting the view that free radical reactions may induce vascular injury leading to endothelial disruption [35]. But it is less well established that such reactions are responsible for the damage affecting neurons or glial cells. However, the situation may change since recent evidence shows that drugs which inhibit specific reactions involved in the generation of oxygen free radical (i.e., indomethacin, allopurinol) ameliorate hypoxic – ischemic brain damage.

Intracellular Calcium and Neuronal Damage

Under physiologic conditions, intracellular Ca^{2+} serves as a key regulator of a multitude of cellular functions. Specific ion channels regulate calcium homeostasis, thus, any rise in intracellular Ca^{2+} is rapidly reversed by its extrusion through the plasma membrane or by its sequestration into subcellular organelles such as mitochondria and endoplasmic reticulum.

During hypoxia–ischemia the opening of voltage-sensitive calcium channels and agonist-operated calcium channels allow intracellular Ca^{2+} to rise. Simultaneously, the lack of ATP and oxygen prevent Ca^{2+} extrusion and the sequestration of Ca^{2+} by mitochondria and endoplasmic reticulum. The result is an accumulation of cytosolic calcium [2, 29, 33].

The mechanisms by which Ca^{2+} accumulation induces cell damage are still under investigation. When these mechanisms occur, the activation of phospholipase A_2 seems of considerable importance. It promotes a breakdown in the phospholipid components of the cell membrane which releases free fatty acids and produces harmful agents such as prostaglandins, leukotriene, and free radicals. There are numerous other biochemical pathways triggered by calcium entry that could play a role in neuronal damage [11]. They include the activation of protease, which may degrade neurofilaments, the disruption of mitochondria with further interference to cellular energy homeostasis, and the activation of phospholipase C with the generation of diacylglycerol and subsequent translocation of protein kinase C.

Glutamate Neurotoxicity and Hypoxia–Ischemia

Many of the aforementioned biochemical perturbations produced by hypoxia–ischemia are not unique to the brain. Thus, elevations in lactic acid, a decrease in ATP and phosphocreatine, and an elevated glycolytic rate are common to virtually all hypoxic–ischemic tissue. This section will focus on those changes that appear to be unique to the brain. This new information suggests that the sensitivity of the immature brain to hypoxia–ischemia is partly a result of an excessive release of the endogenous transmitter glutamate. In 1957 Lucus and Newhouse [14] first described glutamate neurotoxicity. These authors showed that systemically administered glutamate is capable of damaging the neuronal layers of the retina in newborn mice.

It is now evident that glutamate is released from the neuron terminals in the mammalian brain and mediates most fast excitatory synaptic transmission [4, 15]. Three major receptor types for glutamate have been identified on the cell bodies and neurodendrites. They are the kainic acid receptor, the *N*-methyl-D-aspartic (NMDA) acid receptor, and the quisqualic acid receptor. These receptors subserve agonist-operated channels through which ions can pass across the cell membrane independent of the electrochemical gradient [25]. The ion channels gated by quisqualic acid are permeable to Na^+ and K^+ and depolarize neurons when they are activated. The receptors that bind NMDA also gate a channel that depolarizes neurons, but these channels differ from those opened by quisqualic acid by having an extremely high permeability to calcium.

Under conditions of hypoxia–ischemia, neurons depolarize and release glutamate which cannot be removed rapidly from the extracellular space. Several observations suggest that this toxic level of glutamate may be partially responsible for neuronal damage. First, glutamate induces excessive swelling and neuronal death in vitro in neuronal cultures subjected to hypoxic conditions [1]. Second, different glutamate antagonists can attenuate the damage produced by hypoxia–ischemia in vivo in mature and immature brain [13, 16]. Third, destroying glutamatergic axons that project to specific brain regions (i.e., hippocampus) significantly reduces the damage produced by hypoxia–ischemia [21]. Finally, there is a correlation between brain regions that are especially vulnerable to hypoxic–ischemic injury and the location of glutamate receptors, specifically those for NMDA (i.e., CA1 sector of the hippocampus).

The mechanisms by which glutamate and related excitatory amino acids exert their toxic effect are still obscure, but altered ion fluxes across the cell membrane play an important role [17]. Rothman [25] has proposed two mechanisms of ion-mediated neuronal injury. The first, early toxicity, relates to glutamate action on the quisqualate receptor inducing an influx of Na^+ into the neurons. This influx of Na^+ forces Cl^- to enter through normal or activated anion channels and, for osmotic reasons, water passes into the cell. The neuronal edema that ensues is predominantly localized in cell bodies and dentrites. This may eventually lead to neuronal lysis, although much evidence indicates that cell swelling alone does not usually result in cell death. The second mechanism, delayed toxicity, relates to glutamate interaction with the NMDA receptor inducing an influx of Ca^{2+}, which in turn triggers a "cascade" of biochemical events that could lead to cell death [12].

Resistance of the Immature Brain to Hypoxia–Ischemia

It has been well established that the newborn brain is less sensitive to hypoxia–ischemia than the adult brain. The biochemical and physiologic bases for this relative resistance are not entirely understood, but animal experiments indicate three factors that play an important role in the developing animal. First a lower cerebral rate of energy utilization. Thus, with ischemia, brain ATP levels declined progressively among newborn and more rapidly in adult animals [9]. Second, immature synaptic connections and a lack of fully developed glutaminergic innervation characterized the developing brain [24]. Third, a relative resistance of the cardiovascular system, specifically the myocardium, to hypoxia–ischemia. This resistance of the myocardium appears to be related to a greater dependency on energy supplied by glycolysis [5].

Conclusion

Several mechanisms can cause hypoxic brain damage. Those previously described are clearly not the only ones relevant in cerebral hypoxia–ischemia. For example, the mechanisms responsible for the death of astrocytes and oligodendrocytes cannot be explained by the activation of glutamate receptors. Glutamate is not toxic for these cell types at the usual concentrations observed under conditions of prolonged hypoxia–ischemia.

Further research is required to confirm the importance of each of the previously described mechanisms involved in hypoxic–ischemic injury. Based on these findings, we can then hope to prevent or at least minimize brain damage by therapeutic intervention.

References

1. Choi DW (1985) Glutamate neurotoxicity in cortical cell culture is calcium-dependent. Neurosci Lett 58:293–297
2. Diessel GA (1984) Regional accumulation of calcium in postischemic rat brain. J Neurochem 43:913–925
3. Ergander U, Eriksson M, Zetterström R (1983) Severe neonatal asphyxia. Acta Paediatr Scand 72:321–325
4. Foster AC, Fagg GE (1984) Acidic amino acid binding sites in mammalian neuronal membranes: their characteristics and relationship to synaptic receptors. Brain Res Rev 7:103
5. Friedman WF, Kirkpatrick SE (1977) Fetal cardiovascular adaptation to asphyxia. In: Gluck L (ed) Intrauterine asphyxia and the developing fetal brain. Year Book Medical Publishes, Chicago, p 149
6. Ginsberg MD, Frank AW, Budd WW (1980) Deleterious effect of glucose pretreatment on recovery from diffuse cerebral ischemia in the cat. Stroke 11:347–354
7. Halliwell B, Gutteridge JMC (1985) The importance of free radicals and catalytic metal ion in human diseases. Mol Aspects Med 8:89–193
8. Hillered L, Ernster L, Siesjö BK (1984) Influence of in vitro lactic acidosis and hypercapnia on respiratory activity of isolated brain mitochondria. J Cereb Blood Flow Metab 4: 430–437
9. Holowach-Thurston J, McDougal DB Jr (1969) Effect of ischemia on metabolism of the brain of the newborn mouse. Am J Physiol 216:348
10. Kalimo H, Rehncrona S, Söderfelt B, Olsson Y, Siesjö BK (1981) Brain lactic acidosis and ischemic cell damage. II. Histopathology. J Cereb Blood Flow Metab 1:313–327
11. Kennedy MB (1989) Regulation of neuronal function by calcium. Trends Neurosci 12:417–424
12. Kirino T (1982) Delayed neuronal death in the gerbil hippocampus following ischemia. Brain Res 239:57–69
13. Kochlar A, Zivin JA, Lyden PD, Mazzarella V (1988) Glutamate antagonist therapy reduces neurologic deficits produced by focal central nervous system ischemia. Arch Neurol 45:148–153

14. Lucus DR, Newhouse JP (1957) The toxic effect of sodium-L-glutamate on the inner layers of retina. AMA Arch Ophtalmol 58:193–201
15. Mayer ML, Westbrook GL (1987) The physiology of excitatory amino acids in the vertebrate central nervous system. Prog Neurobiol 18:197
16. McDonald JW, Silverstein FS, Johnston MV (1987) MK-801 protects the neonatal brain from hypoxic-ischemic damage. Eur J Pharmacol 140:359–361
17. Meldrum B (1985) Excitatory amino acids and anoxic-ischemic brain damage. Trends Neurosci 8:47–48
18. Myers RE (1979) A unitary theory of causation of anoxic and hypoxic brain pathology. Adv Neurol 26:195–213
19. Myers RE, Yamaguchi S (1977) Nervous system effects of cardiac arrest in monkeys. Arch Neurol 24:65–74
20. Nordström CH, Rehncrona S, Siesjö BK (1976) Restitution of cerebral energy state after complete and incomplete ischemia of 30-minute duration. Acta Physiol Scand 97:270–272
21. Pulsinelli WA (1985) De-afferentation of the hippocampus protects CA1 pyramidal neurons against ischemic injury. Stroke 16:144
22. Pulsinelli WA, Waldman S, Rawlinson D, Plum F (1982) Moderate hyperglycemia augments ischemic brain damage: A neuropathological study in the rat. Neurology (NY) 32:1239–1246
23. Pulsinelli WA, Levy DE, Sigsbee B, Sherer P, Plum F (1983) Increased damage after ischemic stroke in patients with hyperglycemia with or without diabetes mellitus. Am J Med 74:540–544
24. Rothman SM (1992) Biochemistry of hypoxic-ischemic brain injury. In: Polin RA, Fox WW (eds) Fetal and neonatal physiology. Saunders, Philadelphia, pp 1608–1613
25. Rothman SM, Olney JW (1986) Glutamate and the pathophysiology of hypoxic-ischemic brain damage. Ann Neurol 19:105–111
26. Siemkowicz E, Hansen AJ (1978) Clinical restitution following cerebral ischemia in hypo-, normo-, and hyperglycemic rats. Acta Neurol Scand 58:1–8
27. Siesjö BK (1988) Mechanisms of ischemic brain damage. Crit Care Med 16:954–963
28. Siesjö BK, Bengtsson F (1989) Calcium fluxes, calcium antagonists, and calcium-related pathology in brain ischemia, hypoglycemia, and spreading depression: a unifying hypothesis. J Cereb Blood Flow Metab 9:127–140
29. Stein DT, Vannucci RC (1988) Calcium accumulation during the evolution of hypoxic-ischemic brain damage in the immature rat. J Cereb Blood Flow Metab 8:834–842
30. Thiringer K, Blomstrand S, Hrbek A, Karlsson K, Kjellmer I (1982) Cerebral arterio-venous differences for hypoxanthine and lactate during graded asphyxia in the fetal lamb. Brain Res 239:107–111
31. Vannuci RC, Vasta F, Vannuci SJ (1987) Cerebral metabolic responses of hyperglycemic immature rats to hypoxia-ischemia. Pediatr Res 21:524–529
32. Vannuci RC, Christensen MA, Stein DT (1989) Regional cerebral glucose utilization in the immature rat: effect of hypoxia-ischemia. Pediatr Res 26:208–214
33. Van Reempts J, Borgers M (1985) Ischemic brain injury and cell calcium: morphologic and therapeutic aspects. Ann Emerg Med 14:736–742
34. Voorhies JM, Rawlinson D, Vannuci RC (1986) Glucose and perinatal hypoxic-ischemic brain damage in the rat. Neurology 36:1115–1118
35. Wei EP, Kontos HA, Dietrich WD (1981) Inhibition by free radical scavengers and by cyclooxygenase inhibitors of pial arteriolar abnormalities from concussive brain injury in cats. Circ Res 48:95–103
36. Welsh FA, Vannuci R, Brierly JB (1982) Columnar alterations of NADH fluorescence during hypoxia-ischemia in immature rat brain. J Cereb Blood Flow Metab 2:221–228

Clinical Aspects

A. Calame and C.L. Fawer

Introduction

Despite the better identification of prenatal causes associated with subsequent neurodevelopmental impairments (Naeye and Peters 1988; Naeye et al. 1989), perinatal asphyxia remains one of the most important causes of brain injury in neonatal medicine (Levene et al. 1985). However, there is still much controversy as to how perinatal asphyxia should be defined (Clark and Quirk 1990; Levene 1987). Ideally, modern indicators of fetal well-being and sensitive clinical and paraclinical criteria of the infant's condition at birth should be taken into account (Jacobs and Phibbs 1989; Sarnat and Sarnat 1976). Unfortunatly, it is often difficult to routinely collect all these parameters. Furthermore, Apgar score relates poorly to preceding events, is partially gestational-age-dependent, and is only a weak predictor of further neurodevelopmental deficits. Nevertheless, it remains a very effective and reliable system for describing the infant's condition shortly after birth and is still widely used for diagnosing neonatal asphyxia (Nelson and Ellenberg 1981; Ergander et al. 1983). In this chapter, we refer to asphyxia when the Apgar score is $\leqslant 4$ at 1 min and/or $\leqslant 6$ at 5 min (Swiss Neonatal Group definition) in association with the development of neurological signs.

In the late 1970s (Nelson and Ellenberg 1981) incidence rates for very low Apgar scores (0–3) at 1 min were reported in 15.5% of babies with a birth weight below 2500 g and in 4.5% of babies with a birth weight above 2500 g. More recently, an incidence of 2.0% of Apgar scores less than 4 at 1 min was reported in a country-wide survey of neonatal asphyxia (Palme and Ericsson 1986). The incidence of neonatal asphyxia in sick, premature babies is very much higher. In the Neonatal Intensive Care Unit (NICU) Lausanne, Switzerland, among 197 preterm infants of 34 weeks' gestation or less surviving in the neonatal period, 44% had experienced perinatal asphyxia according to the Swiss Neonatal Group definition.

The cerebral effects of asphyxia will depend mainly on brain maturation (Calame et al. 1985). The neuropathological findings reported in premature infants are predominantly hemorrhagic lesions and periventricular leukomalacia (PVL), while a wider spectrum of hypoxic–ischemic injuries were described in term infants (Pape and Wigglesworth 1979). Because of these fundamental differences, clinical aspects of perinatal asphyxia among preterm and term infants will be presented separately in this chapter.

Disturbances in the infant's behavior after birth give a sensitive indicator of significant asphyxia. It is most likely that perinatal asphyxia severe enough to cause neurodevelopmental sequelae will be associated with neonatal clinical neurological dysfunctions. These are several methods for assessing neurological function in both premature and full-term infants (Dubowitz and Dubowitz 1981; Prechtl 1977; Amiel-Tison and Grenier 1980). Using these methods, it is possible to clinically diagnose a variety of neurological insults, including intraventricular hemorrhage and hypoxic–ischemic encephalopathy. In practice, however, most severely asphyxiated preterm and full-term infants are critically ill and need intensive care. Thus, neurological examinations during the first days of life are extremely difficult to perform, or even of no value if the newborn is mechanically ventilated and paralyzed. It is therefore necessary to regularly examine the neurological status up to at least 40 weeks' postmenstrual age in preterm infants and up to hospital discharge in term newborns. In general, the rate of clinical recovery from neurological insult is one of the predictors of neurodevelopmental outcome (Dubowitz et al. 1984).

The use in NICU of several paraclinical investigations, especially cranial ultrasound, has made the correlation between lesions and dysfunctions possible and has enlightened our understanding of clinical signs (Dubowitz et al. 1981). Nowadays, a comprehensive assessment of perinatal asphyxia should include several sequential neurological examinations, repeated cranial ultrasound, and electroencephalographies, in order to assign the most precise prognosis in the early neonatal period. Doppler recordings, other imaging techniques (computed tomography, magnetic resonance imaging), spectroscopy, cerebral blood flow measurements, or evoked potentials may also provide additional information (Haddad et al. 1991).

Preterm Infants

If the hypoxic insult has occurred during fetal life, the infant might show only some degree of hypotonia and mild lethargy at birth. In severe perinatal asphyxia, the preterm newborn is most often apneic, extremely hypotonic and lethargic. After the initial resuscitation, neonatal intensive care is usually required for days or even weeks. In our experience, 80% of asphyxiated preterm infants present respiratory distress syndrome and 44% need mechanical ventilation for more than 24 h (Table 1). In the early neonatal period, a comprehensive neurological assessment in an attempt to detect the effects of hypoxia is of very limited value in tiny babies. The continuous clinical observation of the preterm infant remains, however, essential in order to detect abnormal movements, partial or tonoclonic seizures, and to recognize the different clinical patterns of intracranial hemorrhage.

There are two classic characteristic syndromes associated with intracranial hemorrhage (Volpe 1978). When a large ventricular or parenchymal hemorrhage occurs, the preterm infant becomes comatose, presents generalized tonic seizures, pupils fixed to light, and flaccid quadriparesis. Untreat-

Table 1. Comparison of asphyxiated and nonasphyxiated preterm infants

	Asphyxia (n = 87)	No asphyxia (n = 110)	p Value
Birth weight in g ±1 SD	1538 ±437	1564 ± 373	NS
Gestational age in weeks ±1 SD	31.3 ± 1.96	31.8 ± 1.57	NS
Smallness for date	13 (15%)	16 (15%)	NS
Outborn deliveries	41%	30%	
Mechanical ventilation >24 h	38 (44%)	31 (28%)	0.03
Respiratory distress syndrome	70 (80%)	72 (65%)	0.03
Seizures	4 (5%)	4 (4%)	NS
Abnormal neurological examination at discharge	22 (25%)	17 (15%)	NS

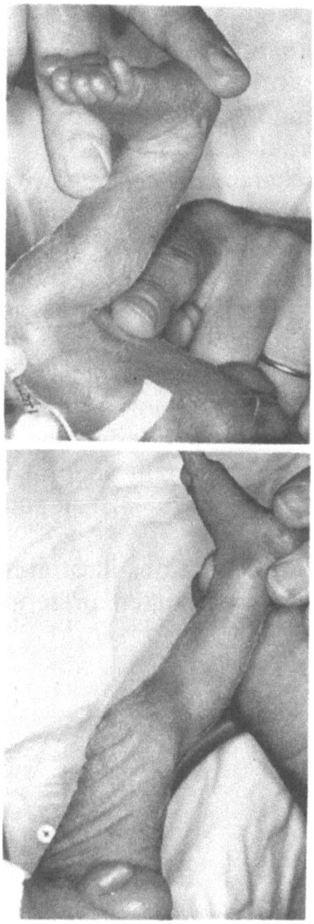

Fig. 1. Preterm infant of 27 weeks' gestation who has sustained a hemorrhage. The tone is poor in the limb on traction (*bottom*) but the popliteal angle is only 100° compared to the expected 150°–170°

able cardiovascular collapse usually occurs and the baby does not survive. Volpe described a second clinical presentation (saltatory syndrome) which consists of an alteration in the level of consciousness, a change in the quantity and quality of spontaneous movements, and aberrations of eye position and movement. Deterioration and improvement continued over many hours (Volpe 1978). Furthermore, with the use of sequential cranial ultrasound and the correlation with repeated clinical evaluations, three clinical stages in infants with germinal-layer intraventricular hemorrhage could be described (Dubowitz 1985). Preceding the hemorrhage or at the time of onset, hypertonicity (more marked in the arms), excessive motility with tremors, and startles may be observed. Tendon reflexes are brisk, the Moro is abnormal, visual and auditory orientations are absent, and the infant is usually irritable. Then, when hemorrhage is evident, tone and motility are decreased but the popliteal angle is relatively tighter (Fig. 1). Tremors and startles are absent, there is generally poor reactivity, and visual orientation is absent. Stage 3 is the recovery phase: limb tone becomes normal first, including popliteal angle, motility improves next, and auditory and visual orientation recover. Head and trunk control are the last to normalize. During this phase, roving eye movements are often observed. The speed of recovery and the presence of persistent deviant signs during this phase are important to be noted. In our experience, isolated hemorrhage occurred in 12% of asphyxiated preterm infants and was associated with PVL in 21%. PVL lesions without hemorrhage were detected in 19%.

The clinical evolution of PVL in infants is characteristic (Dubowitz 1988). After an initial phase of hypotonia and lethargy, the neurological status gradually improves and for a period, some infants may appear normal. However, after several weeks, preterm infants with postasphyxic encephalopathy become more and more irritable. They exhibit markedly abnormal tone patterns with an increased flexor tone in the arms and extensive tone in the legs. The big toe is spontaneously dorsiflexed, marked neck extensor hypertonia is usually present (Fig. 2), and abnormal finger posture may be observed. There is a flexion of the thumb and index finger with the other fingers extended (Fig. 3). The Moro reflex is usually abnormal and there are frequent tremors and startles. At this stage, visual and auditory functions are usually normal.

The evolution of abnormal signs in infants with brain lesions illustrates the importance of repeated evaluations. Among 87 asphyxiated preterm

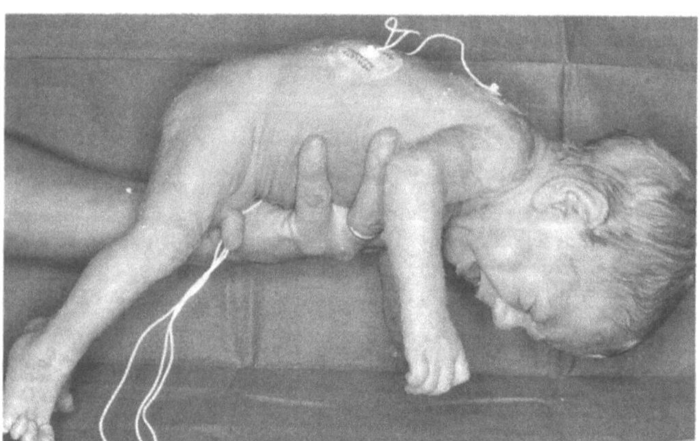

Fig. 2. Infant with PVL lesions on ultrasound exhibiting increased flexor tone in the arms and extensive tone in the legs. The big toe is dorsiflexed

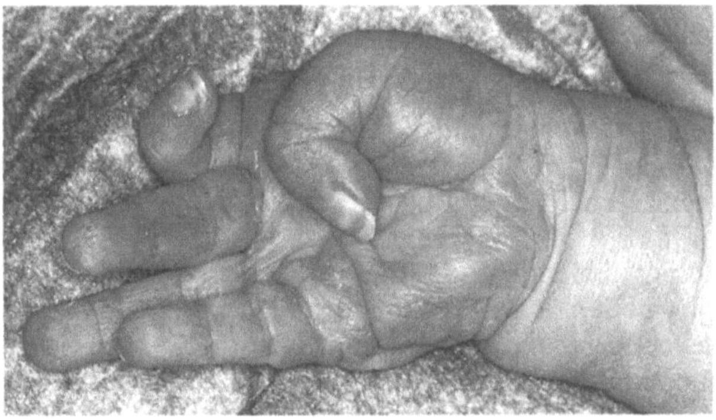

Fig. 3. Abnormal finger posture in a preterm infant with PVL lesions: flexion of the thumb and index finger with the other fingers extended

Table 2. Relationship between neurological examination and ultrasound findings

Neurological examination according to Dubowitz	Normal scans (*n*)	%	Isolated hemorrhage (*n*)	%	Periventricular leukomalacia (*n*)	%
Normal neurological examination	37	92.5	7	64.0	21	58.4
Two abnormal neurological signs	2	5.0	4	36.0	9	25.0
More than three abnormal neurological signs	1	2.5	0		6	16.6
	40	100	11	100	36	100

infants of 34 weeks' gestation or less, 22 (25%) were found to have abnormal neurological signs at 40 weeks' postmenstrual age. Table 2 shows the relationship between neurological examination and ultrasound findings. Almost all infants with normal scans had a normal neurological examination at discharge. A higher incidence of abnormal neurological signs was found in infants with isolated hemorrhage. Of those preterm neonates with PVL 41.6% presented abnormal neurological examinations. Dubowitz (1988) pointed out the differences in the clinical status at 40 weeks' postmenstrual age between infants with intraventricular hemorrhage and those with cystic leukomalacia (Dubowitz 1988). Infants with large intraventricular hemorrhage had decreased head and trunk tone with variable abnormalities in limb tone and motility, while those infants with cystic PVL were usually hypertonic with an extension of the neck and were irritable. Similar patterns of neurological examinations were found in our study. It is, however, important to realize that several infants with brain lesions may nevertheless appear normal at the time of discharge from the unit. This implies that a systematic and standardized neurodevelopmental follow-up is mandatory at least up to 2 years in order to ensure an early detection of handicaps.

Term Infants

Asphyxia is the most important cause of perinatal brain lesions among term infants. Hypoxia together with cerebral hypoperfusion are the two fundamental pathophysiological processes leading to postasphyxial brain injury. Hypoxic–ischemic encephalopathy is then the usual term adopted in these conditions as it describes both facets of functional abnormality. Moderate and severe hypoxic–ischemic encephalopathy are reported in approximately two per 1000 term newborn infants (Levene et al. 1985). One quarter of infants with postasphyxial encephalopathy present intrauterine growth retardation (Finer et al. 1981).

The pathological findings depend on a variety of factors including the duration of asphyxia and whether it was complete or incomplete. Various patterns of damage have been recognized. Edema has usually both vasogenic and cytotoxic components. On macroscopy, the brain appears swollen with a slit-like compression of the lateral ventricles. In some cases, however, there is no macroscopical evidence of lesions. Most often the gray matter is predominently affected. The concept of the vulnerability of the brain of term infants in the region of anastomosing arterial supply is now well accepted (Volpe et al. 1985). Cortical injury is frequent. The visual cortex and hippocampus are particularly affected and the depths of the sulci are more involved than the convolutions. Lesions in the basal nuclei (status marmaratus) may be associated with infants surviving birth asphyxia by many months. Injuries to the brainstem (inferior colliculus and reticular formation) are frequent in infants dying after perinatal asphyxia. On the contrary, pathological involvement of the cerebellum following perinatal asphyxia has been reported infrequently. In term infants, neuronal death is the prominent feature following hypoxic–ischemic insult. The biochemical mechanisms involved have now been well described, leading to a better understanding of the causes of lesions in mature neonatal brain (Vannucci 1990).

Asphyxiated term newborn infants usually show a definite sequence of clinical features. The infant becomes progressively more abnormal over the first 12–24 h of life. This first stage is followed by a period of nonprogressive neurological dysfunction and then, in the most favorable cases, by signs of clinical recovery. The spectrum of neurological symptoms following asphyxia depends on the severity of asphyxia. The hypoxic–ischemic encephalopathy has been graded into mild, moderate, or severe encephalopathy (Sarnat and Sarnat 1976; Fenichel 1983; Amiel-Tison 1979).

Mild Encephalopathy. These infants appear to be "hyperalert," often with staring eyes. They are jittery in response to stimulation and present an exaggerated Moro reflex. Frank hypotonia and head lag are usually present with sometimes an increased tone in the neck extensors. The sucking reflex is often weak. Clinically apparent seizures do not occur. Complete resolution of the abnormal signs is usually noticeable within 48 h.

Moderate Encephalopathy. The main features of this condition are seizures and lethargy with a reduction in spontaneous movements. The infants lie in a hypotonic posture with abducted arms and legs. The arms appear to be relatively more hypotonic than the legs. The pattern of tone abnormality may change from initial hypotonia to hypertonia. The thumbs are often adducted to lie in the palm with hands clenched. The sucking reflex is very poor and tube feeding is usually necessary. Moro reflex may be absent and tendon reflexes may vary from being increased to absent and they may change over the first few days of life. Convulsions occur during the first or second day of life; they may be subtle or fragmentary. Abnormal eye signs are often present and pupils are usually pinpoint. Signs of recovery are seen by the end of the first week but complete recovery, if it occurs, may take several weeks.

Severe Encephylopathy. These infants are comatose with severe hypotonia and usually require respiratory support from birth. The lie in a hypotonic posture with no spontaneous movements and sometimes no response, even to painful stimuli. They show no reflexes but pupillary and doll's eye movements may remain. Decerebrate movements may be seen. Seizure usually occurs within the first hours of life and may be prolonged. In most severely affected infants, pupils are fixed and dilated and there is a complete absence of respiratory efforts. In less severely affected infants a slow progression from hypotonia to extensive hypertonicity may be observed. Later, disturbances in sucking and swallowing are common and the infant may remain extremely hypotonic. Some infants can recover fully, but this may take up to several weeks. In infants who require mechanical ventilation and paralysis, it is necessary to interrupt the paralysis after 48 h in order to assess the neurological status, to adjust the therapy, and even to consider the withdrawal of intensive care in the most severe cases.

Investigations. Together with the sequential neurological assessment, repeated cranial ultrasound, electroencephalography, and cerebral Doppler during the neonatal period are essential to determine the long-term neurodevelopmental outcome. Other investigations such as magnetic resonance imaging and spectroscopy, evoked potentials, and techniques measuring cerebral circulation may also help. Nevertheless, at the present time, the early determination of a long-term prognosis of perinatal asphyxia in term neonates remains a difficult task for all neonatologists. Follow-up studies of these infants are necessary to specify their final outcome and to assess the quality of our intensive care as well as our ongoing policy.

Acknowledgements. We thank Mrs M. Forcada and L. Jaunin for their collaboration, and M. Marion for preparing the manuscript. This work was supported by the Swiss National Science Foundation no 32-30215.90.

References

1. Amiel-Tison C (1979) Birth injury as a cause of brain dysfunction in full-term newborns. In: Korobkin R, Guilleminault C (eds) Advances in perinatal neurology, vol 1. Spectrum, New York, pp 57–83
2. Amiel-Tison C, Grenier A (1980) Evaluation neurologique du nouveau-né et du nourrisson. Masson, Paris
3. Calame A, Fawer CL, Anderegg A, Perentes E (1985) Interaction between perinatal brain damage and processes of normal brain development. Ultrasonographic and neurodevelopmental study in the first year of life. Dev Neurosci 7:1–11

4. Clark RB, Quirk JG (1990) What is birth asphyxia (Letter)? Am J Obstet Gynecol 163:1367–1368
5. Dubowitz LMS (1985) Neurological assessment of the full-term and preterm newborn infant. In: Harel S, Anastolsiow NY (eds) The at-risk infant: psycho/social/medical aspects. Brooks, Baltimore, pp 185–196
6. Dubowitz LMS (1988) Clinical assessment of the infant nervous system. In: Levene MI, Bennett MJ, Punt J (eds) Fetal and neonatal neurology and neurosurgery. Churchill Livingstone, Edinburgh, pp 41–58
7. Dubowitz LMS, Dubowitz V (1981) The neurological assessment of the preterm and full-term newborn infant. SIMP/Heinemann, London (Clinics in developmental medicine, vol 79)
8. Dubowitz LMS, Levene MI, Morante A, Palmer P, Dubowitz V (1981) Neurological signs in neonatal intraventricular hemorrhage: correlation with real-time ultrasound. J Pediatr 99:127–133
9. Dubowitz LMS, Dubowitz V, Palmer PG, Miller G, Fawer CL, Levene MI (1984) Correlation of neurologic assessment in the preterm newborn infant with outcome at 1 year. J Pediatr 105:452–456
10. Ergander U, Ericksson M, Zetterstrom R (1983) Severe neonatal asphyxia. Incidence and prediction of outcome in the Stockholm area. Acta Paediatr Scand 72:321–325
11. Fenichel GM (1983) Hypoxic-ischaemic encephalopathy in the newborn. Arch Neurol 40:261–266
12. Finer NN, Robertson CM, Richards RT, Pinnell LE, Peters KL (1981) Hypoxic-ischaemic encephalopathy in term neonates: perinatal factors and outcome. J Pediatr 98:112–117
13. Haddad J, Christmann D, Messer J (1991) Imaging techniques of the CNS of the neonates. Springer, Berlin Heidelberg New York
14. Jacobs MM, Phibbs RH (1989) Prevention, recognition, and treatment of perinatal asphyxia. Clin Perinatol 16:785–807
15. Levene MI (1987) Asphyxia. In: Aynsley-Green A, Chambers TL (eds) Neonatal neurology. Churchill Livingstone, Edinburgh, pp 157–200
16. Levene MI, Kornberg J, Williams THC (1985) The incidence and severity of post-asphyxial encephalopathy in full-term infants. Early Hum Dev 11:21–28
17. Naeye RL, Peters EC (1988) Antepartum events and cerebral handicap. In: Kubli F, Patel N, Schmidt W, Linderkamp (eds) perinatal events and brain damage in surviving children. Springer, Berlin Heidelberg New York, pp 83–91
18. Naeye RL, Peters EC, Bartholomew M, Landis JR (1989) Origins of cerebral palsy. Am J Dis Child 143:1154–1161
19. Nelson KB, Ellenberg JH (1981) Apgar scores as predictors of chronic neurological disability. Pediatrics 68:36–44
20. Palme C, Ericsson A (1986) What happened to asphyxic Swedish neonates in 1985 (Abstr)? Early Hum Dev 14:145
21. Pape KE, Wigglesworth JS (1979) Haemorrhage, ischemia and the perinatal brain. SIMP/Heinemann, London (Clinics in developmental medicine, vol 69/70)
22. Prechtl HFR (1977) The neurological examination of the full-term newborn infant, 2nd edn. SIMP/Heinemann, London (Clinics in developmental medicine, vol 63)
23. Sarnat HB, Sarnat MS (1976) Neonatal encephalopathy following fetal distress. Arch Neurol 33:696–705
24. Vannucci RC (1990) Experimental biology of cerebral hypoxia-ischemia: relation to perinatal brain damage. Pediatr Res 27:317–326
25. Volpe JJ (1978) Neonatal periventricular hemorrhage, past, present and future. Pediatrics 92:693–696
26. Volpe JJ, Herscovitch P, Perlman JM et al. (1985) Positron emission tomography in the asphyxiated term infant: parasagittal impairment of cerebral blood flow. Ann Neurol 17:287–296

II. Fetal Assessment

Clinical and Biophysical Profile

F. Pierre

For a quarter of a century, obstetricians have been fighting perinatal mortality on two fronts: prenatal diagnosis and care in specific pathologies (diabetes, hypertension, etc.) and fetal safety in the labor ward (fetal monitoring, cesarean section). With this progress, antenatal deaths account for more than half the perinatal mortalities and more than two thirds of these occur before labor. When morphological anomalies and catastrophic accidents like cord prolapse or placental abruption are excluded, the main cause of fetal death is chronic asphyxia (usually called utero-placental insufficiency), whatever its etiology. It may occur where there is no risk factor and screening tests and investigations are needed for diagnoses in which clinical and biophysical fetal profiles can be included.

Clinical Fetal Asphyxia Assessment

Maternal Perception of Fetal Movement

It has been known for a long time that a reduction in fetal movements may announce fetal death (Sadovsky and Yaffe 1973). With the diminishing perinatal mortality rate, the proportional incidence of unexplained antepartum late fetal deaths is increasing, and explains why there is a renewed interest in fetal movement assessment using a widespread method applicable to all pregnancies and not only to high-risk women. In a minority of cases the risk of fetal death increases postterm, with most antenatal fetal deaths occurring at any time during the third trimester. This is the reason why fetal movement counts should be applied not only during the last weeks but also during the last months of pregnancy.

The perception of fetal movements by an expectant mother in late pregnancy varies greatly from one day to the next both in an individual woman (Sadovsky 1985) and between individuals (Benhamou and Tournaire 1981; Valentin and Marsal 1986). The average number of movements perceived each hour is 40 ± 20 (Rayburn 1982; Valentin et al. 1984). The average number of perceived movements per 12-h period varies from ten to 1000 for most normal fetuses (Pearson and Weaver 1976; Rayburn 1982; Sadovsky et al. 1977).

Major fluctuations are seen:

1. Between Different Fetuses, even if one excludes fetuses with malformations (reduced activity for most: renal agenesis, spina bifida, etc.; increased activity in anencephaly).

2. For the Same Fetus with alternating active and quiet phases during a 24-h period. Average "rest periods" last around 20 min. There is no demonstrable evidence of a pathological limit for these quiet phases. A lot of environmental factors modify this activity during the day: eating and maternal blood glucose level, medications (Rayburn 1982), maternal posture or physical activity, time of day (Ehrstrom 1984; Rayburn 1982), sound and light stimulations which were supposed to add efficacy to the test (Arulkumaran et al. 1989), and uterine stimulations. This is why it is important to repeat the monitoring of fetal movements under the same conditions at regular intervals. Variations also depend upon gestational age (Ehrstrom 1979, 1984; Pearson and Weaver 1976; Sadovsky et al. 1977).

3. In Perception of Movements by One Woman Many studies have compared movements as detected with external tocometry or ultrasonography; 82%–87% are detected by an attentive mother (Sadovsky et al. 1984).

Two factors are important for maternal perception (Sorokin et al. 1982): (a) the number of body segments participating in the movements: the mother only perceived 8%–50% of movements when one segment only was moving versus 63%–73% when the entire body was moving and (b) the duration of movement: only 51% were perceived when it was less than 1 s and 84% when it was more than 3 s.

Rayburn (1982) established a correlation between mother perception and the type of movement detected by ultrasonography. About 25% of women show great variations when comparing each day's count (Grant and Hepburn 1984). Attentive counts are needed and this is why it is important to repeat them at the same time each day, under the same conditions. Prolonged counting periods are thought to reduce the perception of movement and it seems better to count two or three times a day for a short period than for only one prolonged period.

4. In Perception Between Different Women There is a great variation in the perception of movement, with some women reporting more or less all their babies' movements and others none, even though most of the women are consistent in the way they feel and record fetal activity. Valentin and Marsal (1986) found that nulliparas and women between 20 and 35 years old had a better perception than multiparas (distraction from other children), teenagers, and women older than 35 years. In a french population Benhamou and Tournaire (1981) found important variations between the women's motivation for counting: 10% were anxious women who even counted the uterine contractions as movements, 60% were unaware of movement, and only 30% perceived and counted fetal movements correctly.

That is why it seems important to use an individualized approach, with each fetus/mother pair having their own average control count and looking for changes in fetal movement rates within its defined norms.

There are two different types of approach to fetal movement counts: (1) Count for a determined length of time with a minimum number of movements counted in this period. These counts have been based on 1-h periods for most authors, or shorter (half an hour for Valentin et al. 1984). For some authors, this counting period could be repeated two or three times a day (Sadovsky et al. 1974). This method has the advantage of moderating the possibility that a reduction in movement is due to a prolonged quiet phase. But even with this method, a high false-positive alarm rate remains the main problem for this form of counting (Grant and Hepburn 1984); (2) The mother has to count until she reaches a predetermined number of movements, with the normal limit defined by a fixed maximum time in which to record them. The method most used is the "Cardiff count to ten". Its interest is that the mother with a healthy fetus, i.e., most women counts for a very short length of time in order to add the fixed number of movements. On the other hand, this method is dreadful for women who do not feel fetal movements very well, with the movements becoming less acute over longer periods of counting, and the mothers more anxious, most of the time for no reason (Grant and Hepburn 1984). An adaptation of this method could be a fixed number for each individual.

Whatever the approach is, the predictive value of the test is considered to be poor for a positive result (more than 50% give a false diagnosis), but is good for negative results and the specificity of the test is more than 90% in most studies (Thacker and Berkelman 1986). A result below the normal limit could help to detect an altered fetus, but it would need more investigations to be certain. Even if some deaths are not preceded by a reduction in fetal movements, there is a clear association between reduced counts and antepartum deaths.

There is no reason why organizing a fetal movement count in pregnant women should permit a reduction in the fetal mortality rate as there are great difficulties in not only applying the count but also in providing the correct complementary investigations. Even though some authors (Moore and Piacquadio 1989) with limited study groups concluded that counts were a real benefit, others in a large randomized controlled trial came to the opposite conclusion (Grant et al. 1989).

Even if a reduction in fetal movements predicts an imminent fatal outcome, it does not mean that fetal extraction will improve neonatal morbidity, based on the fact that when a change in movement is due to a pathology, in most cases it could be antenatal damages with the risk of severe handicap for surviving babies. This aspect has to be and will be difficult to demonstrate.

In conclusion, even if this method of evaluation has to be explored, not all antepartum late fetal deaths are potentially preventable by fetal movement monitoring, but only 50%–60% had no obvious cause found.

If screening by fetal-movement counting seems easy because it can be performed everyday by the mother herself, we do not agree with Jackson's affirmation (Jackson et al. 1991) that it is most applicable "because of near universal applicability, ease of assay, lack of associated expense, and specificity as a screening test for good fetal health."

Compliance seems not to be so easy (70%–80%; Valentin et al. 1984): women are not monitored as regularly as advised and a negligible group of them do not contact those responsible for their antenatal care when fetal movement decreases. Valentin and Marsal (1986) found that a third of the patients failed to report a decrease in fetal-movement count and that many women (representing 30% of all calls) with normal movement patterns were calling the hospital about fetal movements (Valentin and Marsal 1986).

Contrary to Jackson's affirmation, we think the cost of complementary investigations and the psychological impact on women and pregnancy evolution is not negligible and should be evaluated at the same time.

Amniotic Fluid Volume

Reduced or absent amniotic fluid with intact membranes (called oligohydramnios) are found in 3%–5% of all pregnancies. It is associated with an increased risk of intrauterine growth retardation, fetal asphyxia, and stillbirth. Even if its diagnosis can only be proved by ultrasonography and it takes an important place in the biophysical profile, it can be detected clinically. Abdominal palpation and fundal-height measurements should not be neglected. These have been well evaluated for the prediction of a low birth weight and several studies have shown a good sensibility and specificity (Pearce and Campbell 1987), especially when the same observer is present at successive examinations.

Abdominal palpation and fundal-height measurements have not been evaluated for amniotic fluid volume measurement. It always should be interpreted in the context.

If one excludes postterms (progressive decreasing of volume with gestational age), ruptured membranes, and maternal dehydration, oligohydramnios should be considered as a possible marker of fetal asphyxia. Clinical evaluation should, in the right context, be considered an inexpensive screening device in places where ultrasound is not widely available. Clinical suspicion of oligohydramnios should always lead to ultrasonographic confirmation.

Amniotic Fluid Aspects

Meconium-stained liquor occurs in 11%–17% of all pregnancies. There is conflicting evidence as to whether a meconium-stained liquor is associated with asphyxia (Mitchell et al. 1985) or not (Yeamans et al. 1985).

It is easy to diagnosis when membranes are ruptured, but in all other cases amnioscopy is required. It is not acceptable and cannot be considered as a means of fetal assessment, except in the management of postdate or following admittance to the labor ward at term.

Biophysical Profile

Biophysical methods have identified means of evaluation of fetal function that provide indirect information about the integrity and functional state of neurological, neuromuscular, and excretory functions. Contraction stress tests and nonstress tests are studied in another chapter of this book and we will only discuss the use of fetal biophysical profile.

In a prospective study, Manning et al. (1980) identified five biophysical "variables": fetal movement, tone, reactivity, breathing, and qualitative amniotic fluid volume (Table 1). A combination of these assessments was used in an attempt to reduce false-positive and false-negative results obtained when using nonstress tests alone (Table 2).

Table 1. Biophysical profile scoring: variables (Manning et al. 1980)

Biophysical variable	Normal (score = 2)	Abnormal (score = 0)
Fetal breathing movements	The presence of at least one episode of fetal breathing of at least 60-s duration with a 30-min observation period	The absence of fetal breathing movement or the absence of such an episode of at least 60-s duration during a 30-min observation period
Gross fetal body movements	The presence of at least three discrete episodes of fetal movements with a 30-min period. Simultaneous limb and trunk movements were counted as a single movement	Two or discrete fetal movements in a 30-min observation period
Fetal tone	Upper and lower extremities in position of full flexion. Trunk in position of flexion and head flexed on chest. At least one episode of extension of extremities with return to position of flexion and/or extension of spine with return to position of flexion	Extremities in position of extension or partial flexion. Spine in position of extension. Fetal movement not followed by return to position of flexion
Nonstress test	Two or more fetal heart rate accelerations of a least 15 bpm in amplitude and at least 30-s duration associated with fetal movement(s) in a 20-min period	One or less fetal heart rate accelerations of at least 15 bpm and 30-s duration associated with fetal movement in 40 min
Amniotic fluid volume	Fluid evident throughout the uterine cavity. Largest pocket of fluid greater than 1 cm in vertical diameter	Fluid absent in most areas of uterine cavity. Largest pocket of fluid measures 1 cm or less in vertical axis. Crowding of fetal small parts

Table 2. Comparison of controlled trials of biophysical profile versus nonstress test

		Sensitivity		Specificity		Predictive positive value		Predictive negative value	
		NST	BP	NST	BP	NST	BP	NST	BP
Perinatal death	Manning et al. (1984)	75	50	84	94	–	–	–	–
	Platt et al. (1985)	33.3*	66.7	95	97	6.7*	25	99.3*	99.5
Low Apgar score at 5 min	Manning et al. (1984)	57	77	85	97	–	–	–	–
	Platt et al. (1985)	33.3*	25	96*	97	20*	25	97.7*	97

NST, Nonstress test; BP, biophysical profile.
* NS

Reasons Biophysical profile reproducibility could be discussed:

- Its five variables are interdependent: the movements are influenced by the volume of amniotic fluid which is the fifth variable. Variations in amniotic fluid volume induced by physiological or pathological processes directly affect the score of other variables. Its significance may be different in ruptured membranes and intrauterine growth retardation.
- Movements could be influenced by many factors including maternal blood glucose levels (Boylan and Lewis 1980; Devoe et al. 1987), maternal stress, and ingestion of some sedative medications (Rayburn 1982).
- The evaluation of fetal movements is subjective and surely subject to major variations depending on the experience of the operator, their concentration, and availability.

It is a great problem in units where one sees major emphasis on the tests performed and this is surely a reason to evaluate the place it should take in different investigations: good indications for the test will allow its better quality.

Supposed Interest in the Biophysical Profile

In the literature, for most outcomes (perinatal death, fetal distress in labor, low Apgar score, low umbilical cord pH, neonatal morbidity with its various definitions), the biophysical profile score seems to have a higher specificity and a better positive predictive value (Manning et al. 1980). Its specificity is higher than 95% and its predictive positive value is higher than 85% for all abnormal outcomes. If the specificity of nonstress tests is similar to the biophysical profile, its predictive positive value would seem to be significantly worth while (33% for Platt et al. 1985). But this needs to be confirmed by comparison to a well-analyzed nonstress test.

Comparative Analysis of Biophysical Profile Versus a Nonstress Test

To our knowledge, only two controlled trials were performed comparing nonstress tests to biophysical profiles (Manning et al. 1984; Platt et al. 1985). In Manning et al. (1984) a population of high-risk pregnancies referred for antepartum testing were randomly assigned to either fetal-biophysical profile scoring (375 patients) or a nonstress test (360 patients). As Mohide and Keirse (1991) pointed out, Platt et al. (1985) could have a statistical bias in enrolling two very different populations by odd and even selection. In addition, there is a difference for test indication with 43% postdate in the biophysical profile group and 34% in the nonstress test group, with a significant difference. The management decision was taken on the test results, whatever the group was for Manning et al. (1984), whereas it was followed by contraction stress tests for Platt et al. (1985). There were no great differences between the two groups (Table 2).

Recent Publications

Ribbert et al. 1990; Vintzileos et al. 1987, 1991 try to evaluate fetal hypoxia through biophysical profile. They examined the relationship between the biophysical profile and fetal gasometry, on samples obtained by cordocentesis. Even if their populations were small, the biophysical profile score was significantly correlated to Δ-pH ($p < 0.001$) but not to ΔpO_2, ΔpCO_2, or Δ oxygen saturation (Ribbert et al. 1990).

They confirm the feelings of Vintzileos et al. (1987) that fetal heart-rate reactivity and fetal breathing movements are the first to be compromised during acidemia. As the fetal gross body movements and tone are the last biophysical activities to be modified in these cases, it seems to be more useful to make a partial analysis of each individual component of the biophysical profile than of the total score. Absence of tone and fetal gross body movements may identify severely acidemic fetus, but subsequent sonographic examination for fetal tone and activity could help to distinguish between acidemic and nonacidemic fetus.

It is impossible to draw conclusions from these few studies with a very limited population and it is important to note that in Ribbert et al.'s (1990) patients there was very severe intrauterine growth retardation.

Conclusion

It is important when considering these fetal assessment tests to remember that there is some subjectivity in their resulting analyses (either for mother or doctor), they are not for one specific fetal pathology, but are used to evaluate many various pathologic phenomena of different gravity, and they can only currently be considered as a complementary exploration of fetal health, and should therefore be compared to other clinical and paraclinical investigations and considered pathologies.

References

1. Arulkumaran S, Anandakumar C, Wong YC, Ratnam SS (1989) Evaluation of maternal perception of sound-provoked fetal movement as a test of antenatal fetal health. Obstet Gynecol 73:182–186
2. Benhamou M, Tournaire M (1981) Le compte des mouvements actifs du foetus. Son intérêt dans le diagnostic de la souffrance foetale chronique. J Gynecol Obstet Biol Reprod 10:449–458
3. Boylan P, Lewis PJ (1980) Fetal breathing in labor. Obstet Gynecol 56:35–38
4. Devoe LD, Searle N, Castillo RA, Searle J (1987) Fetal biophysical testing. The effects of prolonged maternal fasting and the oral glucose tolerance test. J Reprod Med 32:563–568
5. Ehrstrom C (1979) Fetal movement monitoring in normal and high-risk pregnancies. Acta Obstet Gynecol Scand Suppl 80:315–325
6. Ehrstrom C (1984) Circadian rythm of fetal movements. Acta Obstet Gynecol Scand 63:539–541
7. Elbourne D, Grant A (1990) Study results vary in count-to-10 method of fetal movement screening. Am J Obstet Gynecol 163:264–265
8. Grant A, Hepburn M (1984) Merits of an individualized approach to fetal movement counting compared with fixed-time and fixed-number methods. Br J Obstet Gynecol 91:1087–1090
9. Grant A, Elbourne D, Valentin L, Alexander S (1989) Routine formal fetal movement counting and risk of antepartum late death in normally formed singletons. Lancet 2: 345–349
10. Jackson GM, Forouzan I, Cohen AW (1991) Fetal well-being: non-imaging assessment and the biophysical profile. Semin Roentgenol 26:21–31
11. Manning FA, Platt LD, Sipos L (1980) Antepartum fetal evaluation: development of a fetal biophysical profile. Am J Obstet Gynecol 136:787–795
12. Manning FA, Lange IR, Morrison I, Harman CR (1984) Fetal biophysical profile score and the non-stress test: a comparative trial. Obstet Gynecol 64:326–331
13. Mitchell J, Schulman J, Fleisher A (1985) Meconium aspiration and fetal acidosis. Obstet Gynecol 65:352–355
14. Mohide P, Keirse M (1991) Biophysical assessment of fetal well-being. In: Chalmers I, Enkin M, Keirse M (eds) Effective care in childbirth, vol 2. Oxford University Press, Oxford, pp 477–492
15. Moore TR, Piacquadio K (1989) A prospective evaluation of fetal movement screening to reduce the incidence of antepartum fetal death. Am J Obstet Gynecol 160:1075–1080
16. Pearce JM, Campbell S (1987) A comparison of symphysis-fundal height and ultrasound as screening tests for light for gestational age infants. Br J Obstet 94:100–104
17. Pearson JF, Weaver JB (1976) Fetal activity and fetal well-being: an evaluation. Br Med J 1:1305–1307
18. Platt LD, Walla CA, Paul RH, Trujillo ME, Loesser CV, Jacobs ND, Broussard PM (1985) A prospective trial of the fetal biophysical profile versus the nonstress test in the management of high-risk pregnancies. Am J Obstet Gynecol 153:624–633
19. Rayburn WF (1982) Antepartum fetal assessment. Monitoring fetal activity. Clin Perinatol 9:231–232
20. Ribbert LS, Snijders RJ, Nicolaides KM, Visser GH (1990) Relationship of fetal biophysical profile and blood gas values at cordocentesis in severely growth-retarded fetuses. Am J Obstet Gynecol 163:569–571
21. Sadovsky E, Yaffe H (1973) Daily fetal movement recording and fetal prognosis. Obstet Gynecol 41:845–850
22. Sadovsky E, Yaffe H, Polishuk WZ (1974) Fetal movements monitoring in normal and pathologic pregnancy. Int J Gynaecol Obstet 12:75–79
23. Sadovsky E, Polishuk WZ, Yaffe H et al. (1977) Fetal movements recorder use and indications. Int J Gynaecol Obstet 15:20–24
24. Sadovsky E, Rabinowits R, Freeman A, Yarkoni S (1984) The relationship between fetal heart rate accelerations, fetal movements and uterine contractions. Am J Obstet Gynecol 149:187–189

25. Sorokin Y, Leroy J, Dierker LJ (1982) Fetal movements. Clin Obstet Gynecol 25:719–734
26. Thacker SB, Berkelman RL (1986) Assessing the diagnosis accuracy and efficacy of selected antepartum fetal surveillance techniques. Obstet Gynecol Surv 41:121–141
27. Valentin L, Marsal K (1986) Subjective recording of fetal movements. II. Screening of a pregnant population; methodological aspects. Acta Obstet Gynecol Scand 65:639–644
28. Valentin L, Lofgren O, Marsal K, Gullberg B (1984) Subjective recording of fetal movements. I. Limits and acceptability in normal pregnancies. Acta Obstet Gynecol Scand 63:223–228
29. Vintzileos AM, Gaffney SE, Saliger LM, Campbell A, Nochimson DJ (1987) The relationship among the fetal biophysical profile, umbilical cord pH and Apgar score. Am J Obstet Gynecol 157:627–631
30. Vintzileos AM, Campbell WA, Rodis J, McLean DA, Fleming AD, Scorza WE (1991) The relationship between fetal biophysical assessment, umbilical artery velocimetry and fetal acidosis. Obstet Gynecol 77:622–626
31. Yeamans ER, Houth JC, Gilstgrap LC (1985) Umbilical cord pH, pCO_2 and bicarbonates following uncomplicated term vaginal deliveries. Am J Obstet Gynecol 151:798–800

Umbilical, Cerebral, and Renal Circulations in Growth-Retarded and Hypoxic Fetuses

Ph. Arbeille, A. Fignon, and P. Leguyader

Introduction

Until now most clinical applications of the fetal Doppler method have concerned the investigation of the placental and cerebral hemodynamics. The amplitude of the end-diastolic flow in the fetal vessels is directly related to the vascular resistance in the area supplied by these vessels. In order to quantify the vascular resistance, various indices which measure the proportion of systolic flow within the total forward flow during one cardiac cycle have been proposed. (Adamson et al. 1990; Gosling and King 1975; Maulik et al. 1989; Pourcelot 1974; Stuart et al. 1980).

Several studies have already demonstrated the possibilities and limits of the umbilical Doppler for the assessment of fetal growth (Arbeille et al. 1983; Cameron et al. 1988; Eiknes et al. 1984; Erskine et al. 1985; Fleisher et al. 1985; Gudmundsson and Marsal 1988; McCullum et al. 1978; Maulik et al. 1990; Reuwer et al. 1984; Schulman et al. 1984; Trudinger et al. 1985, 1987). The sensitivity of this method in this application is generally accepted to be 65%–70%. When used as a predictor of fetal well-being, it seems that only strong disturbances in the umbilical flow, such as absent end-diastolic flow, are frequently associated with acute fetal patency, but in this case it is still difficult to evaluate the degree of hypoxia according to the reduction in placental perfusion (Brar and Platt 1988; Divon et al. 1989; Nicolaides et al. 1988; Rochelson et al. 1987).

Several reasons may explain these results, but we have to keep in mind that the umbilical Doppler only detects hemodynamic abnormalities which are not in all cases associated with a deterioration in placental function. Therefore one can easily imagine that placental vascular disease exists without inducing any fetal hypoxia. Conversely the alteration in the mother-to-fetus oxygen exchange may be altered without any placental hemodynamic lesion. The fetal cerebral circulation was then investigated with the objective of studying the brain response to placental dysfunction and to test the sensitivity of the cerebral Doppler as a predictor of fetal outcome.

Fetal Hemodynamics – Intrauterine Growth Retardation – Hypoxia

Cerebral and Umbilical Flows in Intrauterine Growth Retardation

In 1984, the main difficulty was to locate the brain vessels without the help of color Doppler, so only arteries easy to identify from the well-known anatomical structures such as the internal carotid artery (Wladimiroff et al. 1986) and the anterior cerebral artery (Arbeille et al. 1986; Woo et al. 1987) were explored. These studies demonstrate the absence of diastolic flow at the beginning of the second half of the pregnancy (20–25 weeks) and the development of a diastolic circulation later on. The increase in the diastolic component with the age of the pregnancy was interpreted as a decrease in cerebral resistance, and the same Doppler indices as for the placenta were used for the evaluation of cerebral vascular resistance.

The first clinical application of the cerebral Doppler index based on the comparison of the cerebral and the placental vascular resistance was proposed in 1986 (Arbeille et al. 1986) and confirmed in 1987 (Arbeille et al. 1987; Wladimiroff et al. 1987). This application concerned the relation between the cerebro-placental ratio (CPR, cerebral resistance index/umbilical resistance index) and the fetal growth. In normal pregnancies the diastolic component in the cerebral arteries is lower than in the umbilical arteries at any gestational age. Therefore the cerebral vascular resistance remains higher than the placental one and the CPR is greater than one (Fig. 1).

In the case of pathological pregnancies (hypertensive pregnancies, for example) with intrauterine growth retardation (IUGR), we frequently observe a reduction in placental perfusion and an increase in the flow towards the brain. This phenomenon, so-called brain-sparing effect, is supposed to compensate the fetal hypoxia and it is reasonable to consider that this flow redistribution is associated most of the time with both fetal hypoxia and fetal growth retardation. The main advantage of comparing the placental and the cerebral vascular resistance is that we take into account, first the existence of placental disease that can be responsible for an alteration in the maternal-to-fetal exchanges and second the cerebral hemodynamic consequences of these abnormalities. The CPR can be pathological for different reasons – there is an increase in the placental resistance but no hypoxia and a normal cerebral perfusion, the placental resistance is normal but hypoxia exists and so the cerebral resistance has decreased abnormally, both placental and cerebral resistance are abnormal, and both indices are within their normal range but the cerebral index is lower than the placental one (Fig. 2c). All these combinations describe a fetal flow redistribution and are associated, in most of the cases, with IUGR. Several studies based on the same philosophy, using either the ratio between the anterior cerebral A index and the umbilical index CPR = cerebral resistance/placental vascular resistance (Rc/Rp) which is pathological when

a
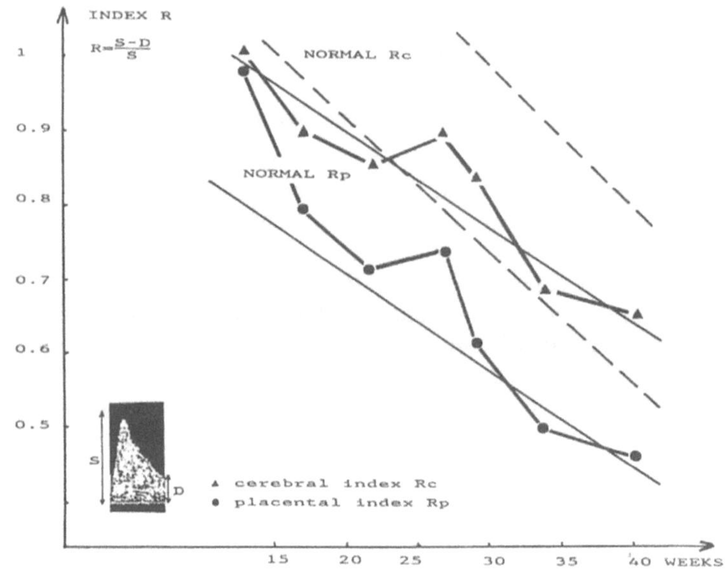

b
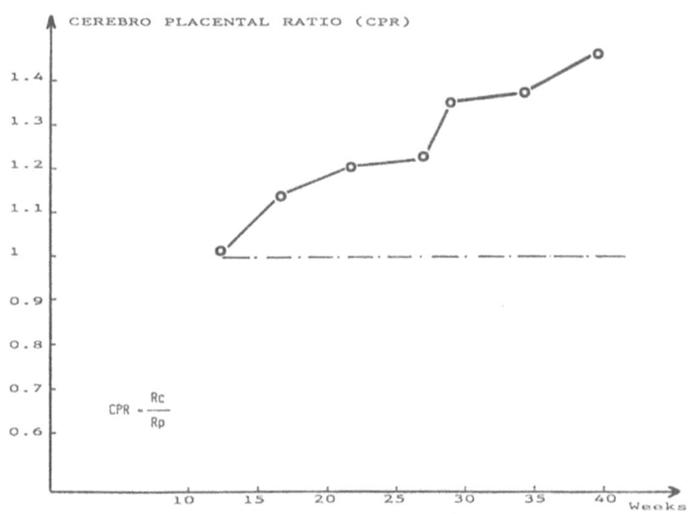

Fig. 1. a Evolution of the cerebral (*Rc*) and placental vascular resistance (*Rp*) on the same fetus during a normal gestation. The fluctuations in parallel of the two indices are related to the variations in fetal heart rate. Note that the cerebral resistance are superior to the placental resistance at any gestational age. **b** Evolution of the cerebro-placental ratio (*CPR*) during a normal gestation (Rc > Rp → Rc/Rp always greater than one). *R*, resistance; *S*, systolic amplitude; *D*, diastolic amplitude

less than one (Fig. 2, and Arbeille et al. 1987; Brar et al. 1989), or the ratio between the umbilical index and the internal carotid index I = Rp/Rc which is pathological when greater than one (Wladimiroff et al. 1987; Woo et al. 1987) demonstrated a sensitivity of approximately 86% and a specificity of about 98%.

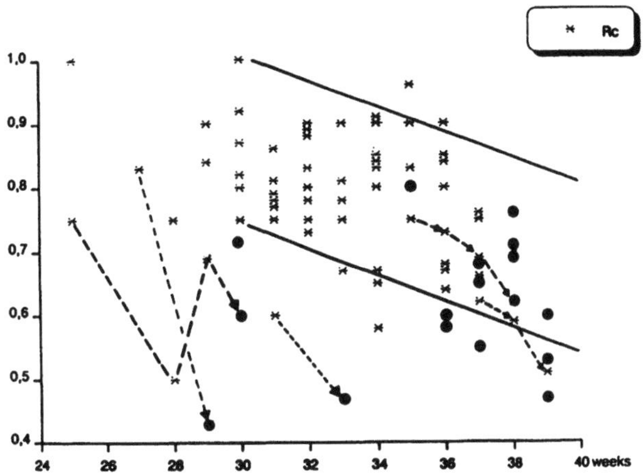

a

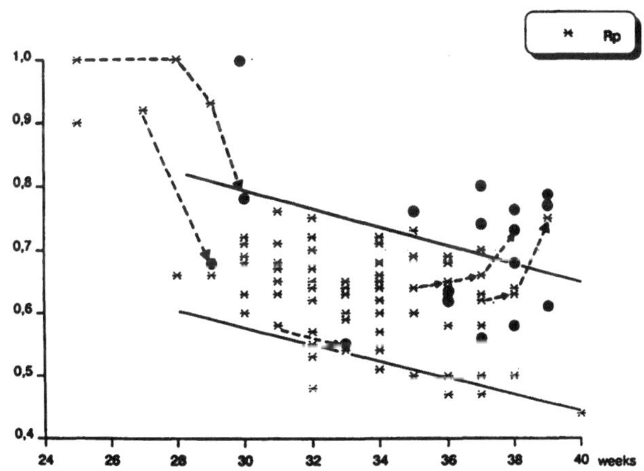

b

Fig. 2. **a** Evolution of the cerebral vascular resistance (*Rc*) in normal, and in growth-retarded fetuses (•). **b** Evolution of the umbilical vascular resistance (*Rp*) in normal, and in growth-retarded fetuses (•). The population consists in 90 hypertensive pregnancies, with 19% of moderate IUGR. The sensitivity of these indices for the detection of IUGR is about 60%. **c** Different *Rc* and *Rp* values associated either with a normal fetus (CPR = Rc/Rp > 1) or with an IUGR (CPR < 1). *1* and *4*, normal Rp and Rc (CPR > 1): normal fetus; *2*, abnormal Rc and Rp (CPR < 1): IUGR; *3*, abnormal Rc, normal Rp (CPR < 1): IUGR; *5*, normal Rc, abnormal Rp (CPR < 1): IUGR; and *6*, normal Rc, normal Rp (CPR < 1): IUGR. **d** Relationship between the fetal growth and the cerebro-placental ratio (*CPR*) in hypertensives pregnancies. The sensitivity of this index for the detection of IUGR is about 85%, (•), hypotrophic fetuses. *R*, resistance; *S*, systolic; *D*, diastolic

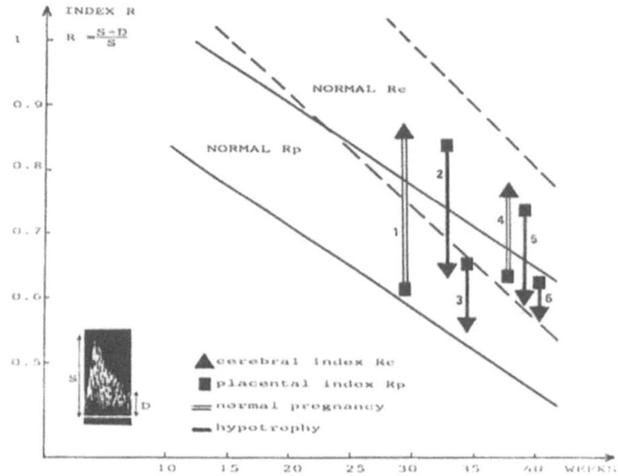

c

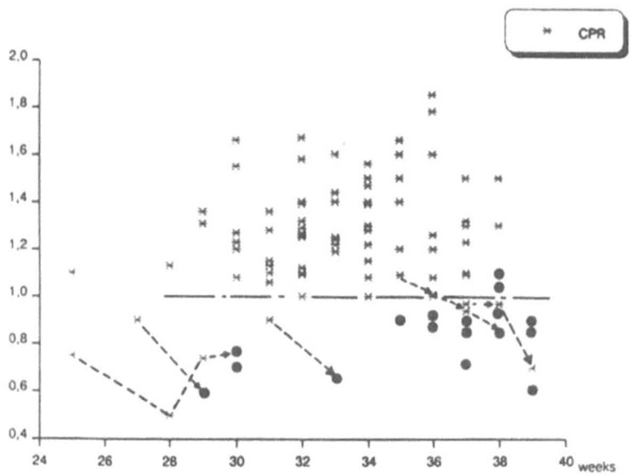

d

Fig. 2. *Continued*

The CPR has been tested on hypertensive pregnancies, idiopathic IUGR, or twin pregnancies and shows in all these cases the same accuracy. At the present time this parameter is the most widely used in clinical practice.

Another advantage of the CPR index is that this parameter does not depend on the heart rate. It is well known that an elevation in the fetal heart rate increases the end-diastolic velocity and therefore decreases the resistance index. On the other hand, the diminution in the heart rate increases the value of the index. This effect of the heart rate is partially eliminated by the use of the CPR. Both indices, Rc and Rp, are measured on the same fetus and the ratio is supposed to be nonheart-rate dependent.

Cerebral Flow and Hypoxia

The Doppler indices measured on the spectrum of the main fetal cerebral arteries are sensitive to any vasoconstriction or vasodilatation of the brain vessels. The increase in the diastolic cerebral flow is interpreted as a vasomotor response (vasodilation) to the hypoxia and even though the hypoxia is confirmed in most of the cases, it remains difficult to quantify and to follow up (Archer et al. 1986; Laurin et al. 1987). The comparison between the cerebral Doppler index and the measurement of PO_2, PCO_2, pH, and O_2 content by cordocentesis have demonstrated a good correlation between PO_2 and the cerebral Doppler index during the early development of hypoxia (the cerebral index decreases with the PO_2). Nevertheless, when the acidosis appears we observe an increase in the cerebral vascular resistance index due to a decrease in the diastolic flow (Bilardo et al. 1990).

On a first approach it looks difficult to use the cerebral index to quantify hypoxia, but if in fetuses with abnormal CPR (brain sparing) or decreased cerebral resistance, we measure the cerebral index every 2 days, we can expect to follow the evolution of the hypoxia: (1) the cerebral vascular index decreases progressively as in normal fetuses, so the hypoxia seems to be compensated by the brain hyperperfusion, (2) the cerebral index decreases significantly and becomes more and more pathological – the hypoxia develops but the fetus is probably not acidemic –, and (3) the cerebral index, which was lower than the normal limit, increases and enters the normal range again. In this case the capability of the brain vessels to vasodilate has been overloaded, the hypoxia is decompensated, and the fetus becomes acidemic.

These hypotheses are still under clinical evaluation, but now it is clear that it is hazardous to use only the absolute value of the cerebral Doppler index for the assessment of hypoxia and to take the decision of extraction. Only the evolution of the cerebral index or the CPR over several days may provide information on the development of fetal hypoxia. Nevertheless, a good correlation between the existence of a significantly decreased cerebral resistance (−2 SD) and the development of postasphyxial encephalopathy on the neonate (Rizzo et al. 1989) has been found. The specificity and the sensitivity of the cerebral Doppler as a predictor of neonatal outcome were about 75% and 87%, respectively.

Cerebral/Renal Flow and Hypoxia

Hypoxia does not affect just the brain circulation. The change in PO_2, pH, and O_2 content may affect several vascular systems such as the renal or the splanchnic circulation. In order to make the Doppler method more sensitive in the evaluation of fetal hypoxia, several groups have carried out studies on the fetal renal circulation (Moretti et al. 1990; Vyas et al. 1989).

Some studies concerning severe IUGR and hypoxic fetuses have demonstrated a reduction in renal perfusion which expresses itself as an increase in the pulsatility index (Vyas et al. 1989). Conversely, another study carried out by our group on a population of moderate IUGR with no sign of severe hypoxia or fetal distress, has shown that the renal response could be vasoconstriction or vasodilation (Fig. 3). These contradictory findings may be explained by the fact that the renal flow is sensitive to many factors other than hypoxia (Moretti et al. 1990). One can suggest that in the case of severe IUGR and hypoxia or acidemia, the fetus adaptation consists of a cerebral vasodilation and a vasoconstriction of the rest of the vascular bed. Conversely, during the early phase of development of the pathology, the renal vasculature still adapts to PO_2 changes. Moreover, in the case of oligoamnios or hydramnios for instance, the renal resistance increase or decrease and probably the renal flow is also affected by any change in fetal volemia.

Therefore it seems that the renal index needs to be evaluated in a larger population in order to conclude whether or not it is a reliable parameter for the evaluation and follow-up of fetal hypoxia.

Regional Cerebral Flow

With the color Doppler technique it is now possible to investigate the main cerebral arteries and to evaluate the vascular resistance in the different brain vascular areas supplied by these arteries (Arbeille et al. 1989). In normal fetuses the resistance index was significantly higher in the middle cerebral artery than in the anterior and posterior cerebral arteries. In pathological pregnancies with cerebral vasodilation (decreased cerebral index), the sensibility of the cerebral Doppler was not dependent on the choice of the cerebral artery explored.

Cerebral Flow Reactivity

When the cerebral Doppler index is lower than the normal range or when the CPR point out a fetal flow redistribution, the fetus is considered to be hypoxic, but it is difficult to evaluate the consequences of this exposure to hypoxia on the brain structures and cerebral functions (Archer et al. 1986; Rizzo et al. 1989).

The oxygen test (maternal oxygenation administration) was used to test fetal brain reactivity (Arduini et al. 1989; Edelstone et al. 1985; Nicolaides et al. 1987). During the oxygen treatment the cerebral index was measured at the level of the internal carotid. On fetuses with brain sparing (cerebral resistance below normal), but which did not develop fetal distress, the oxygen treatment induced an increase in cerebral resistance. On the other hand, those fetuses with cerebral vasodilation, which did not respond to the oxygen test (no increase in cerebral resistance), developed fetal distress (Fig. 4). The sensitivity of the oxygen test when used as a predictor of

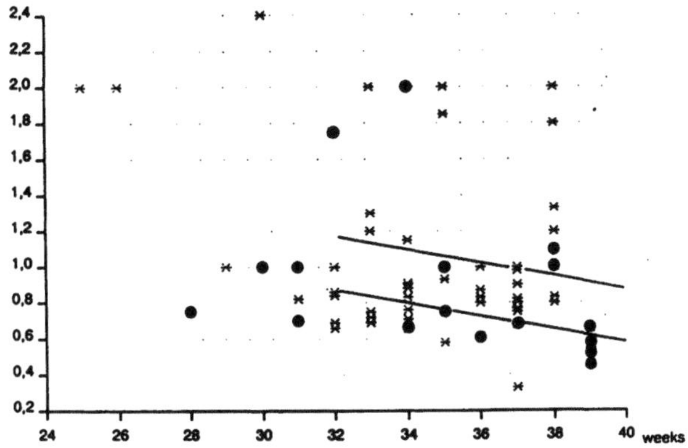

Fig. 3. Renal resistance index on eutrophic and hypotrophic (•) fetuses. On this population of 90 hypertension pregnancy with 19% moderate growth-retarded fetuses we observe three different renal vascular responses

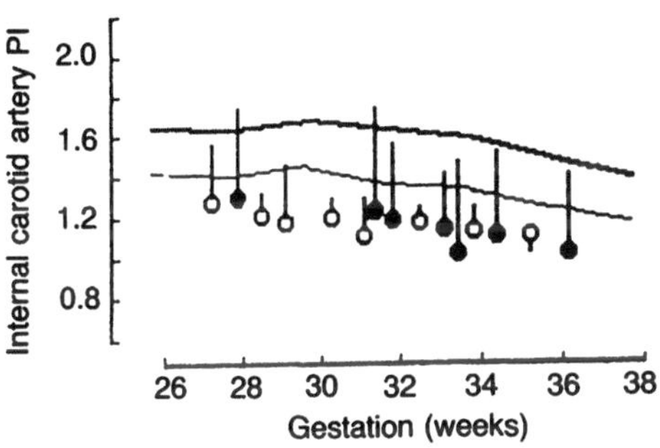

Fig. 4. Pulsatility index (*PI*) from the internal carotid artery before (*circles*) and 20 min after (*vertical lines*) maternal administration of 60% humidified oxygen. The *open circles* indicate those fetuses that developed acute fetal distress. Normal range for gestation is shown as the mean −1 SD

imminent fetal distress is about 70% (Arduini et al. 1989). The positive cerebral response proves a maintained placental transfer, this condition could justify the intrauterine treatment of the fetuses with long-term maternal oxygen therapy. Conversely, the absence of vascular response to oxygen test means that either the placental transfer and/or cerebral reactivity are impaired.

Fetal cerebral vascular resistance is also sensitive to the variation in CO_2 content in the air inspired by the mother. A recent study (Richardson et al. 1990) demonstrated that a mixture with 2% CO_2 induces a decrease in fetal cerebral resistance, but does not affect either the fetal heart rate or umbilical flow.

Cerebral Flow Response to the Administration of Drugs and to Induced Pathologies on Animals

These kinds of studies are usually performed on animal fetuses and particularly on lamb fetuses. The most frequent drug abuse during pregnancies concerns nicotine (Miller et al. 1976) and cocaine. The effect of nicotine on the fetal flows has been studied in pregnant ewes receiving a daily dose

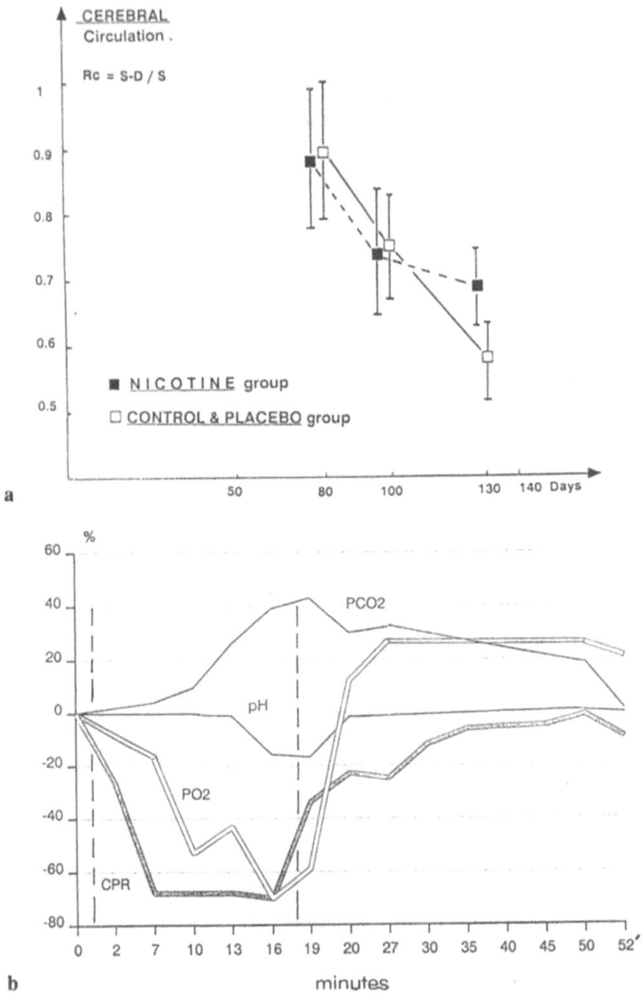

Fig. 5. **a** Variation in the fetal cerebral resistance index (*Rc*) in lamb fetuses during gestation (140 days). In the "control – placebo" group the cerebral index decreases progressively until the end of gestation, but not in the "nicotine group" which had been submitted to repeated maternal injections of nicotine from day 30. **b** Variations in the cerebro-placental ratio (*CPR*), PO_2, PCO_2, and pH in lamb fetus during the period of cord compression (between the two *vertical dotted lines*). *S*, systolic; *D*, diastolic

of nicotine equivalent to 20 cigarettes (Arbeille et al. 1990). The results showed an absence of the normal decrease in vascular resistance at the end of the gestation (Fig. 5a). This phenomenon (more apparent on the cerebral than on the placental vessels) could be interpreted as a cerebral vasoconstriction or as a reduction in brain development. In addition to these hemodynamic findings, a higher percentage of stillborns were observed in the nicotine group (63%) than in the control group (13%). The newborns of the nicotine group showed an abnormal cerebro-vascular response when submitted to a CO_2 test (no cerebral vasodilation). The conclusion of the study was that the repeated administration of nicotine is responsible for abnormal brain vascular development, loss of cerebral reactivity, and poor fetal outcome. These results lead to the hypothesis that if fetal hypoxia develops, the ability of the brain vessels to adapt through vasodilation will be reduced because of the nicotine "vasoconstrictive" effect.

Induced hypoxia in lamb fetuses has demonstrated a very sensitive and rapid brain vasodilation. The fetal cerebral and umbilical flows were assessed by Doppler sensors implanted on the fetus during fetal hypoxia induced by cord compression of about 15 min (Arbeille et al. 1992). In this experiment the umbilical flow was reduced by 80%, however, the cerebral flow was maintained. The simultaneous measurements of the cerebral and umbilical indices, together with the PO_2 showed that the CPR decreases proportionally with the PO_2 and precedes the decrease in the PO_2. The pH becomes abnormal later on, as does the heart rate (Fig. 5b). These results confirm that the Doppler parameters can accurately detect the development of hypoxia; the remaining objective will be to use them to quantify the degree of hypoxia.

On the other hand, during simulated fetal hypoxia in lamb fetuses, a reverse diastolic flow into the aortic arch has been observed. This confirms the presence of hypoxic blood coming from the right ventricle in the flow moving towards the brain (Fouron et al. 1991). This phenomenon will probably limit the beneficial effect of the brain-sparing reflex.

At last a significant redistribution of fetal flows during the acute administration of drugs commonly used in pregnancies like propanolol (Arbeille et al. 1992) has been reported. This hemodynamic adaptation, observed during an approximately 40-min period after the injection, favors the existence of fetal hypoxia due to the placental flow reduction induced by the drug during this period.

Discussion

The recent developments in Doppler technology now allow us to explore most of the fetal vessels. The cerebral flow changes in relation to hypoxia and fetal distress remain one of the most interesting areas to be investigated.

Nevertheless, although many studies have already demonstrated good correlations between cerebral Doppler data and fetal hypoxia or fetal well-being, it is too early to decide how to use this cerebral Doppler method in routine practice for the management of fetal distress and to take the decision to terminate the pregnancy.

The cerebro-placental index is now the most widely used parameter for the assessment of IUGR, probably because it takes into account the causes and consequences of the placental insufficiency responsible for IUGR and hypoxia. It is also clear that only several successive daily measurements of the Doppler indices may lead to a more accurate evaluation of fetal well-being.

Finally, even if the Doppler measurements help the obstetrician more, the objectives for the near future are first to test in routine practice the true possibilities of the Doppler indices in large randomized studies and second to point out the physiopathological phenomenon responsible for fetal flow disturbances, and for that the animal experiments will be of great interest.

References

1. Adamson SL, Morrow RJ, Langille BL et al. (1990) Site-dependent effects of increases in placental vascular resistance on the umbilical arterial velocity waveform in fetal sheep. Ultrasound Med Biol 16(1):19–27
2. Arbeille P, Asquier E, Moxhon E et al. (1983) Nouvelle technique dans la surveillance de la grossesse: l'étude de la circulation foetale et placentaire par les ultrasons. J Gynecol Obstet Biol Reprod (Paris) 12:851–859
3. Arbeille P, Tranquart F, Body G et al. (1986) Evolution de la circulation artérielle ombilicale et cérébrale du foetus au cours de la grossesse. In: Relier JP Progrès en néonatologie, vol 6. Karger, Basel, pp 30–37
4. Arbeille P, Roncin A, Berson M et al. (1987) Exploration of the fetal cerebral blood flow by Doppler ultrasound in normal and pathological pregnancies. Ultrasound Med Biol 13(6):329–337
5. Arbeille P, Collet M, Fignon A et al. (1989) Cerebral flow assessment by conventional and color-coded Doppler in human fetuses during pregnancies with hypertension. Echocardiogr J 6(3):265–270
6. Arbeille P, Bosc M, Vaillant MC, Bodard S (1990) Effect of repeated nicotine injection on the cerebral and placental circulation in ovine fetuses. Am J Perinat 9(4):268–272
7. Arbeille P, Berson M, Maulik D et al. (1992) New implanted Doppler sensors for the assessment of the main fetal hemodynamics. Ultrasound Med Biol 18(1):97–103
8. Archer L, Levene MI, Evans DH (1986) Cerebral artery Doppler ultra sonography for prediction of outcome after perinatal asphyxia. Lancet 2:1116
9. Arduini D, Rizzo G, Romanini C, Mancuso S (1989) Hemodynamic changes in growth-retarded fetuses during maternal oxygen administration as predictors of fetal outcome. J Ultrasound Med 8:193–196
10. Bilardo CM, Nicolaides KH, Campbell S (1990) Doppler measurements of fetal and uteroplacental circulations: relationship with umbilical venous blood gases measured at cordocentesis. Am J Obstet Gynecol 162:115–120
11. Brar HS, Platt LD (1988) Reverse end-diastolic flow on umbilical artery velocimetry in high-risk pregnancies: an ominous finding with adverse pregnancy outcome. Am J Obstet Gynecol 159:559–561
12. Brar HS, Horenstein J, Medearis AL, Platt LD et al. (1989) Cerebral, umbilical and uterine resistance using Doppler velocimetry in postterm pregnancy. J Ultrasound Med 8:187–191

13. Cameron AD, Nicholson SF, Nimrod CA, Harder JR, Davies DM (1988) Doppler waveforms in the fetal aorta and umbilical artery in patients with hypertension in pregnancy. Am J Obstet Gynecol 158:339–345
14. Divon MY, Girz BA, Lieblich R, Langer O (1989) Clinical management of the fetus with markedly diminished umbilical artery end-diastolic flow. Am J Obstet Gynecol 161:1523–1527
15. Edelstone DI, Peticca BB, Goldblum LJ (1985) Effects of maternal oxygen administration on fetal oxygenation during reductions in umbilical blood flow in fetal lambs. Am J Obstet Gynecol 152:351–358
16. Eiknes SH, Marsal K, Kristoffersen K (1984) Methodology and basic problems related to blood-flow studies in the human fetus. Ultrasound Med Biol 10(3):329–337
17. Erskine RLA, Ritchie JWK (1985) Quantitative measurement of fetal blood flow using Doppler ultrasound. Br J Obstet Gynaecol 92:600–604
18. Fleisher A, Schulmann M, Farmakides G et al. (1985) Umbilical artery velocity waveform and intrauterine growth retardation. Am J Obstet Gynecol 151:502–505
19. Fouron JC, Teyssier G, Maroto E, Lessart M, Marquette G (1991) Diastolic circulatory dynamics in the presence of elevated placental resistance and retrograde diastolic flow in the umbilical artery: Doppler echographic study in lambs. Am J Obstet Gynecol 164:195–203
20. Gosling RG, King DH (1975) Ultrasonic angiology. In: Harcus AW, Adamsson L (eds) Arteries and veins. Churchill Livingstone, Edinburgh, pp 61–98
21. Gudmundsson S, Marsal K (1988) Umbilical artery and uteroplacental blood-flow velocity waveforms in normal pregnancy – a cross-sectional study. Acta Obstet Gynecol Scand 67:347–354
22. Laurin J, Marsal K, Persson P et al. (1987) Ultrasound measurement of fetal blood flow in predicting fetal outcome. Br J Obstet Gynecol 94:940–948
23. Maulik D, Yarlagadda P, Nathanielsz PW et al. (1989) Hemodynamic validation of Doppler assessment of fetoplacental circulation in a sheep model system. J Ultrasound Med 8:177–181
24. Maulik D, Yarlagadda P, Downing G (1990) Doppler velocimetry in obstetrics. Obstet Gynecol Clin North Am 17(1):163–186
25. McCallum WD, Williams CS, Daigle RE (1978) Fetal blood velocity waveforms. Am J Obstet Gynecol 132:425–429
26. Miller HC, Hassanein K, Hensleigh PA (1976) Fetal growth retardation in relation to maternal smoking and weight gain in pregnancy. Am J Obstet Gynecol 125:55
27. Moretti M, Mercer B, Cartier M et al. (1990) Pulsatility index of the fetal renal artery in postdates pregnancies and relation to amniotic fluid. Proceedings IIIrd International Perinatal Doppler Society, 27–29 Sept 90 – Los Angeles (USA), p 48
28. Nicolaides KH, Bradley RJ, Soothill PW et al. (1987) Maternal oxygen therapy for intrauterine growth retardation. Lancet 1:942
29. Nicolaides KH, Bilardo CM, Soothill PW, Campbell S (1988) Absence of end-diastolic frequencies in umbilical artery: a sign of fetal hypoxia and acidosis. Br Med J 297:1026–1027
30. Pourcelot L (1974) Applications cliniques de l'examen Doppler transcutané: vélocimétrie ultrasonore Doppler. Semin INSERM 34:213–240
31. Reuwer PJHM, Nuyen WC, Beijer HJM et al. (1984) Characteristics of flow velocities in the umbilical arteries, assessed by Doppler ultrasound. Eur J Obstet Gynecol Reprool Biol (Paris) 17:397–408
32. Richardson B, Potts P, Connors G, Gillis S, Hunse C (1990) The effect of carbon dioxide on cerebral flow velocity waveforms in the human fetus. Proceeding of IIIrd International Perinatal Doppler Society, 27–29 Sept 90 – Los Angeles (USA), p 56
33. Rizzo G, Arduini D, Luciano R et al. (1989) Prenatal cerebral Doppler ultrasonography and neonatal neurologic outcome. J Ultrasound Med 8:237–240
34. Rochelson B, Schulman H, Farmakides G et al. (1987) The significance of absent end-diastolic velocity in umbilical artery velocity waveforms. Am J Obstet Gynecol 156:1213–1218

35. Schulman H, Fleischer A, Stern W et al. (1984) Umbilical velocity wave ratios in human pregnancy. Am J Obstet Gynecol 148:985–990
36. Stuart B, Drumm J, Fitzgerald DE, Duignan NM (1980) Fetal blood velocity waveforms in normal pregnancies. Br J Obstet Gynaecol 87:780–786
37. Trudinger BJ, Giles WB, Cook CM et al. (1985) Fetal umbilical artery flow velocity waveforms and placental resistance: clinical significance. Br J Obstet Gynaecol 92:20–23
38. Trudinger BJ, Cook CM, Giles WB et al. (1987) Umbilical artery flow velocity waveforms in high-risk pregnancies. Lancet 1:188–190
39. Wladimiroff JW, Tonge HM, Stewart PA (1986) Doppler ultrasound assessment of cerebral blood flow in the human fetus. Br J Obstet Gynaecol 93:471–475
40. Wladimiroff JW, van de Wijngaard JA, Degani S et al. (1987) Cerebral and umbilical arterial blood-flow velocity waveform in normal and growth-retarded pregnancies. Obstet Gynaecol 69:705–709
41. Vyas S, Nicolaides KH, Campbell S (1989) Renal artery flow velocity waveforms in normal and hypoxemic fetuses. Am J Obstet Gynecol 161:168–172
42. Woo JSK, Liang ST, Lo RLS, Chang FY (1987) Middle cerebral artery Doppler flow velocity waveform. Obstet Gynaecol 70:613–616

Heart and Fetal Asphyxia

J. Le Bidois and L. Fermont

Life is dependent on a permanent supply of oxygen. Cardiovascular function is particularly important in fetal oxygenation since low arterial O_2 saturation must be compensated for by high blood flows. Oxygen is delivered to the fetus from the maternal circulation across the placenta. Umbilical blood flow, and thus the fetal heart performance, is one of the most important factors influencing oxygen uptake (Longo et al. 1972). The microsphere method, combined with the sampling of arterial and venous blood from these organs, has allowed the measurement of oxygen uptake in fetal brain, heart, liver, gastrointestinal tract, and kidney (Rudolph 1984). In the fetal lamb, the heart has the highest oxygen consumption in relation to tissue weight and accounts for about 12% of total oxygen consumption. Therefore there should be an appropriate response of the cardiovascular system to fetal asphyxia.

The Normal Fetal Cardiovascular System

Most of what is known of the fetal circulation and its distribution results from studies in chronically instrumented fetal lambs (Rudolph 1974). Blood is oxygenated in the placenta and returns to the heart through the umbilical vein, the ductus venosus, and the inferior vena cava. Inferior vena caval blood represents about two thirds of total venous return; it is made up of lower body venous blood (O_2 saturation 35%), umbilical venous blood (O_2 saturation 85%), and right (O_2 saturation 50%–55%) and left (O_2 saturation 70%–75%) hepatic venous blood (Rudolph 1985). Distal inferior vena caval and right hepatic venous blood preferentially flows through the tricuspid valve, whereas blood with a higher oxygen content (umbilical and left hepatic venous blood) is preferentially directed through the foramen ovale to the left atrium where it mixes with pulmonary venous blood. The left ventricular blood has an O_2 saturation of about 65% and is distributed to the myocardium, the head, and the forelimbs; only a small proportion crosses the isthmus.

Superior vena caval blood represents about 25% of total venous return. It has an O_2 saturation of about 40%. It passes almost totally through the tricuspid valve to the right ventricle where it mixes with inferior vena caval and coronary sinus blood, so that O_2 saturation in the right ventricle is about 50%–55%. Less than 10% of the blood ejected by the right ventricle enters the pulmonary circulation. Most of the blood ejected by the right ventricle crosses the ductus arteriosus to the descending aorta where it mixes with the blood ejected by the left ventricle that crosses the isthmus. O_2 saturation in the descending aorta is about 55%; its blood is distributed to the placenta, the abdomen, and the lower body. Thus the blood with the lower O_2 saturation is delivered to the site of oxygenation, whereas the blood with the higher O_2 saturation is distributed to the heart and brain.

Fetal cardiac output varies with heart rate. In fetal lambs, resting heart rate is about 170/min. Increasing the heart rate by pacing the left atrium does increase ventricular output up to rates of about 300/min; the maximum increase in output is about 15%. Above this rate, a fall in ventricular output occurs. Conversely, if heart rate is decreased by stimulating the vagus nerve, ventricular output falls (Rudolph and Heymann 1976). An increase in ventricular filling pressure increases the ventricular output only slightly (Heymann and Rudolph 1973), but a fall in filling pressure causes a marked decrease in output (Gilbert 1980). Cardiac output depends also on afterload: an increase in afterload, whether acute or chronic, causes a decrease in stroke volume. These studies show that the fetal heart has a limited capacity for increasing output as compared with the adult, mainly by increasing the heart rate.

Myocardial Oxygenation in the Fetus and Cardiovascular Responses to Asphyxia

A reduction in umbilical vein oxygen content or umbilical blood flow results in a decreased oxygen delivery to the fetus. Oxygen consumption can be maintained by an increased oxygen extraction until oxygen delivery is reduced by 50% (Istkovitz et al. 1983). However, in the fetal lamb with induced sustained hypoxemia, acidemia secondary to lactic acid accumulation occurs despite maintained oxygen consumption (Rurak et al. 1990). During fetal hypoxia, redistribution of flow allows the maintenance of myocardial (Fisher et al. 1982) and cerebral blood flow and oxygen delivery, whereas flow to other organs is reduced. In the hypoxic fetal lamb, there is an increased umbilical venous return through the ductus venosus and a preferential streaming of this highly oxygenated blood through the foramen ovale to the left heart and thus to the coronary arteries and the myocardium (Rudolph et al. 1981).

The redistribution of cardiac output can be maintained for up to 48 h in fetal lambs with acutely induced hypoxemia. When fetal oxygenation is deeply decreased, oxygen delivery to the heart becomes insufficient despite a maximal blood flow.

Left ventricular oxygen consumption per weight of myocardium is similar in fetal and adult ovines (Fisher et al. 1980), despite a lower oxygen content of the coronary arterial blood. This is due to the higher resting myocardial blood flow (about twice the adult values) and oxygen extraction in the fetus. In fetal lambs with acute or chronic hypoxemia, coronary artery vasodilation results in an increase in myocardial blood flow (two- to threefold) and myocardial oxygen consumption is maintained (Fisher et al. 1982a, b). When hypoxemia is deeper, acidemia and myocardial dysfunction appear, although fetuses can maintain a myocardial contraction function during periods of very severe hypoxemia.

Myocardial Alterations Secondary to Fetal Asphyxia

Fetal heart rate, as well as combined ventricular output decrease in experimental acute fetal hypoxemia, but these changes do not appear to be related to myocardial ischemia; the decrease in cardiac output may be due to increased afterload.

Acidosis, as observed in perinatal asphyxia, alters myocardial function (Fisher 1983). Even mild acidemia combined with hypoxemia can result in myocardial ischemia and decreased cardiac output, whereas the same degree of hypoxemia with no acidemia would not produce these changes. It is not clear whether acidemia has a direct effect on the myocardium or acts through its effect on the oxygen dissociation curve.

Perinatal asphyxia may result in myocardial dysfunction, low cardiac output, and/or tricuspid insufficiency in the neonate (Rowe and Hoffman 1972; Bucciarelli et al. 1977). Typical ECG changes (ischemia ± necrosis), elevated serum creatine kinase, and abnormal myocardial thallium uptake show that these manifestations are associated with myocardial ischemia. In a population of 84 infants with no congenital heart malformation who died before the end of the first postnatal month, autopsy showed papillary muscle necrosis in 29% (Setzer et al. 1980), suggesting myocardial ischemia occurring pre- or postnatally. It has been demonstrated in some cases that transient tricuspid insufficiency of the newborn is associated with fetal asphyxia and tricuspid papillary muscle necrosis occurring prenatally (Bucciarelli et al. 1977). It is likely that these changes can occur at any moment during gestation in hypoxic fetuses, resulting in tricuspid valve insufficiency and/or ventricular dysfunction.

In 1989, we retrospectively studied 33 fetuses with tricuspid insufficiency with no congenital heart malformation (Fermont et al. 1989). Six fetuses died in utero and eight in the early postnatal period, i.e., a global mortality of 42%. Among the 19 survivors, nine (45%) needed intensive care for more than 24 h. Thus fetal tricuspid insufficiency seems to indicate a poor prognosis which may be related to the hemodynamic consequences of valvar insufficiency, and/or to the associated ventricular dysfunction due to myocardial ischemia, and/or to multiorgan failure due to fetal asphyxia.

Conclusion

Experimental data show that the fetal myocardium has little potential to increase its performance. However, there is a considerable blood flow redistribution in fetal asphyxia, so that the heart is "protected" even in moderate to severe hypoxemia. When, despite an increased coronary flow, oxygen delivery to the myocardium becomes insufficient, myocardial ischemia occurs resulting in tricuspid insufficiency and/or ventricular dysfunction. Little echo Doppler data regarding this issue are available in the fetus because fetal echocardiography has primarily focused on malformations or arrythmias. Assessment of fetal myocardial function should be undertaken whenever fetal asphyxia is present or suspected. This may lead in the next future to a better understanding of fetal hemodynamics in this instance and possibly to the improved management of these high-risk fetuses.

References

1. Bucciarelli RL, Nelson RM, Egan EA, Eitzman DV, Gessner IH (1977) Transient tricuspid insufficiency of the newborn: a form of myocardial dysfunction in stressed newborns. Pediatrics 59:330–337
2. Fermont L, Le Bidois J, Sidi D (1989) Functional tricuspid insufficiency in the fetus: a good index of fetal distress (Abstr)? 19th International Congress of Pediatrics, 23–28 July, p 142
3. Fisher DJ (1983) Acidemia depresses cardiac contractility and output in newborn lambs (Abstr). Circulation 68(3):122
4. Fisher DJ, Heymann MA, Rudolph AM (1980) Myocardial oxygen and carbohydrate consumption in fetal lambs in utero and in adult sheep. Am J Physiol 238:H399–H405
5. Fisher DJ, Heymann MA, Rudolph AM (1982a) Fetal myocardial oxygen and carbohydrate consumption during acutely induced hypoxemia. Am J Physiol 242:H657–H661
6. Fisher DJ, Heymann MA, Rudolph AM (1982b) Fetal myocardial oxygen and carbohydrate metabolism in sustained hypoxemia in utero. Am J Physiol 243:H959–H963
7. Gilbert RD (1980) Control of fetal cardiac output during changes in blood volume. Am J Physiol 238:H80–H86
8. Heymann MA, Rudolph AM (1973) Effects of increasing preload on right ventricular output in fetal lambs in utero (Abstr). Circulation 48 Suppl 4:37
9. Istkovitz J, La Gamma EF, Rudolph AM (1983) The effect of reducing umbilical blood flow on fetal oxygenation. Am J Obstet Gynecol 145:813–818
10. Longo LD, Hill EP, Power GG (1972) Theoretical analysis of factors affecting placental O_2 transfer. Am J Physiol 222:730–739
11. Rowe RK, Hoffman T (1972) Transient myocardial ischemia of the newborn infant: a form of severe cardiorespiratory distress in full-term infants. J Pediatr 81:243–250
12. Rudolph AM (1974) The fetal circulation. In: Rudolph AM (ed) Congenital diseases of the heart. Yearbook Medical Publishers, Chicago, pp 1–16
13. Rudolph AM (1984) Oxygenation in the fetus and neonate. A perspective. Semin Perinatol 8:158–167
14. Rudolph AM (1985) Distribution and regulation of blood flow in the fetal and neonatal lamb. Circ Res 57:811–821
15. Rudolph AM, Heymann MA (1976) Cardiac output in the fetal lamb: the effects of spontaneous and induced changes of heart rate on right and left ventricular output. Am J Obstet Gynecol 124:183–192
16. Rudolph AM, Istkovitz J, Iwamoto H, Reuss ML, Heymann MA (1981) Fetal cardiovascular responses to stress. Semin Perinatol 5:109–121
17. Rurak DW, Richardson BS, Patrick JE, Carmichael L, Homan J (1990) Oxygen consumption in the fetal lamb during sustained hypoxemia with progressive acidemia. Am J Physiol 258:R1116–R1122
18. Setzer E, Ermocilla R, Tonkin I, John E, Sansa M, Cassady G (1980) Papillary muscle necrosis in a neonatal autopsy population: incidence and associated clinical manifestations. J Pediatr 96:289–294

Antepartum Fetal Heart-Rate Monitoring and Fetal Asphyxia

H. Schulman, E. Schneider, S. Schulman, P. Lai, and G. Farmakides

Introduction

Mayor is credited by Lejumeau de Kergardec as the first to recognize the fetal heart-rate (FHR) sounds [26]. It occurred by accident while he was employing direct auscultation for the purpose of listening for fetal movements. In 1923, in the text by Kerr et al. [23] there is mention of the use of FHR in clinical medicine: "The fetal heart sounds can be heard from the fifth month onward. The rate averages between 120 and 150 to the minute; the most common rates are from 130 to 140. When they sink persistently below 100 or are increased to 160, this is evidence that the child's life being in danger owing to some circulatory disturbance."

The development of the nervous control of the circulation and the heart has been extensively studied [8]. The earliest and simplest experiments used vagal stimulation to see whether the heart slowed. In fetal puppies and newborn kittens it did slow. In fetal lambs early in the second trimester, stimulation of the vagus nerve slows the heart and stimulation of the left cardiac nerves accelerates it. Bauer, in 1937 (cited in [8]) demonstrated in the goat that occlusion of the umbilical cord resulted in an immediate slowing of the pulse rate. However, if the vagi were cut, there was a 10-s delay in the response. A series of elegant experiments in rabbits showed that responses were gestational and age-dependent. For example, after birth the infant did not tolerate hypoxia as long as the fetus. Barcroft [4] showed that vagal tone abruptly rose in the third trimester. The response to adrenaline did not seem to be affected by gestational age.

Pharmacologic studies using various autonomic nervous system blockers and stimulants have confirmed the earlier experiments of the regulation of pulse rate. Central nervous system stimuli to the heart diminish during sleep and increase during awake states. Fetal movement is associated with an acceleration in pulse rate and breathing is associated with sinus arrhythmia. The denervated heart also has an intrinsic slow-beating rhythm. Interestingly, infusion of acid or alkali into the fetal circulation did not produce clear changes. Making the fetus hypoxic by having the mother breathe low oxygen–gas mixtures also lacked a clear effect on pulse rate.

Introduction of Continuous Electronic Monitoring of the FHR

Hon and Hess [21] introduced continuous electronic monitoring of the fetal heart into clinical medicine in 1959. Their studies were primarily carried out in labor, using transabdominal and direct fetal scalp electrodes. These types of studies were not done in the animal model. He showed that compression of the umbilical cord produced the same effect that Barcroft [4] and Bauer had shown in the goat, lamb, and rabbit. Hon and Hess [21] showed also that compression of the fetal head during a uterine contraction elicited a vagal response while the pressure was being applied. They identified a third type of change that was characterized by a transient smooth slowing of the heart rate that began during the relaxation phase of the contraction cycle of the uterus. These observations stimulated considerable interest and further clinical and basic animal experiments ensued.

Animal studies in the monkey and sheep showed that the smooth late slowing of the heart rate was associated with a drop in oxygen in the fetal blood, mild hypoxemia [23–25]. An explosion of clinical studies followed when the FHR signal could be obtained by transabdominal Doppler ultrasound. Clinicians hoped that this new technology would lead to a decrease in fetal deaths during labor, particularly from cord accidents. Secondly, a great pursuit was on to eliminate the asphyxia that was thought to be causing brain damage during labor [2, 6, 12, 13, 18, 20, 36, 38–40, 46, 50, 56].

Shortly after the appearance of Hon's work, Saling [44] showed the feasibility of obtaining blood from the fetal scalp during labor. These micro samples could be analyzed for pH, and blood O_2 and CO_2.

Seven varieties of variable decelerations were described, many of which correlated with some decrease in cord-blood pH. Head compressions or early decelerations rarely were associated with changes in pH. Late decelerations were the most serious and were frequently associated with significant falls in pH. Low Apgar scores were more likely to be present when pH was low.

Clinical Application of Antepartum FHR Monitoring

In the United States, antepartum FHR monitoring became the principle mode of fetal surveillance. The demonstration of pathophysiologic correlates between FHR patterns and hypoxia in labor, stimulated an active search for a surveillance program that might reduce the incidence of fetal deaths and brain-damaged children. A vigorous investigational program began at the Southern California University in Los Angeles, the home of Hon's disciples [38, 40].

The first major thrust was the development of the contraction stress test or CST [18]. The hypothesis of this approach was that a uterine contraction reduces uterine blood flow, therefore a compromised fetus would be uncovered because a compensatory response would be evoked. Most studies confirmed the utility of this approach because (1) compromised fetuses were discovered, particularly those with growth retardation and (2) women who had the test had a lower perinatal mortality than those not having it. These studies were not controlled, randomized, or prospective.

The issue of whom to select for testing was resolved by the adoption of risking systems. These systems are composed of a compilation of epidemiologic-derived factors that were associated with adverse perinatal outcomes. In other words, a panorama of categories in which fetal death or perinatal morbidity–mortality are more likely to happen. The beneficial likelihood was concluded from statistical associations and odds ratios, not sensitivities or predictive values. Therefore, the exact number of women needing testing was loosely defined [1].

Rochard [42] introduced the idea of assessing the antepartum FHR without superimposing the uterine contractions as part of the testing scheme. This became known as the "nonstress test" or NST. (Europeans more correctly have called this cardiotocography or CTG). They also discovered an important pathologic variant, the sinusoidal rhythm. In this FHR pattern, there are rapid smooth oscillations of the FHR approximately five times per minute in a pathologic basal rate range, e.g., more than 160 or less than 120 beats per minute.

Visser and Huisjes in 1977 [57] described the preagonal or preterminal tracing. It was characterized by a normal baseline heart rate and an absence of accelerations and short-term oscillations of the baseline. When shallow late decelerations appear, the fetus is close to death, the maximal latent period being about a week.

The focus on accelerations and the variability of the baseline heart rate led to the observation that these events are affected by fetal sleep and awake states. Therefore, some of the principle important pathologic observations might also occur when the fetus was sleeping. Similarly, when the mother was taking drugs, there might be a sedative effect on the fetus producing an absence of accelerations and decreased heart-rate variability.

The NST was attractive because the CST was time consuming and contaminated with too many unexplained false positives. Shortcuts were devised, such as using a window of a 10-min tracing that contained two accelerations that could serve as an adequate sample. The schemes for analyzing the NST range from a simple search for fetal activity or accelerations to more complex scoring tabulations. Points were scored based on the baseline heart rate, presence or absence of accelerations or decelerations, and the frequency and amplitude of the oscillations.

Several problems arose for users of the NST. If accelerations were the most important criteria, 15%–20% of the fetuses might be sleeping during the entire test period, creating the dilemma of whether this was distress or disease. A battery of stimuli were proposed to differentiate sleep from depression, such as heat, light, food, and noise.

The most comprehensive method proposed to investigate the suspicious FHR tracing was the "biophysical profile." Advances in ultrasound allowed the physician, nurse, or technician to look at the fetus and semiquantitate its muscular tone, the vigor of its movements, breathing pattern, and quantity of amniotic fluid [33–35]. This proved useful for approximately 90% of fetuses having an FHR tracing without accelerations. If the fetus appeared normal, it was probable that the FHR tracing was a false positive.

These surveillance programs appeared to be successful. Corrected perinatal outcomes in those tested reached a low value of two per thousand. The interval for which the test was considered reassuring was 1 week for the CST and 2–3 days for the NST.

Critical Analysis of Antepartum FHR Monitoring

As mentioned above, women having antepartum FHR testing had a very low perinatal mortality. In women not having the testing, approximately 70% of the pregnant population, the perinatal mortality rate continued to be 6%–13%, depending on the population. Could extending the testing to a low-risk population be of benefit?

Randomized clinical trials began to appear in which a tested population was compared to a group closely followed but not having an antepartum NST. No benefit could be shown for the tested group and even more damaging, the tested group appeared to have more operative interventions. If there were no demonstrable benefits in a high-risk population, it is not likely that it would be helpful in a low-risk population. The four randomized studies revealed no advantages in using antepartum FHR monitoring but paradoxically showed more perinatal deaths. In other words, the test was good for identifying dying fetuses but not for saving lives [32, 48].

The analyses of FHR tracings have not been uniform [43]. The American College of Obstetrics and Gynecology classified tracings as reactive and nonreactive. When reactive it is called "reassuring" and if it is abnormal it may be called "ominous." The choice of the latter adjective may have contributed to increasing the anxiety of United States' obstetricians faced with an increasing litiginous environment and to provoking premature interventions. The International Federation of Obstetricians and Gynecologists proposed a classification of normal, suspicious, and abnormal. This classification combined many elements available on a tracing, e.g., accelerations, decelerations, baseline heart rate, and baseline variability. A more detailed type analysis was offered by several investigators. These systems made a sum of points given for each component of the tracing [20].

How reliable are the interpretations? The results of the studies asking this question are uniform: all showed that there is considerable subjectivity. Every study, at least nine, revealed that there is poor intra- and interobserver reliability [5, 48].

What is the meaning of a tracing? The search for pathophysiologic correlates have revealed many end points that were associated with an abnormal tracing when compared to a normal one [12]. These included pH, Apgar scores, meconium passage, fetal distress in labor, cesarean sections, and the need for neonatal intensive care [41, 58]. Although multiple studies showed associations between abnormal tracings and these end points, the predictive values were low. All end points are indirectly related to heart rate, thus not optimum.

When newer statistical analytic techniques were applied to fetal monitoring studies, the results were quite damaging. The sensitivity of the test or its ability to identify disease (true positive/true positive plus false negative) was low. Specificity was better, meaning that a normal tracing implied a healthy fetus. The positive predictive value was also low because of the large number of false positives [32, 48].

The most objective end point for the consequences of an abnormal FHR tracing would be fetal death or the demonstration of acidosis–hypoxia in fetal blood at the time of the tracing but before the onset of labor [53]. Again, such studies have shown good correlations but sensitivities and positive predictive values low enough to suggest that the cause of the tracing abnormality may be more complex than a simple matter of a change in acid base status.

The Search for Alternative Methods of Analysis

Since most fetuses have an abnormal tracing prior to death, a search continues to find a more effective way to analyze these tracings. One approach to improving the predictability and objectivity of the analysis might be to develop a numerical technique of quantitating the basic signal. Signal processing is a problem because of the type of techniques involved. Fetal ECG potentials must be secured across the maternal abdomen and considerable filtering is necessary to exclude maternal signals from the heart and rectus abdominis muscles. Doppler ultrasound is more easily secured but there are multiple sources of reflection in a single cardiac cycle [54].

One method proposed was that of spectral density analysis [28, 45]. In this method, the frequency domain of the individual beat-to-beat intervals is plotted. The method appears useful because peaks emerge that seem to correlate with the relative influence of the sympathetic and parasympathetic nervous system (Fig. 1). The component associated with parasympathomimetic activity, respiratory arrhythmia, is approximately 0.15 0.35 Hz. A component centered at 0.1 Hz is thought to be associated with oscillations of the baroreflex system. Components at a lower frequency, 0–0.05 H_z, are attributed to the influences of temperature control systems. In the fetus, the spectral analysis reveals a pattern with predominately mid frequency distribution and small activity in the vagal or parasympathomimetic range. After 36 weeks, a prominent peak appears in the 0.6–1 Hz range [17]. The higher frequency domain in the fetus may be caused by its higher respiratory rate. The latter signals were analyzed via ECG signals obtained across the maternal abdomen. It is generally believed that a respiratory arrhythmia commences around 24–26 weeks and better signal retrieval might reveal this component [14, 15, 22, 29, 59]. These data coincide with those using other methodology that suggest that the sick fetus has decreased respirations. On-line spectral analysis has been initiated in the adult. These studies are an evolution from the continuous monitoring that is

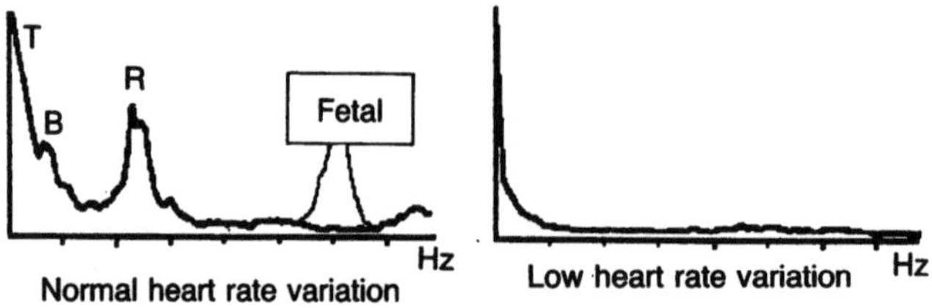

Fig. 1. A plot of a power spectrum analysis of the adult heart rate variation. The *ordinate* represents the power spectral density. The *abscissa* is Hz, or cycles per second where beat-to-beat intervals are located. *T* is <0.05 Hz and is associated with temperature regulation. *B* is 0.05–0.15 Hz and is associated with blood pressure regulation. *R* is 0.15–0.35 Hz and is associated with respiration. In the fetus, there is a peak at 0.6 Hz is seen after 35 weeks, presumably from respirations

used to monitor individuals with arrhythmias at risk of developing cardiac arrest [49]. Continuous spectral-analysis monitoring in the adult heart reveals the expected changes in the two main low (0.1 Hz) and mid (0.3 Hz) peaks, the latter predominating in the evening hours. In the adult with a diseased heart, the frequency is dramatically reduced and the mid peak is small to absent [37].

The Dawes-Redman Computerized Analysis

Geoffrey Dawes, the distinguished professor of fetal physiology at Oxford University, in association with a clinician counterpart Christopher Redman, developed a computerized on-line analysis of the FHR that is now commercially available (the system 8000, Sonicaid Company of Oxford, England [9–11, 51, 52]). Working with a dedicated group of engineers, they first defined an external Doppler ultrasound signal that could be electronically recognized as a standard for a computer-counting analysis. A definable signal resolves two important issues, identification of artifact or signal loss and the ability to count accurately beat-to-beat variation. They determined a time frame in which acceptable analyses could be done and used 3.75 s, or the range of 16 intervals per minute. The maximum minus the minimum was the number used to compare and be averaged with succeeding minutes. The long-term variation (average of range of differences in epoch intervals) within the test period is calculated. Using standard statistical methodology, approximately 95% of the population showed a variation more than 30 ms; these are classified as normal. A little more than 4% show a variation of 20–30 ms; these are classified as questionable. Less than 1%, three standard deviations, show a variation of less than 20 ms. As expected, the variation is stable after 30 weeks but may be diminished before that because of the absence of parasympathomimetic input. The program provides a descriptive analysis and summary of the random events that clinicians use, e.g., accelerations, decelerations, contractions, and fetal movements. The essential departure of the program is the emphasis on variation and the periods that the fetus spends in low and high variation. The perspective changes from an emphasis on random events to an in-depth examination of the basic rhythm of the regulatory factors of heart rate.

The reproducibility of the System 8000 was demonstrated by trials at three other clinical centers, two in England and one in our hospital. The predominant fetal disease with low variation was proteinuric hypertension and growth retardation. The Dawes group reported their findings on 78 fetuses that had at least one episode of low variation. There were five fetal and four neonatal deaths. The average gestational age at delivery was 230 days. Eight were acidemic, arterial pH less than 7.12 and base deficit greater than 12 mmol. They further refined their data to look at short-term variation, the average of the each successive 3.75 s epoch. This became necessary for the detection of sinusoidal patterns. Four of the five fetuses dying in utero had short-term variations of less than 2.6 ms, suggesting that there was a profound loss of variation. Of the 11 live births, seven were severely growth retarded and two thirds were acidotic.

Our experience in the first 1000 women using the System 8000 has been excellent [47]. We use Doppler ultrasound as a screening tool and follow closely those with abnormal umbilical artery flow velocity. Using this combination, we achieved a two per 1000-corrected fetal mortality rate, comparable to the best published studies [16, 55]. A randomized study is under way in Europe to compare physician diagnosis versus computer.

Chaos, a Summary Analysis of Nonrepetitive Phenomena

A new mathematical theory for the analysis of the flow of material through systems has captured the imagination of several investigators [3, 7, 19]. All structures have some sort of relation among its atoms, therefore are coherent and exist in space and time. For example, the periodic cycle of populations can be called a structure. It possesses a coherence in time; the population at any future instant can be estimated. These reactions can give spatial patterns. A simple example is a pendulum, with and without resistance (Fig. 2). Such phase–space plots can be carried out on the heart rate. The plot consists of the current beat-to-beat interval versus some predetermined previous interval. Such a plot reveals that it has a central attractor and many sporadic, apparently random projections or emissions. When the heart-rate variation diminishes the central attractor becomes smaller and the random emissions disappear. This precedes cardiac arrest in the adult.

We have done fractal analyses on FHR signals processed through the Dawes-Redman computer program. Such an analysis is shown in Fig. 3. As the fetus changes from healthy to terminal, an interesting pattern emerges. The healthy fetus shows a large central attractor with lots of random pro-

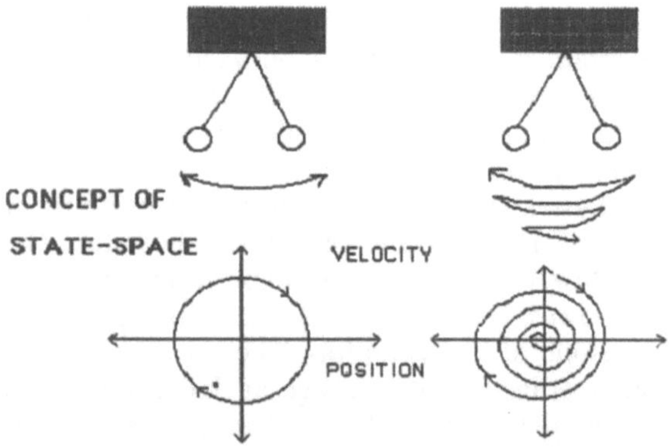

Fig. 2. The concept of state–space represents behavior in a geometric form. On the *left*, the pendulum always returns to its original position. The state or phase–space plot is a *circle*. On the *right* the pendulum gradually comes to a central attraction point

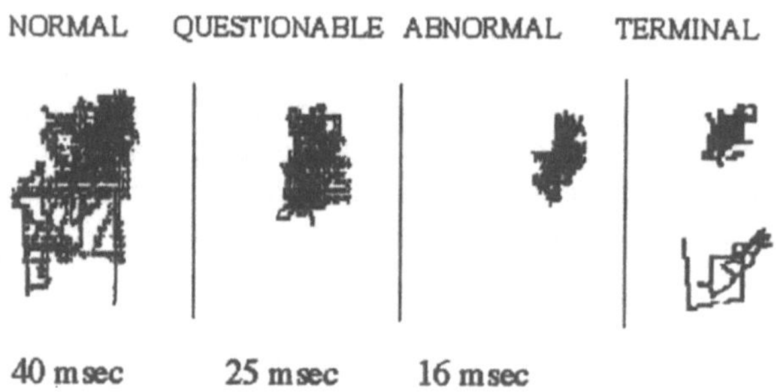

Fig. 3. State–space or chaos plot of the fetal heart rate. This is done by plotting current R-R interval versus a previously defined interval, e.g., −1, −2, −3, −5 etc. Normal chaos plot on the *left* reveals a central attractor and random projectiles. When variation diminishes, projectiles diminish and the attractor becomes smaller. Preterminal fetus shows small or no attractor

jections or energy. When the tracing becomes questionable, the attractor becomes smaller and the emissions diminish in number. With further deterioration, the attractor becomes very small and there are no random projections. In one fetus, the central attractor was lost.

The phase–space plot suggests a new approach to the analysis of FHR patterns, a system analysis. The output is the variation. The input is the myriad of factors that contribute to the regulation of the FHR. The structure of the system is the fractal plot. It is a dynamic process similar to others in nature in which there is a central bundle of energy with random emissions of energy projections. Diminution of energy is manifest by a smaller bundle and the disappearance of the random emissions. When death approaches, the energy bundle may not be recognizable. Of interest is that nuclear magnetic resonance studies on the fetal brain energy systems suggest that we may never be able to recognize the point of no recovery in a system based upon nervous regulation [30, 31].

We believe that a chaos analysis may be an improved way of thinking about FHR tracings. Is the system working or not? The system is the basic energy; it is not simply a matter of acid-base regulation or hypoxia. The regulatory inputs are too numerous to quantitate and therefore the systems approach may be more reasonable. On-line studies will be necessary to find how much difference there is in a fractal plot of a sleeping fetus versus one whose system regulation has gone awry.

References

1. ACOG Technical Bulletin (1987) Antepartum fetal surveillance
2. Ahn MO, Phelan JP, Smith CV et al. (1987) Antepartum fetal surveillance in the patient with decreased movement. Am J Obstet Gynecol 157:360–364
3. Atkins PW (1984) The second law. Scientific American Books, New York
4. Barcroft J (1946) Researches on pre-natal life. Blackwell, Oxford
5. Borgatta L, Shrout PE, Divon MY et al. (1988) Reliability and reproducibility of nonstress test readings. Am J Obstet Gynecol 159:554–558
6. Bracero LA, Schulman H, Baxi LV et al. (1986) Fetal heart rate characteristics that provide confidence in the diagnosis of fetal well being. Clin Obstet Gynecol 29:3–11
7. Chaffin DG, Goldberg CG, Reed KL et al. (1991) The dimension of chaos in the fetal heart rate. Am J Obstet Gynecol 165:1425–1429
8. Dawes GS (1964) Physiologic changes in the circulation after birth. In: Fishman A, Dickinson WR (eds) Circulation of the blood. Men and ideas. Oxford University Press, Oxford
9. Dawes GS, Visser GHA, Goodman JDS, Redman CWG (1981) Numerical analysis of the human fetal heart rate: the quality of ultrasound records. Am J Obstet Gynecol 141:43–53
10. Dawes GS, Houghton CRS, Redman CWG, Visser GHA (1982) Pattern of the normal fetal heart rate. Br J Obstet Gynaecol 89:276–284
11. Dawes GS, Moulden M, Redman CWG (1990) Limitations of antenatal fetal heart rate monitors. Am J Obstet Gynecol 162:170–173
12. Devoe LD, Castillo RA, Sherline DM (1985) The nonstress test as a diagnostic test: a critical reappraisal. Am J Obstet Gynecol 152:1047–1053
13. Devoe LD, McKeuzie J, Searle N, Sherline D (1985) Nonstress test: dimensions of normal reactivity. Obstet Gynecol 66:617–620
14. Divon MY, Yeh SY, Zimmer EZ et al. (1985) Respiratory sinus arrythmia in the human fetus. Am J Obstet Gynecol 151:425–428
15. Donchin Y, Caton D, Porges SW (1984) Spectral analysis of fetal heart rate in sheep: the occurrence of respiratory sinus arrhythmia. Am J Obstet Gynecol 148:1130–1135
16. Farmakides G, Schulman H, Winter D et al. (1988) Prenatal surveillance using nonstress testing and Doppler velocimetry. Obstet Gynecol 71:184–187
17. Ferrazzi E, Pardi G, Setti PL et al. (1989) Power spectral analysis of the heart rate of the human fetus at 26 and 36 weeks of gestation. Clin Phys Physial Meas 10 Suppl B:57–60
18. Freeman RK, Anderson G, Dorchester W (1982) A prospective multi-institutional study of antepartum fetal heart-rate monitoring. I. Risk of perinatal mortality and morbidity according to antepartum fetal heart-rate test results. Am J Obstet Gynecol 143:771–777
19. Goldberger AL, Rigney DR, West BJ (1990) Chaos and fractals in human physiology. Sci Am 264:43–49
20. Hammacher K (1969) The clinical significance of cardiotocography. In: Huntingford PJ, Huter KA, Saling E (eds) Perinatal medicine. Academic, New York

21. Hon EH, Hess OW (1960) The clinical value of fetal ECG. Am J Obstet Gynecol 79:1012–1017
22. Hrushesky WJM, Schmitt O, Gilbertsen V (1984) The respiratory sinus arrhythmia: a measure of cardiac age. Science 224:1001–1004
23. Itskovitz J, LaGamma EF, Rudolph A (1983) Heart rate and blood pressure responses to umbilical cord compression in fetal lambs with special reference to the mechanism of variable deceleration. Am J Obstet Gynecol 147:451–457
24. James SL, Morishima HO, Daniel SS et al. (1972) Mechanism of late deceleration of the fetal heart rate. Am J Obstet Gynecol 113:578–593
25. James SL, Yeh MN, Morishima HO et al. (1976) Umbilical vein occlusion and transient acceleration of fetal heart rate. Am J Obstet Gynecol 126:276–283
26. Kennedy E (1834) Dublin hospital report. Dublin J Med Sci 4–5:89, 231
27. Kerr JJM, Ferguson JH, Young J, Hendry J (1923) A combined textbook of obstetrics and gynaecology. Livingstone, Edinburgh
28. Kitney RI, Rompelman O (1980) The study of heart-rate variability. Clarendon, Oxford
29. Koyanga T, Nakahara H, Nakano H (1989) A new method of assessing fetal heart rates with potential for analytical quantification. Frontiers Med Biol Eng 1:229–236
30. Kurth CD, Steven JM, Wagerle LC et al. (1989) Bedside measurement of cerebral oxygenation by reflectance spectroscopy (NIRS). Pediatr Res 25(4):357A
31. Pane CR, Gwiazdowski S, Misra OP et al. (1989) Effect of graded hypoxia on brain-cell membrane injury of newborn piglets. Pediatric Res 25(4):225A
32. Lumley J, Lester A, Anderson I et al. (1983) A randomized trial of weekly cardiotocography in high-risk obstetric patients. Br J Obstet Gynecol 90:1018–1026
33. Manning FA (1985) Assessment of fetal condition and risk: analysis of single and combined biophysical variable monitoring. Semin Perinatol 9:168–183
34. Manning FA (1987) Fetal assessment based on fetal biophysical profile scoring: experience in 19221 referred high-risk pregnancies. Am J Obstet Gynecol 157:880–884
35. Manning FA (1990) Fetal assessment based on fetal biophysical profile scoring. Am J Obstet Gynecol 162:703–709
36. Myers RE, Mueller-Heubach E, Adamonx K (1973) Predictability of the state of fetal oxygenation from a quantitative analysis of components of late deceleration. Am J Obstet Gynecol 115:1083–1090
37. Myers GA, Martin GJ, Magrid NM et al. (1986) Power-spectral analysis of heart-rate variability in sudden cardiac death: comparison to other methods. IEEE Trans Biomed Eng 33:1149–1156
38. Paul RH, Hon EH (1974) Clinical fetal monitoring. V. Effect on perinatal outcome. Am J Obstet 118:529–533
39. Pello LC, Dawes GS, Smith J, Redman CWG (1988) Screening of the fetal heart rate in early labor. Br J Obstet Gynaecol 95:1128–1136
40. Platt LD (1987) Fifteen years of experience with antepartum fetal testing. Am J Obstet Gynecol 156:1509–1515
41. Ribbert LM (1991) Relation of fetal blood gases and data from computer-assisted analysis of fetal heart-rate patterns in small for gestation fetuses. Br J Obstet Gynaecol 98:820–823
42. Rochard F (1976) Nonstressed fetal heart-rate monitoring in the antepartum period. Am J Obstet Gynecol 126:699–706
43. Rooth G, Huch A, Huch R (1987) Guidelines for the use of fetal monitoring. FIGO news. Int J Gynaecol Obstet 25:159–167
44. Saling EP (1969) Fetal blood studies. In: Huntingford PJ, Huter KA, Saling EP (eds) Perinatal medicine. Academic, New York, pp 113–119
45. Sayers BM (1973) Analysis of heart-rate variability. Ergonomics 16:17–32
46. Schneider EP, Hutson JM, Petrie RH (1988) An assessment of the first decade's experience with antepartum fetal heart-rate testing. Am J Perinatol 5:5–11
47. Schneider EP, Schulman H, Farmakides G et al. (1992) Clinical comparison of the interpretation of antepartum fetal heart rate tracings between a computer program and experts. J Maternal Fetal Invest 1:205–208
48. Shy KK (1987) Evaluating a new technology: the effectiveness of electronic fetal heart-rate monitoring. Annu Rev Public Health 8:165–190

49. Singer DH, Martin GJ, Magrid N et al. (1988) Low-heart-rate variability and sudden cardiac death. J Electrocardiogr Suppl: S46–S55
50. Smith CV, Nguyen HN, Kovacs B et al. (1987) Fetal death following antepartum fetal heart-rate testing: a review of 65 cases. Obstet Gynecol 70:18–20
51. Snijders RJM, Ribbert LSM, Fransens M et al. (1989) Heart-rate variability in small for date fetuses with abnormal umbilical artery velocity wave forms. Fetal Neonatal Meas 3:259–263
52. Snijders RJM et al. (1990) Computer-assisted analysis of fetal heart-rate patterns at 20–41 weeks gestation. Fetal Diagn Ther 79–83
53. Street P, Dawes GS, Moulden M et al. (1991) Short-term variation in abnormal antenatal fetal heart-rate records. Am J Obstet Gynecol 165:515–523
54. Suidan JS, Young BR, Hochberg HM, George MED et al. (1985) Observations on perinatal heart-rate monitoring. II. Quantitative unreliability of Doppler fetal heart-rate variability. J Reprod Med 30:519–522
55. Trudinger BJ, Cook CM, Jones L, Giles WB et al. (1986) A comparison of fetal heart-rate monitoring and umbilical artery waveforms in the recognition of fetal compromise. Br J Obstet Gynecol 93:171–175
56. Vintzileos AM, Campbell WA, Nochimson DJ, Weinbaum PJ et al. (1986) The use of the nonstress test in patients with premature rupture of the membranes. Am J Obstet Gynecol 155:149–153
57. Visser GHA, Huisjes HJ (1977) Diagnostic value of the nonstressed antepartum cardiotocogram. Br J Obstet Gynaecol 84:321–326
58. Visser GHA, Sadovsky G, Nicolaides KH et al. (1990) Antepartum heart-rate patterns in small for gestational age third trimester fetuses: correlations with blood gas values obtained at cordocentesis. Am J Obstet Gynecol 162:698–703
59. Zho Y, Szeto HH (1987) Cyclic variation in fetal heart rate and sympathetic activity. Am J Obstet Gynecol 156:1001–1005

Physiopathological Basis of Doppler Indices

D. Maulik

Introduction

Doppler ultrasound offers a unique noninvasive technology for assessing circulatory dynamics. Doppler frequency shift waves represent velocity pulse waves at the vascular site of ultrasound interrogation. The Doppler waveform is, therefore, shaped by hemodynamic phenomena both proximal and distal to the point of measurement (Fig. 1). The main technique for analyzing the waveform has been the determination of descriptor indices of the waveform. Most of these indices are virtually angle-independent ratios which are derived from various combinations of the peak systolic, end-diastolic, and average values of the Doppler maximum frequency shift envelope. Most of these indices express wave pulsatility although their formulations are not identical. The indices that are commonly used include resistance index (RI), pulsatility index (PI), and systolic-diastolic (S/D) ratio (Fig. 2). The indices have been used widely for noninvasively investigating regional circulations based on the assumption that they are capable of reflecting downstream flow impedance. Obviously, such an assumption requires experimental verification as the contour of an arterial Doppler wave, being representative of the arterial flow velocity waveform, is expected to be affected not only by peripheral but also by central circulatory dynamics. Relatively recently, a number of studies have sought such experimental verification of the hemodynamic basis of Doppler waveform analysis. Investigations have also been conducted to elucidate the changes in Doppler waveform in relation to homeostatic challenges to the fetus, such as asphyxia or chronic nicotine exposure. These investigations have utilized a variety of experimental approaches using a wide spectrum of models encompassing in vitro circulatory simulation, in vivo models involving fetuses and neonates, and vascular morphological studies on human placentae (Table 1). This paper critically reviews these studies and also presents basic concepts regarding opposition to flow in a pulsatile circulation.

Investigations on Hemodynamic Validity of Doppler Waveform Analysis

Studies in this area have, in general, looked for any correlation between the commonly used Doppler indices and independently measured parameters of

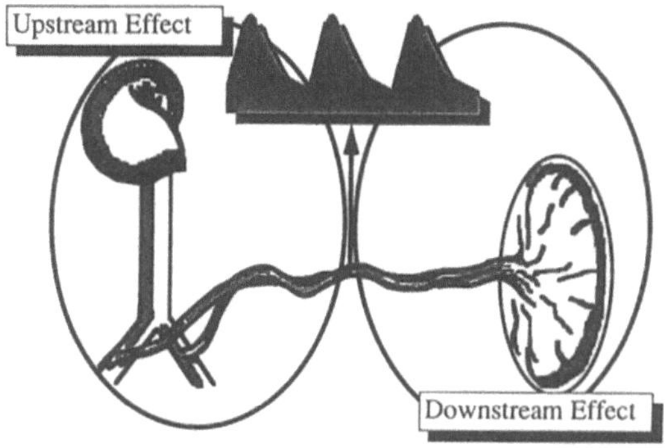

Fig. 1. Up- and downstream effects on arterial velocity waveform. The figure illustrates the concept that an arterial velocity waveform is shaped by hemodynamic factors both from the central and peripheral components of the circulation. As the Doppler waveform represents the arterial velocity waveform, the descriptor indices of the Doppler waveform are also influenced by these hemodynamic factors

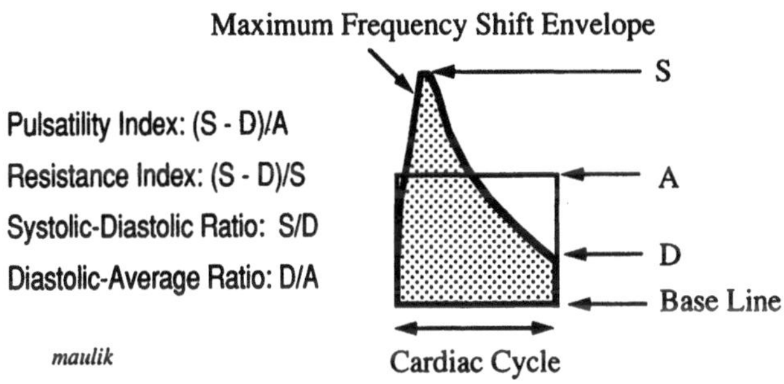

Fig. 2. Doppler indices estimated from the maximum frequency shift envelope. *S*, peak systolic value; *D*, end-diastolic value; *A*, the time-averaged value of the envelope

central and peripheral hemodynamics. The latter included the heart rate, the peripheral resistance, and the components of arterial input impedance. In addition, association between placental angiomorphologic changes and the Doppler indices has been investigated by several workers. This section summarizes these findings and also presents a brief description of basic hemodynamic concepts relevant for Doppler validation studies.

Arterial Input Impedance: Basic Concepts and Relevance to Doppler Waveform Analysis

In this section, the relevant aspects of peripheral circulatory dynamics are briefly reviewed. Specifically, the basic principle of arterial input impedance and its relevance to Doppler waveform analysis are discussed. Traditionally,

Table 1. Hemodynamic studies of the Doppler indices

- Investigations on hemodynamic validity of Doppler waveform analysis
  - In vitro hemodynamic validation
    - Doppler indices and peripheral resistance
    - Doppler indices and arterial input impedance
  - In vivo hemodynamic validation – fetal lamb model
    - Doppler indices and peripheral resistance
      - Acute umbilical arterial flow occlusion
      - Chronic fetoplacental embolization
    - Doppler indices and heart rate
      - Acute uteroplacental flow occlusion
      - Maternal hypoxia
  - In vivo hemodynamic validation – neonatal lamb model
    - Doppler indices, arterial input impedance, and heart rate
  - Angiomorphologic validation studies
    - Doppler indices and placental vascular morphology
- Investigations on the validity of Doppler waveform analysis in experimental fetal compromise
  - Doppler indices and fetal asphyxia
    - Uteroplacental insufficiency – maternal aortic occlusion
    - Fetoplacental embolization
  - Doppler indices and fetal drug exposure
    - Chronic nicotine administration

opposition to flow has been expressed in terms of peripheral resistance (Zpr) which is the ratio of the mean pressure to mean flow: Zpr = Pm/Qm, where Pm and Qm represent mean pressure and mean flow, respectively. Although the concept of Zpr is extensively used, it is applicable only to steady nonpulsatile flow conditions. Flow of blood in an arterial system, however, is a pulsatile phenomenon driven by myocardiac contractions with a periodic rise and fall in pressure and flow associated with systole and diastole of the ventricle. The pressure and flow pulses thus generated are also profoundly affected by the downstream circulatory conditions which include opposition to flow offered by the peripheral vascular system. The idea of vascular impedance provides the foundation for understanding this complex phenomenon in a pulsatile circulation. Vascular impedance is analogous to electrical impedance in an alternating current system. It has been shown by Womersley [1] that the equations dealing with electrical impedance are applicable for solving the problems of impedance. It should be noted in this context that the idea of vascular resistance is analogous to the principle of electrical resistance in direct current electrical transmissions.

Closely related to impedance is the phenomenon of wave reflection in a vascular tree. Configuration of pressure and flow waves at a specific location results from the interaction of the forward-propagating orthograde waves with the reflected backward-propagating retrograde waves. Current evidence suggests that the arterial–arteriolar junctions serve as the main source of wave reflections in an arterial system. An in-depth discussion of arterial impedance and wave reflection is beyond the scope of this review. However, relevant impedance parameters are briefly described below.

The input impedance of an arterial system (Zi, [2]) at a specific vascular site is the ratio of the sinusoidal harmonic components of pressure waves to those of the corresponding flow waves at that particular location. As the pressure and flow waves are complex in shape, it is necessary to convert them by Fourier analysis into their constituent sinusoidal components. The input impedance is described by the following two parameters (Fig. 3): (a) the Zi modulus which is the amplitude of the pressure-flow harmonic ratios as the function of the frequency; the units of measurement for the modulus are dynes/s cm^3 for velocity and dynes/s cm^5 for volumetric flow, and (b) the phase of the impedance, which expresses the phase-angle lag between the pressure and the flow waves as a function of frequency. The phase angle is measured in radians, negative or positive.

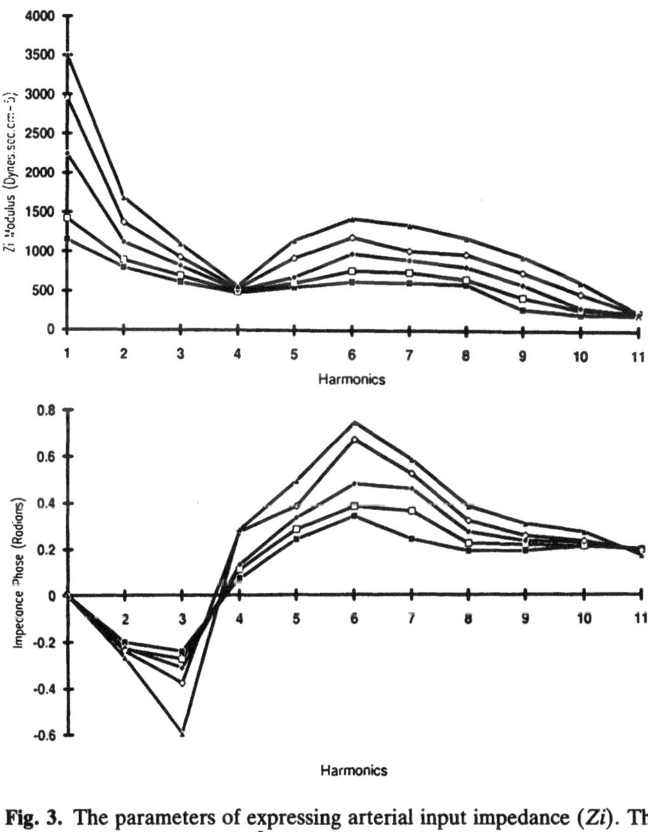

Fig. 3. The parameters of expressing arterial input impedance (*Zi*). The *upper panel* shows the modulus of Zi in dynes/s cm^5. The *vertical axis* shows the magnitude, and the *horizontal axis* the harmonics. The Zi at 0 frequency expresses opposition to flow in the nonpulsatile or the steady state and is therefore equivalent to the peripheral resistance. The *lower panel* shows the changes in impedance phase measured in radians; the *horizontal axis* shows the harmonics. Note the effects of increasing opposition to flow on both these parameters. Five levels of impedance are indicated by the symbols ■, □, ◆, ◇, and ▲, where ■ represents the highest and ▲ the lowest

Other descriptors of vascular impedance are: (a) impedance at the fundamental harmonic (zero frequency) which is the same as the Zpr and represents the steady nonpulsatile component of vascular impedance, (b) characteristic impedance (Zo) which reflects the properties or characteristics of the arterial system such as the cross-sectional area, wall thickness, and vasculo-elasticity; any discontinuity or change in these properties gives rise to reflections of pressure of flow waves, and (c) the reflection coefficient (Rc) which expresses the magnitude of the phenomenon of wave reflection in an arterial tree.

In Vitro Hemodynamic Validation

In vitro circulatory simulation has been extensively used to investigate complex hemodynamic phenomena. In this type of simulation, nonbiological materials are used for prototyping circulatory systems in order to study circulatory phenomena. In an intact organism, it may be difficult to perform comprehensive hemodynamic measurements without profoundly altering the physiological state of the preparation. Indeed, it may be impossible to conduct certain hemodynamic experiments in vivo because of the inherent complexities of modelling. It is also well recognized that the fundamental hydrodynamic principles are equally applicable to explaining circulatory phenomena in a physical simulation and in a biological system. The principles of in vitro simulation for hemodynamic studies have been comprehensively reviewed by Hwang [3]. Hemodynamic parameters for designing an in vitro circulatory system for validating the Doppler indices have been reviewed by Maulik and Yarlagadda [4].

Spencer et al. [5] used an in vitro flow model to validate the Doppler indices in terms of Zpr (mean pressure/mean flow). The in vitro flow model was a modification of a commercially available Doppler-flow phantom system. Two types of perfusates were used: microspheres suspended in water and heparinized sheep blood. The Doppler measurements were performed using an 8-MHz continuous transducer; the flow and pressure were measured using transit-time flowmeter and strain gauge, respectively. The flow was reduced in a stepwise fashion using a clamp. The decline in flow led to increases in RI, PI, and S/D ratio. The pressure and the pressure waveform remained unchanged. The authors noted that the rises in the Doppler indices reflected the increasing Zpr; the RI and PI had a mostly linear increase, whereas the S/D ratio demonstrated an exponential increase.

Maulik and Yarlagadda [6] used an in vitro circulatory simulation system for hydrodynamic validation of the Doppler indices. The in vitro system (Fig. 4) consisted of a ventricular pump, an "arterial" line, a proximal "arterial" compliance unit, a "fetal placental vascular" branching model, a systemic bypass circuit, and a "venous" line for the return of the perfusate into an "atrial" reservoir which was connected to the pump. The following parameters of the input impedance were used for the hemo-

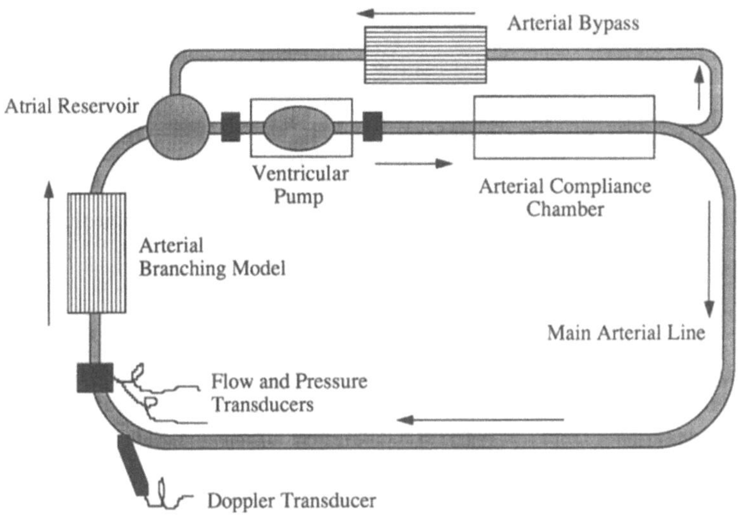

Fig. 4. The in vitro flow circuit

dynamic validation: (a) Zi modulus, (b) Zi at zero frequency representing the Zpr, (c) Zo, and (d) Rc. A continuous wave Doppler system with a 4-MHz transducer was used at a fixed angle of approximately 30°. Flow and pressure were recorded from the same location. The Doppler indices included the S/D ratio, the RI, the PI, and the diastolic/average ratio. The impedance to arterial flow was progressively increased by sequentially occluding the model fetal placental branches. All the indices correlated significantly ($p < 0.01$) with the Zpr and Rc; this indicates that the changes in the Zpr and in the wave reflections from the downstream circulation are expressed by the changes in the Doppler waveform as described by the indices. For Doppler indices vs Zo, the correlations were not significant at $p < 0.05$; this is explained by the fact that in these experiments, no changes occurred in the vascular wall characteristics which is reflected by this impedance parameter.

These in vitro simulation studies indicate that within the general confines of a hydrodynamic model characterized by an increasing downstream opposition to a pulsatile flow and by a constant pump function, the descriptor indices of the Doppler waveform are capable of reflecting the state of the downstream flow impedance.

In Vivo Hemodynamic Validation – Fetal Lamb Model

These investigations aimed at elucidating changes in the Doppler indices with independently measured hemodynamic parameters under acute and chronic experimental conditions. These are briefly reviewed below.

Doppler Indices and Peripheral Resistance

These investigations aimed at establishing a hemodynamic foundation for the Doppler indices by increasing umbilical circulatory resistance and analyzing the changes in the indices and in the calculated vascular resistance. Increases in the resistance were achieved by mechanical occlusion or chronic embolization. One of the limitations of these studies was that resistance rather than impedance was used to assess peripheral circulatory state in a pulsatile flow system.

Maulik et al. [7] investigated the relationship between the various umbilical arterial Doppler indices and the umbilical arterial resistance in fetal lambs. Seven ovine fetuses of Rambouillet-Columbia variety with gestational ages from 100–110 days were used. The experiments were conducted under general anesthesia. Umbilical arterial pressure and flow were measured using a strain-gauge pressure transducer and transit time-flow meters, respectively. A continuous wave Doppler device with a 4 MHz transducer was used. The hemodynamic perturbation was induced by mechanical constriction of the umbilical artery and by fetal administration of vasopressin. The hemodynamic parameters included pressure, volumetric flow, and the Zpr; the latter was calculated from the mean pressure and flow. The Doppler indices measured were S/D ratio, RI, PI, and the diastolic/average ratio. The mechanical constrictions resulted in declines in flow by 74%–89% and increases in pressure by 8%–23%. The correlation between the Doppler indices and the hemodynamic parameters are shown in Table 2, which indicates that in well-defined obstruction to downstream flow the indices correlate highly with the downstream hemodynamic parameters including the Zpr ($p < 0.005$). The vasopressin experiments failed to produce consistent circulatory changes and were not utilized in the validation process.

Trudinger et al. [8] studied the consequences of chronic embolization of the umbilical circulation on umbilical arterial S/D ratio and umbilical circulatory resistance in fetal lambs. Thirteen singleton ovine fetuses were studied from day 120 of gestation. Over a period of 9 days, approximately 9×10^8 microspheres of 15-μm diameter were injected into the fetal placental cotyledons along the umbilical arteries in order to occlude the small vessels of the fetoplacental vascular bed. Doppler measurements were

Table 2. Correlation coefficients, Doppler indices vs hemodynamic parameters, constriction experiments

	S/D	PI	RI
Zpr	0.95	0.89	0.88
Q	−0.91	−0.96	−0.97
P	0.81	0.84	0.82

S/D, systolic-diastolic ratio; PI, pulsatility index; RI, resistance index; Zpr, peripheral resistance; Q, volumetric flow; P, pressure. All significant at $p < 0.005$.

performed with a 4-MHz continuous wave Doppler device; pressure was measured in the fetal aorta, fetal inferior vena cava, and amniotic cavity using pressure catheters and transducers. Umbilical and uterine flows were measured by radiolabelled microsphere technique. Fetoplacental vascular resistance was calculated from fetal aortic-vena caval pressure gradient and total umbilical flow. Chronic embolization resulted in (a) increasing fetoplacental vascular resistance – 0.25 to 0.35 mmHg/ml min, (b) increased umbilical arterial S/D ratios, and (c) significant declines ($p < 0.05$) in the umbilical total flow and in the umbilical to splanchnic flow ratio (3.36 to 1.53). The authors concluded that the umbilical artery flow velocity waveform S/D ratio measures the reflection coefficient at the peripheral vascular bed of the placenta. It should be noted, however, that reflection coefficient, which has a precise hemodynamic definition (see above), was not measured in these experiments and no correlations were performed between the hemodynamic parameters and the indices. Nevertheless the study established a relationship between chronic circulatory alteration and the umbilical arterial Doppler waveform. The above findings were corroborated by Morrow et al. [9] who observed progressive increases in the umbilical arterial pulsatility consequent to fetoplacental arterial embolization. In this study involving seven chronically catheterized sheep fetuses, the umbilical arterial S/D ratio correlated significantly ($r = 0.76$, $p = 0.001$) with placental resistance derived from aortic inferior vena caval pressure gradient and mean peak velocity.

Doppler Indices and the Heart Rate

As stated at the outset, central circulation potentially influences the Doppler waveform and should be an important consideration in interpreting the changes in Doppler indices. The clinical significance of fetal heart-rate effect on the indices has been somewhat controversial [10, 11]. Most of the clinical studies were conducted in normal pregnant populations. It remained unclear whether Doppler indices are of independent value in acute fetal distress associated with significant alterations in the fetal heart rate. This prompted in vivo animal studies which investigated this issue in models of experimentally produced acute fetal compromise.

Maulik et al. [12] investigated the effects of acute uteroplacental flow occlusion on the umbilical arterial Doppler indices and the possible role of the central circulation in modulating this response. Five pregnant ewes of Rambouillet-Columbia variety with gestational ages of 115–120 days were used. Following appropriate surgical procedures, the maternal peritoneal cavity was entered and an electromagnetic flowmeter was placed around the maternal common hypogastric artery. The vascular occluding device was then placed around the artery proximal to the flowmeter using the technique of Clark et al. [13]. After stabilization of the maternal and fetal condition, baseline data on the arterial flow rate and umbilical arterial Doppler waveforms were collected. The Doppler instrumentation consisted of a continuous wave device with a 4-MHz transducer. The artery was then completely

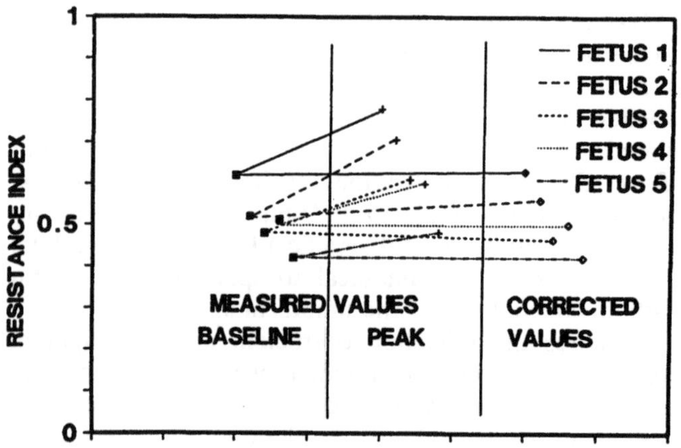

Fig. 5. The effect of variations in the cardiac cycle time correction on the resistance index. The three divisions in this figure correspond to baseline, measured peak, and corrected values, respectively. Please note that the differences between the baseline and the measured peak values (all significant at $p < 0.005$) disappeared when corrected for fetal heart rate (none significant at $p < 0.05$). [From 12]

occluded for an approximate duration of 20 seconds following which the occlusion was released. The hypogastric flow rate and umbilical arterial Doppler waveform were continuously recorded during the occlusion and subsequent recovery. The Doppler indices measured were RI, S/D ratio, and PI. The duration of the cardiac cycle and its systolic and diastolic times were approximated from the Doppler waveform. The occlusion produced transient increases in the Doppler indices and cardiac cycle time. The observed increases ($p < 0.005$) in the Doppler indices disappeared ($p < 0.05$) when the latter were corrected for the changes in the cardiac cycle time (Fig. 5). Furthermore, diastolic time was the main component of the changes in the cardiac cycle time (Fig. 6) and the Doppler indices demonstrated a higher correlation with diastolic time than with systolic time. The results verified that changes in the duration of the fetal cardiac cycle play an important role in shaping the umbilical arterial Doppler waveform and hence need to be taken into account in estimating the Doppler indices.

The effect of acute maternal hypoxemia on umbilical arterial Doppler waveform was assessed by Downing et al. [14] in a chronically instrumented ovine model. Hemodynamic parameters measured in this study included maternal heart rate, blood gases, and fetal heart rate; the indices measured were S/D ratio, PI, and RI. During the periods of hypoxemia, fetal heart rate decreased (<80 beats/min) and umbilical arterial Doppler indices increased ($p < 0.001$). Significant negative correlation ($p < 0.001$) was noted between the Doppler indices and the fetal heart rate. Analysis of the Doppler waveform phase intervals revealed nearly constant systolic intervals, while diastolic intervals varied inversely with the heart-rate alterations. Moreover, the umbilical arterial Doppler indices, when cor-

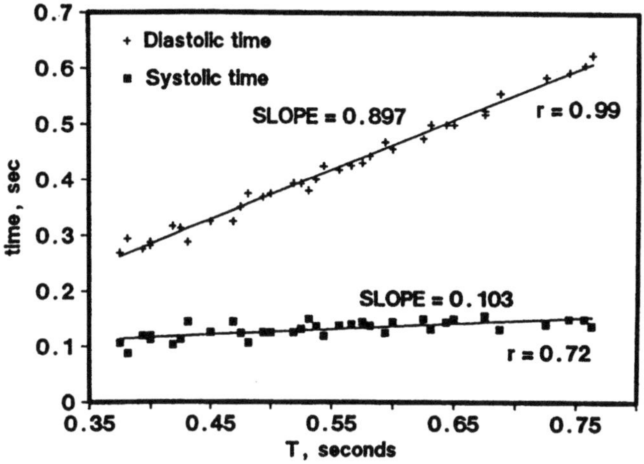

Fig. 6. The correction between the systolic and the diastolic components of the Doppler waveform, and the cardiac cycle time (*T*). Please note that the slope of the cardiac cycle time vs diastolic time is approximately nine times greater than the slope of the cardiac cycle time vs systolic time. [From 12]

rected for fetal heart-rate changes, were relatively unchanged from baseline measurements.

These studies emphasize the importance of the contribution of the central circulatory changes to alterations in the Doppler waveform.

In Vivo Hemodynamic Validation – Neonatal Lamb Model

The above investigations in vivo did not offer hemodynamic validation of the Doppler indices in terms of arterial impedance. The latter, however, is the hemodynamic parameter of choice in assessing obstruction to flow in a pulsatile circulation. In contrast, Zpr is applicable only for the nonpulsatile component of a circulation and as such constitutes one of the parameters of arterial impedance. Our recent studies [15, 16] have addressed this in a neonatal lamb model in which the validity of Doppler waveform analysis was investigated by correlating the descending aortic PI with the measures of aortic input impedance.

Doppler Indices, Input Impedance, and Heart Rate

Downing et al. [15] developed a pharmacological model of hemodynamic alteration for this purpose. Five cross-bred term neonatal lambs were used. General anesthesia was used for the initial surgical preparation which allowed the in situ placement of a 4-MHz continuous-wave Doppler transducer and a transit time-flow transducer on the infra-renal descending aorta. A pressure transducer was retrogradely introduced up to the level of the flow probe by the left femoral artery. The probe lines were exteriorized. Each animal was

Table 3. Alterations of physiologic parameters in response to drug-induced vasodilation and vasoconstriction

Parameter	Baseline	Hydralazine	Norepinephrine
Heart rate (beats/min)	132 ± 15	164 ± 14*	82 ± 10*
Mean arterial blood pressure (mmHg)	78 ± 5	55 ± 5*	122 ± 8*
Mean volumetric flow, descending aorta (ml/min)	319 ± 21	405 ± 28*	138 ± 22*

Data are means ± standard error of mean.
* Significant difference from baseline parameter $p < 0.005$

allowed to recover for 4 days following which experimental intervention was initiated. Following baseline recordings of the blood pressure, heart rate, aortic flow rate, and Doppler waveforms, either vasodilation or vasoconstriction was produced with the administration of hydralazine or norepinephrine, respectively. The alternate drug was administered on the following day. The sequence of drug regimen was randomized.

The following parameters were measured: peripheral vascular resistance, characteristic impedance, reflection coefficient, and Gosling pulsatility index. Basic hemodynamic changes are shown in Table 3.

As evident, vasodilation led to tachycardia, hyperperfusion, and hypotension. Opposite changes were noted with vasoconstriction. Significant increases in PI, Zpr, Zo, and Rc are seen in response to the administration of norepinephrine, and decreases in PI were noted with the administration of hydralazine. Figure 7 depicts the changes in the PI and Rc and Fig. 8 the changes in the input impedance moduli and phase. As evident, the impedance modulus curve showed distinct changes with vasoconstriction and vasodilation. Further analysis of the data showed statistically significant

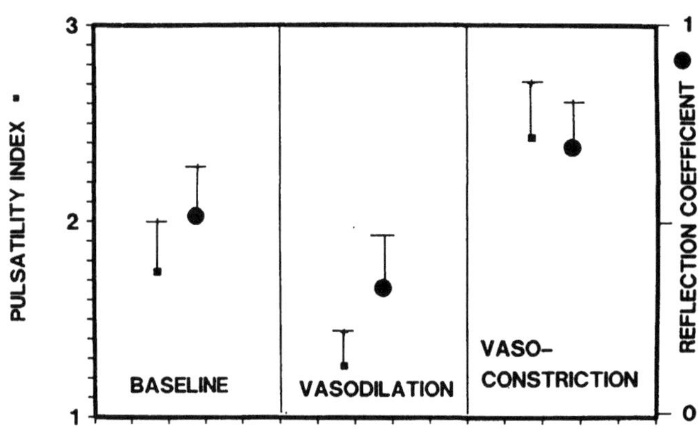

Fig. 7. Responses of the pulsatility index (PI) and reflection coefficient (Rc) to vasodilation with hydralazine and vasoconstriction with norepinephrine. PI and Rc increased significantly from baseline in response to norepinephrine. PI decreased significantly with hydralazine; however, Rc was not significantly affected. [From 15]

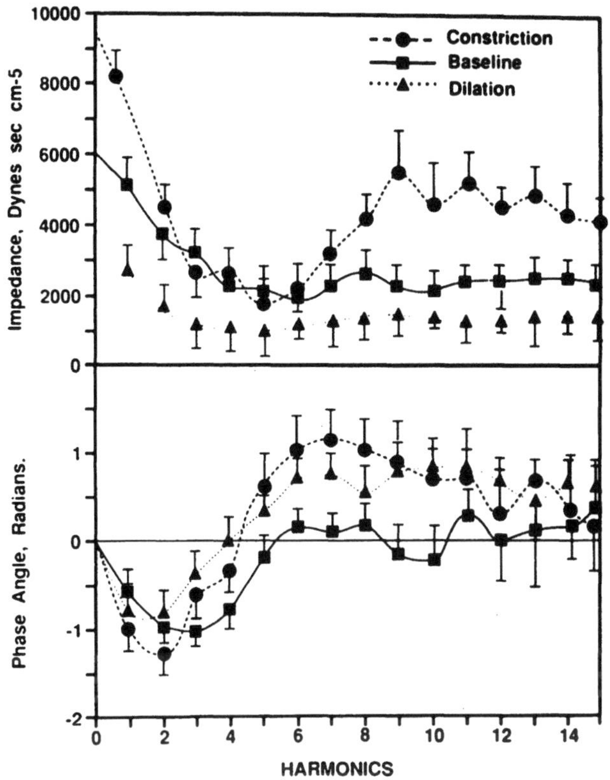

Fig. 8. Impedance moduli and phase angle during baseline, vasodilation, and vasoconstriction with hydralazine and norepinephrine, respectively. Each *curve* represents the mean ± standard error of mean impedance moduli for all study animals. [From 15]

correlation between PI and Zpr (Fig. 9). However, changes in PI did not significantly correlate with the changes in characteristic impedance and reflection coefficient. The next study investigated this issue.

The experimental model was the same as in the previous study. The experimental protocol, however, ensured that heart-rate changes due to vasoactive interventions were pharmacologically suppressed by trimethophan (TM) during vasodilation or atropine methylbromide (AMB) during vasoconstriction. Relative changes in PI were correlated with relative changes in input impedance parameters of characteristic impedance (Zo) and Zpr as corrected for body surface area (dynes/s cm^5/m^2). The results are presented in Table 4.

In response to both hydralazine (HY) and phenylephrine (PE) without controlled reflex heart-rate responses, aortic PI was highly and positively correlated with Zpr ($r = 0.78$, $p < 0.001$), but not with Zo. However, changes in PI correlated well with Zpr and Zo when reflex heart-rate responses were inhibited ($r = 0.92, 0.95$, respectively, $p < 0.001$).

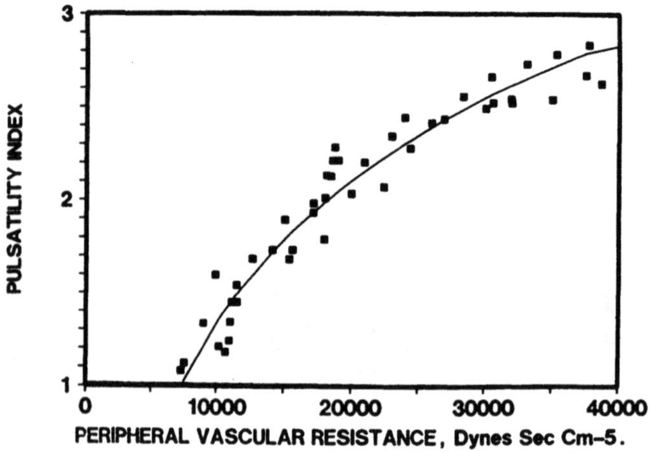

Fig. 9. Correlation between peripheral vascular resistance and pulsatility index during baseline, vasodilation, and vasoconstriction conditions ($r = 0.97$; $p < 0.001$). *Points* represent pooled data from each experiment. [From 15]

Table 4. Hemodyamic parameters and Doppler indices in drug-induced vasodilation and vasoconstriction with and without heart rate changes

	Control	Hydralazine	Hydralazine trimethophan	Phenylephrine	Phenylephrine atropine
HR (beats/min)	128 ± 10	186 ± 12	130 ± 8	65 ± 7	126 ± 8
PI	1.91 ± 13	1.50 ± 21	1.71 ± 13	2.54 ± 19	2.28 ± 0.21
Zpr	39960 ± 8728	12350 ± 1210	10460 ± 961	65380 ± 8130	60840 ± 7590
Zo	4010 ± 340	1683 ± 350	1333 ± 410	8960 ± 760	7683 ± 630

Data are means ± standard error of mean.
HR, heart rate; PI, pulsatility index; Zpr, peripheral resistance; Zo, characteristic unpedance

It is reasonable to conclude from these findings that PI reflects the state of downstream circulation independent of the changes in the heart rate. However, heart rate does influence PI in the pharmacologically altered hemodynamic state, and should therefore be taken into consideration.

Angiomorphologic Validation of Doppler Indices

A number of investigators reported fetoplacental morphological changes in relation to umbilical arterial velocimetry and pregnancy complications including fetal growth compromise. Giles et al. [17] correlated fetoplacental histopathological changes with umbilical arterial A/B (systolic-diastolic) ratio in three groups of pregnant women: (a) 38 normal pregnancies, (b) 33 pregnancies with normal A/B ratio, and with maternal or fetal risk factors such as clinical suspicion of intrauterine growth retardation (IUGR), hypertension, renal disease, or antepartum hemorrhage, and (c) 35 pregnancies with abnormal A/B ratios (>3) and with the above high-risk factors. A statistically significant decrease ($p < 0.01$) in the model small arterial count (arteries in the tertiary stem villi measuring less than 90 μm in diameter) was observed in abnormal pregnancies with an abnormal Doppler index (the third group) in comparison with the first and the second groups. Moreover, there was a statistically significant ($p < 0.001$) increase in the prevalence of small for gestational age infants in the third group. No differences were observed between the first and the second groups. Furthermore, no significant differences were noted in the tertiary villi count between the three groups. The authors suggested fetoplacental vaso-obliterative pathologic processes as the underlying cause of an abnormal Doppler waveform.

It is also possible that under certain circumstances, a failure of angiogenesis may lead to a subnormal vascularity and an abnormal Doppler waveform. This was suggested by Schulman's group. Bracero et al. [18] performed morphological examinations of placentae from 47 women in whom measurements of umbilical arterial S/D ratios were also conducted. Those who had an S/D ratio value of three or more had increased perinatal complications, including higher incidence of low-birth-weight infants, Cesarean deliveries for fetal distress, and neonatal intensive care admissions. Consistent with the findings of Giles et al., a statistically significant ($p < 0.0005$) decrease in the number of small muscular fetoplacental arteries was noted. Interestingly, the investigators also observed a continuing decline in the S/D ratios although they remained in the abnormal range. Furthermore, vascular obliteration was seen in association with both normal and abnormal S/D ratios. These findings prompted Bracero et al. to suggest that the observed decline in the fetoplacental arteries in the group with elevated S/D ratios probably represents a failure of normal angiogenesis in the fetoplacental vascular system. It is possible that both the mechanisms are responsible for this abnormal vascular phenomenon. The importance of these studies lies with their demonstration of an inverse relationship between fetoplacental vascularity and pulsatility of the umbilical arterial Doppler waveform.

Investigations on the Validity of Doppler Waveform Analysis in Experimental Fetal Compromise

These studies investigated the changes in the Doppler indices in experimentally produced alterations of the fetal state. The latter included various models of producing in utero asphyxia and chronic fetal exposure to nicotine. These are reviewed below.

Doppler Indices and Fetal Asphyxia

Although umbilical arterial velocimetry has been related to fetal asphyxia [19], data from animal experiments do not completely corroborate this.

Aortic vascular state and velocimetry, in contrast, demonstrate changes with fetal asphyxia. These apparently contradictory findings may be explained in terms of circulatory redistribution associated with fetal homeostatic response to asphyxial stress. The following review examines this.

Copel et al. [20] studied four fetal sheep at 114–139 days' gestation to evaluate the effect of acute placental embolization on placental perfusion, fetal oxygenation and acid-base status, and the umbilical arterial S/D ratio. Studies were performed in an acute model under general anesthesia. Serial embolizations with microspheres were performed in the umbilical artery at 20- to 30-min intervals. Data on umbilical flow, S/D ratio, and arterial blood gas levels were measured and collected. In one fetus, umbilical flow decreased by 50% and the pH became 6.91 following the first injection; however, a normal S/D ratio was maintained. This fetus eventually succumbed after the second injection. Significant flow declines were also observed in the remaining three fetuses, although only after the third injection ($p < 0.02$); again, no changes were noted in the S/D ratio. The ratio increased significantly only after the fourth injection ($p < 0.05$) which was associated with an even greater decline in the umbilical flow. Corresponding decreases in PO_2 and pH did not achieve statistical significance. Subsequent embolizations accentuated the magnitude of all changes. This study indicated relative insensitivity of the umbilical arterial S/D ratio to a declining umbilical perfusion and developing fetal asphyxia. Significant changes in the S/D ratio occurred only with severely diminished flow and profound hypoxia with acidosis.

These findings received further support from Morrow et al. [9] who used an ovine umbilical arterial embolization model (see above) and noted that a transient decrease in fetal PO_2 levels had no effect on the umbilical arterial Doppler waveform and even progressive severe hypoxic acidemia did not cause the waveform shape to change until the fetus was moribund.

A clearer understanding of fetal circulatory response to asphyxia may be obtained from a preliminary report by Malcus [21] who evaluated fetal hemodynamic changes during experimentally induced fetal asphyxia in an acute ovine model. Fetal asphyxia was induced by occlusion of the maternal aorta by either stepwise decreasing perfusion pressure to the placenta, or total occlusion of the maternal aorta for 1-, 2-, and 4-min duration. A combined 3.5-MHz real time and a 2-MHz-pulsed Doppler ultrasound device was used. The fetal aortic, common carotid arterial and umbilical arterial Doppler waveforms were recorded. In addition, in six fetuses, the diameter changes of the fetal aorta and the common carotid artery were measured simultaneously with the flow velocity waveform recordings. The authors observed that acute asphyxia altered central circulatory dynamics in the fetal lamb. In the fetal-descending aorta vasoconstriction occurred, resulting in a decline and eventual disappearance of the end-diastolic flow velocity and consequent increase in the PI, and decrease in the mean velocity and blood flow. Concomitantly, carotid vasodilation occurred associated with increases in blood flow and end-diastolic velocity, and with a decrease in the PI. Umbilical artery PI did not change during acute asphyxia.

Schroder and Huneke [22] also found significant correlations ($p < 0.01$) between ovine fetal-descending aortic PI, and fetal pH ($r = 0.85$) and PCO_2 ($r = 0.82$). In this chronically instrumented model with a fetal gestational age of 120–130 days, the authors found that harmonic analysis which yielded the harmonic component index was better correlated with Zpr than the PI.

In summary, the above findings suggest that with acute asphyxia, the fetal circulation undergoes a compensatory redistribution of blood flow with increased perfusion to the brain, decreased perfusion to the nonvital circulations, and maintenance of umbilical flow. As the asphyxia becomes worse , these compensatory homeostatic mechanisms fail, so that as a late phenomenon even the umbilical and other vital circulations become compromised.

Doppler Indices and Fetal Drug Exposure

Arbeille et al. [23] used Doppler ultrasound technology to study the hemodynamic effects of drug exposure in a pregnant ewe model. The study group was exposed to maternal nicotine administration, 10 mg intramuscularly daily for 5 days per week. Five ewes were assigned to the study group, five to the placebo group, and six to the control group. The RI was measured in the umbilical, fetal cerebral, and uterine arteries on days 80, 100, and 130 of gestation. Normal ranges of the Doppler indices were derived from the measurements in the six control ewes. No difference was found on the Doppler indices between the control group and the placebo group. Although the cerebral RI values did not differ in the three groups at days 80 and 100, these values were significantly higher at day 130 in the nicotine group ($p < 0.01$). The umbilical indices were slightly elevated at days 80 and 130 in the nicotine group. No significant differences were noted in the uterine RI values between the groups, at any stage of gestation. In the control and placebo groups the ewes delivered at a normal gestational age and only two out of 20 (10%) died at birth. In contrast, in the nicotine group, two delivered prematurely and eight out of 13 (63%) were stillborns. The authors concluded that repeated nicotine injections were associated with profound hemodynamic changes in the umbilical and fetal cerebral circulations, and also with a poor perinatal outcome.

Conclusion

Doppler velocimetry offers a wide range of hemodynamic information. For perinatal applications, the assessment of the downstream hemodynamic state is of primary importance. Although waveform analytic techniques were based upon the current understanding of hemodynamic phenomena, data on the hemodynamic validation of Doppler indices have only recently been available. As summarized above, Doppler indices may not clearly reflect the impedance to flow in circumstances where acute insults to the fetus may also affect the heart rate. However, in circumstances of fetal compromise where heart rate remains stable, the indices correlate well with the changes in the Zpr. A reduction in the number of fetoplacental small arteries, whether caused by secondary obliteration, failure of angiogenesis, or both, will lead to a decrease in the total cross-sectional area of the vascular bed. This in turn will lead to an increase in the downstream flow impedance, a compromise in the end-diastolic velocity, and to abnormalities of the Doppler waveform, In relation to the cerebral effect, the fetal heart-rate effect on the end-diastolic velocity has been recognized as a confounding factor for the ability of the indices to express changes in the downstream impedance. However, as reviewed above, when the heart rate is stable, the indices do reflect changes in the downstream impedance. It should be emphasized that this finding is based upon the direct measurement of arterial input impedance rather than on extrapolations. It was also demonstrated that Doppler indices may not clearly reflect the impedance to flow in circumstances where acute insults to the fetus may also affect the heart rate. This is not surprising as the effect of the fetal heart rate on the Doppler indices is well recognized [10, 24]. It is clear that within certain limitations, analysis of the Doppler frequency shift waveform can be used to assess downstream hemodynamic state. The response of the waveform to asphyxia, however, appears to be less encouraging; the animal studies suggest that asphyxia-related changes in the Doppler indices occur as a relatively late phenomenon and only when the fetus is seriously compromised. Finally, the use of Doppler technology in an animal model has extended our ability to investigate fetal hemodynamics in a more comprehensive manner. This was exemplified in the nicotine study [23] in which fetal cerebral, umbilical, and uterine circulations were investigated with implanted Doppler probes.

References

1. Womersley JR (1957) The mathematical analysis of the arterial circulation in a state of oscillatory motion. Wright Air Development Centre (Technical report WADC-TR56-614)
2. McDonald D (1974) Blood flow in arteries. William and Wilkins, Baltimore, p 351
3. Hwang NHC (1977) An engineering survey of problems in cardiovascular flow dynamics and measurements. In: Hwang NHC, Norman A (eds) University Park Press, Baltimore
4. Maulik D, Yarlagadda P (1987) In vitro validation of Doppler waveform indices. In: Maulik D, McNellis D (eds) Perinatology Press, Ithaca, pp 257–282
5. Spencer JAD, Giussani DA, Moore PJ, Hanson MA (1991) In vitro validation of Doppler indices using blood and water. J Ultrasound Med 10:305
6. Maulik D, Yarlagadda P (1990) Hemodynamic validation of the Doppler indices: an in vitro study (Abstr). 3rd Congress of the International Perinatal Doppler Society, Malibu
7. Maulik D, Yarlagadda P, Nathanielsz PW, Figueroa JP (1989) Hemodyamic validation of Doppler assessment of fetoplacental circulation in a sheep model system. J Ultrasound Med 8:177
8. Trudinger BJ, Stevens D, Connelly A et al. (1987) Umbilical artery velocity waveform and placental resistance: the effects of embolization of the umbilical circulation. Am J Obstet Gynecol 157:1443
9. Morrow RJ, Adamson SL, Bull SB, Knox Ritchie JW (1989) Effect of placental embolization on the umbilical arterial velocity waveform in fetal sheep. Am J Obstet Gynecol 161:1055

10. Mires G, Dempster J, Patel NB et al. (1987) The effect of fetal heart rate on umbilical artery flow velocity waveform. Br J Obstet Gynaecol 94:665
11. Thompson RS, Trudinger JB, Cook CM (1985) Doppler ultrasound waveforms in the fetal umbilical artery: quantitative analysis technique. Ultrasound Med Biol 11:707
12. Maulik D, Downing GJ, Yarlagadda P (1990) Umbilical arterial Doppler indices in acute uteroplacental flow occlusion. Echocardiography 7:619
13. Clark KE, Durnwald M, Austin JE (1988) A model for studying chronic reduction in uterine blood flow in pregnant sheep. Am J Physiol 2242:H297
14. Downing GJ, Yarlagadda P, Maulik D (1991) Effects of acute hypoxemia on umbilical arterial Doppler indices in a fetal ovine model. Early Hum Dev 25:1
15. Downing GJ, Yarlagadda AP, Maulik D (1991) Comparison of the pulsatility index and input impedance parameters in a model of altered hemodynamics. J Ultrasound Med 10:317
16. Downing GJ, Maulik D (1991) Correlation of the pulsatility index (PI) with input impedance parameters during altered hemodynamics (Abstr). J Maternal Fetal Invest 1(2):114
17. Giles WB, Trudinger JB, Baird PJ (1985) Fetal umbilical artery flow velocity waveforms and placental resistance: pathologic correlation. Br J Obstet Gynaecol 92:31
18. Bracero LA, Beneck D, Kirshenbaum N, Peiffer M, Stalter M, Schulman H (1989) Doppler velocimetry and placental disease. Am J Obstet Gynecol 161:388
19. Nicolaides KH, Bilardo CM, Soothill PW et al. (1988) Absence of end-diastolic frequencies in umbilical artery: a sign of fetal hypoxia and acidosis. Br Med J 297:1026
20. Copel JA, Schlafer D, Wentworth R, Belanger K, Kreitzer L, Hobbins J, Nathanielsz P (1990) Does the umbilical artery systolic/diastolic ratio reflect flow or acidosis? Am J Obstet Gynecol 163:751
21. Malcus P (1991) Changes of the central blood flow during acute asphyxia in the fetal lamb (Abstr). J Maternal Fetal Invest 1:112
22. Schroder H, Huneke B (1991) Doppler and inductive blood flow velocimetry in the descending aorta of chronically instrumented fetal sheep. J Maternal Fetal Invest 1:111
23. Arbeille P, Bosc M, Vaillant MC, Tranquart F, Pourcelot L (1992) Nicotine-induced changes in the cerebral circulation in ovine fetuses. Exp Biophys Med (in press)
24. Yarlagadda P, Willoughby L, Maulik D (1989) Effect of fetal heart rate on umbilical arterial Doppler indices. J Ultrasound Med 8:215

III. Neonatal Assessment

Neuropathological Aspects of Perinatal Asphyxia and Ultrasound Imaging

E. Saliba and J. Laugier

The neuropathological aspects of perinatal asphyxia depend mainly on brain maturation. Germinal matrix intraventricular hemorrhage (IVH) and periventricular leukomalacia (PVL) are the most common neuropathological findings reported in premature infants, while various patterns of cerebral lesions occur in the mature neonate. Thus, lesions affecting gray matter of cortex, basal ganglia, and brain stem increase in importance with maturity.

Because of these fundamental differences, neuropathological aspects of perinatal asphyxia and the features demonstrable by ultrasound (US) will be discussed separately with regards to brain maturation.

Preterm Infants

Neuropathological Aspects of Intraventricular Hemorrhage

The origin of IVH characteristically is the subependymal germinal matrix located in the region of the head of the caudate nucleus. This cellular region produces glial cells which migrate during the midtrimester to the cerebral parenchyma and ganglia. The matrix undergoes a progressive decrease in size with increasing gestational age. The subependymal plate is most active between 24 and 34 weeks' gestation, this being the time when hemorrhage is most likely to occur. In the full-term infant, the subependymal matrix has almost entirely disappeared. It is supplied by a rich matrix of poorly supported fragile thin-walled capillaries, and it is the rupture of these vessels that initiates periventricular hemorrhage (PVH). Hemorrhage from the choroid plexus occurs in nearly 50% of infants with germinal matrix hemorhage and IVH [2]. Hemorrhage may remain confined to the subependymal matrix adjacent to the caudate nucleus, but frequently ruptures into the lateral ventricles (Fig. 1). In approximately 80% of cases with rupture, spread throughout the ventricular system occurs ([9], Fig. 2).

Several neuropathological states may occur as a consequence of IVH. They include germinal matrix destruction, possibly together with its glial precursor and periventricular hemorrhagic infarction (Fig. 3). Later, post-hemorrhagic hydrocephalus may occur. In about 20% of cases with IVH, a relatively large region of hemorrhagic necrosis in the periventricular white matter may occur (Fig. 4). Approximately 80% of periventricular infarctions are associated with large IVH [8]. Microscopic study of periventricular

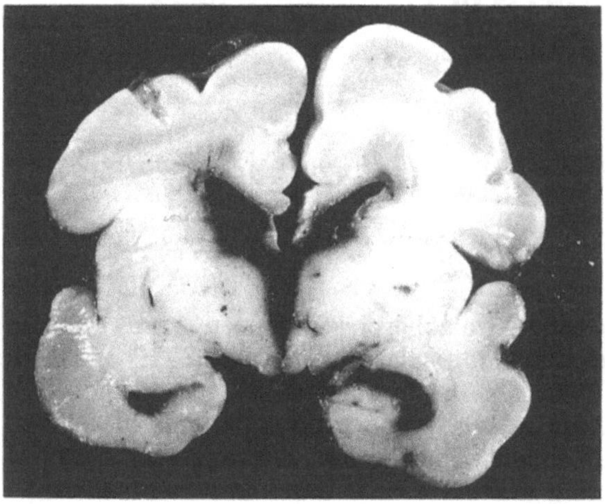

Fig. 1. Bilateral germinal matrix hemorrhage with rupture into the left lateral ventricle which is filled with clot

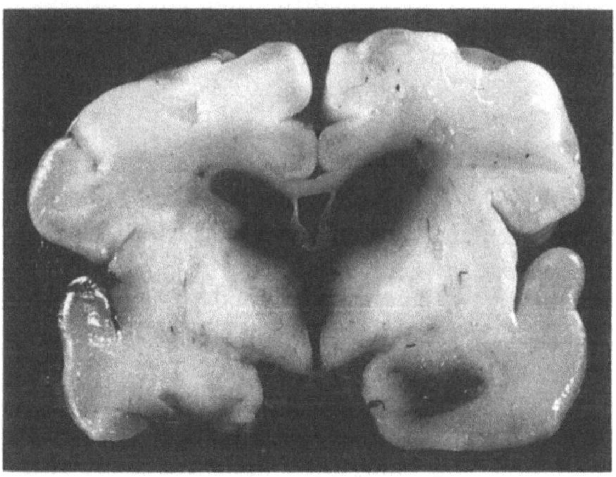

Fig. 2. Spread of intraventricular hemorrhage throughout the ventricular system. Note the small focal PVL corresponding to focal areas of gliosis

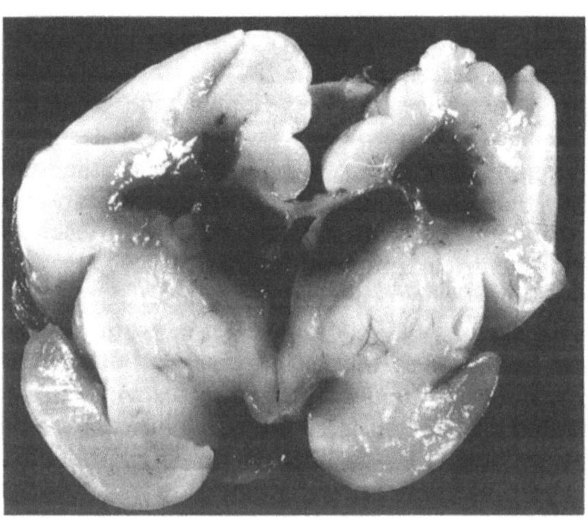

Fig. 3. Periventricular hemorrhagic infarction on both sides as intraventricular hemorrhage

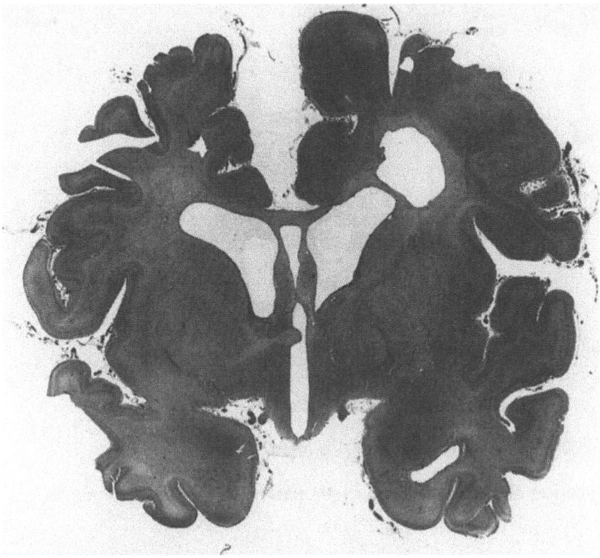

Fig. 4. Hemorrhagic intracerebral involvement with intraventricular hemorrhage: coronal section of cerebrum

hemorrhagic necrosis indicates that the lesion is a venous infarction, probably caused by the obstruction of the terminal veins by the intraventricular blood clot [5]. This lesion is neuropathologically distinguishable from secondary hemorrhage in PVL. However, distinction between these two lesions by US is very difficult.

Two lesions related to hypoxic–ischemic insults, that is, PVL and pontine neuronal necrosis, are common accompaniments of IVH. PVL has been reported in 75% of one series of infants who died with IVH [2]. In the same study approximately 46% of infants with IVH exhibited pontine neuronal necrosis.

Neuropathological Aspects of Periventricular Leukomalacia

Periventricular leukomalacia is an ischemic lesion which occurs primarily in premature infants, although it has occasionally been reported in full-term babies. A major advance in understanding this condition occurred in 1962 when Banker and Larroche [3] described the pathological changes in 51 infants. They introduced the name PVL to describe their observation of the periventricular "white spots" seen macroscopically. Perinatal hypoxia was noted in the clinical history of all their cases and they were the first to point out that the condition was vascular in origin. Macroscopical examination shows a typical distribution of the periventricular lesions involving the

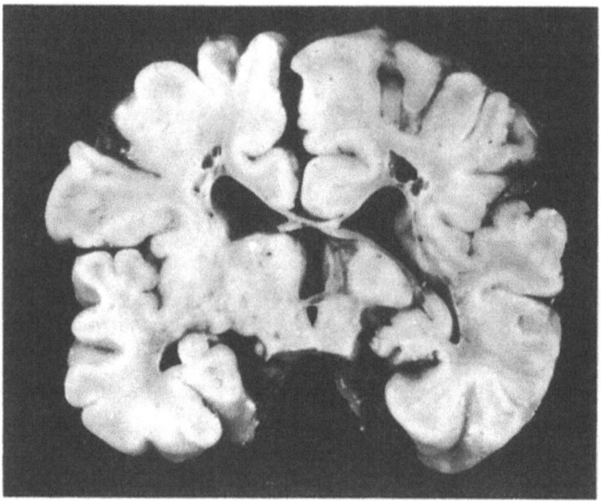

Fig. 5. Postmortem coronal section showing bilateral cystic periventricular leukomalacia

external angle of the lateral ventricles (centrum semiovale), the corona radiata, the optic radiation, and the temporal auditory radiation (Fig. 5).

Term Infants

Neuropathological Aspects Associated with Perinatal Asphyxia

Consistent pathological features associated with perinatal asphyxia in full-term infants have been recognized since the nineteenth century. Ulegyria and gyral scarring were first described by Bressler in 1899. In 1893, Anton reported lesions involving the basal ganglia and named them status marmoratus or *état marbré* [1]. Despite these observations, no single distinct or uniform pathological feature has been recognized following hypoxic–ischemic insult. The brain may be either globally affected with extensive swelling and necrosis or pathological lesions may be more focal and discrete. The majority of brain lesions observed after perinatal asphyxia have been explained either as ischemic injury based on vascular etiology or as selective neuronal necrosis. The latter is due to the vulnerability of certain regions of the brain to metabolic injury such as accumulation of glutamate.

Vascular Injury

Parasagittal cerebral injury refers to a lesion of the cerebral cortex and subcortical white matter. The injury is bilateral and occurs at the junction between the territories of the anterior, middle, and posterior cerebral arteries. The pathogenesis of parasagittal cerebral injury relates principally

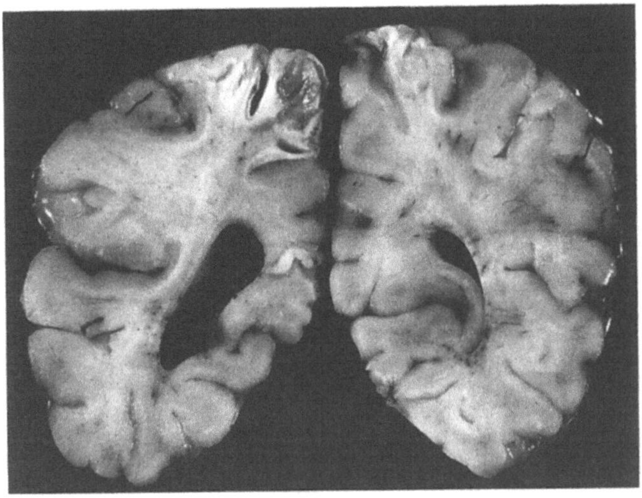

Fig. 6. Postmortem coronal section of cerebrum in an asphyxiated full-term infant showing necrosis in the parasagittal regions

to impaired cerebral perfusion. This injury is the principal ischemic lesion in full-term infants (Fig. 6). Experimental support for this hypothesis was provided in full-term fetal monkeys by Brann and Myers in 1975 [4], by inducing maternal hypotension for 1–5 h. This lesion has been diagnosed in vivo by means of technetium brain scans or positron emission tomography [9]. Seventeen infants with evidence of significant postasphyxial encephalopathy were studied within the first 5 days of life. There was a consistent symmetrical decrease in cerebral blood flow of up to 50% to the parasagittal region. These lesions were more evident posteriorly than anteriorly.

Selective Neuronal Necrosis

Selective neuronal necrosis is the most common variety of injury observed in neonatal hypoxic–ischemic encephalopathy. The major sites of predilection for this neuronal injury are the cerebral cortex, the basal ganglia, the brain stem, and the cerebellum.

Cerebral Cortex. Neurons of the cortex, particularly the pyramid cells of the hippocampus, are very vulnerable to hypoxic insult [7].

Basal Ganglia. Within the basal ganglia, neurons of the caudate, putamen, and globus pallidus are often injured. Status marmoratus is the classic neuropathological lesion affecting the basal ganglia. This lesion has a marble-like appearance and is due to abnormal myelination (Fig. 7).

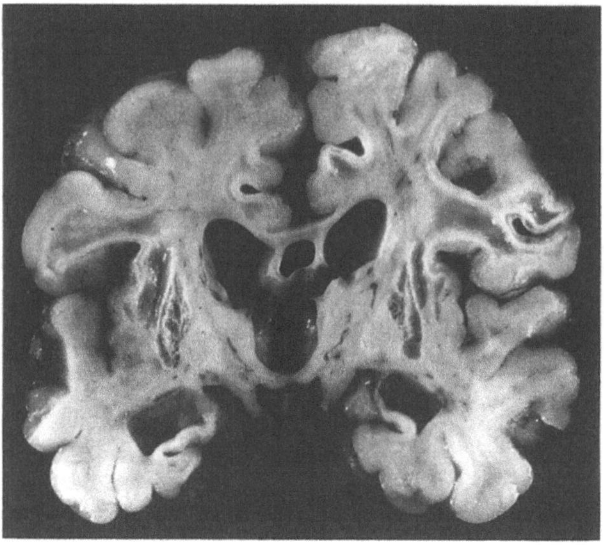

Fig. 7. Necrosis of the basal ganglia after severe perinatal asphyxia

Brain Stem. Lesions are rarely evident on macroscopic examination. The most vulnerable structures are the inferior colliculus formation, the lateral geniculate bodies, and the pontine nuclei.

Cerebellum. The cerebellum is especially vulnerable to hypoxic–ischemic insults. The dentate nuclei, the Purkinje's cells appear to be the most vulnerable.

Ultrasound Imaging

Hemorrhagic and Hypoxic–Ischemic Lesions in Preterm Infants

Subependymal Germinal Matrix Hemorrhage. The subependymal germinal matrix hemorrhage (SEH) appears on US as a highly echogenic lesion adjacent to, and inferior and lateral to, the lateral ventricle (Fig. 8). If the bulging, highly echogenic hematoma is present on both coronal and saggital views, one can be certain of the diagnosis. Typically, on the sagittal view the SEH will appear as a bulge just anterior to the termination of the choroid. However, pitfalls do exist. A hemorrhage may also occur in the anterior part of the choroid plexus, particularly among very immature infants, making it difficult to differentiate it from a SEH. As the hematoma resolves, the small hemorrhages disappear. In moderate-sized hematoma, a subependymal cyst may form.

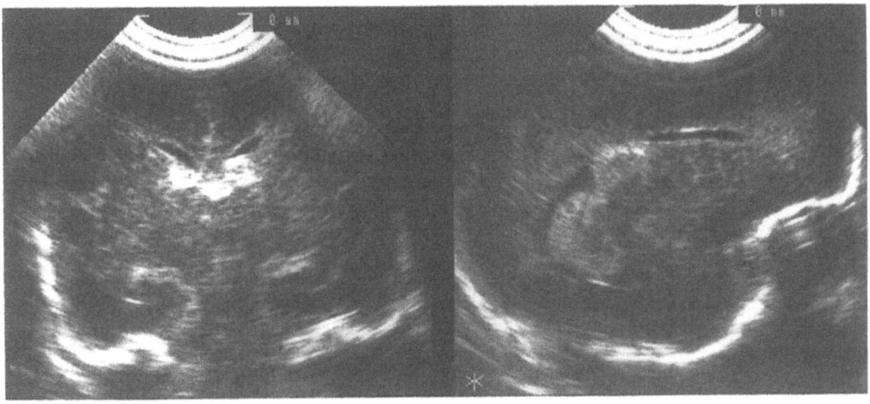

Fig. 8. Coronal scan showing a germinal-layer hemorrhage. *Right panel*, parasagittal scan demonstrating a germinal-layer hemorrhage

Intraventricular Hemorrhage. If clots form in the ventricles, intraventricular echoes will be seen. Initially it may be difficult to identify a ventricle if it is filled with clots that obscure the ventricular margin (Fig. 9). As the ventricles dilate, the diagnosis of IVH becomes easier, especially if echoes are seen within the occipital horns (Fig. 10). In some infants, IVH is suspected when there is a clot attached to an unusually bulky choroid plexus.

Periventricular Hemorrhagic Infarction. The lesion appears first as a homogeneous, highly echogenic intraparenchymal hematoma (Fig. 11). Over a period of 2–3 weeks, resolution of the lesion occurs with a reduction in the density of the echoes. A retraction of clots can be seen around the demarcated hemorrhage. Finally, in 2–3 months' time, a large porencephalic cyst, which is completely anechoic, replaces the parenchymal hematoma (Fig. 12).

Periventricular Leukomalacia. In US examination, the early stage of PVL is characterized by areas of increased echogenicity localized in the periventricular white matter at the external angle of the lateral ventricles. These areas have a characteristic triangular shape with the apex pointing towards the ventricle. The echogenicity should be demonstrated on both coronal and sagittal planes; it may be focal of diffuse (Fig. 13).

Within 1–3 weeks, areas of increased echogenicity resolve into echo-free areas. The final stage is a cyst which does not communicate with the ventricular system. The development of cysts will confirm the initial diagnosis of ischemic lesions (Fig. 14). These cysts remain visible for several weeks and then gradually resolve. Other patterns of PVL are frequently observed and include: (a) transient echogenicity or periventricular flares which refer to subtle, badly delineated hyperechogenicities. They are visible

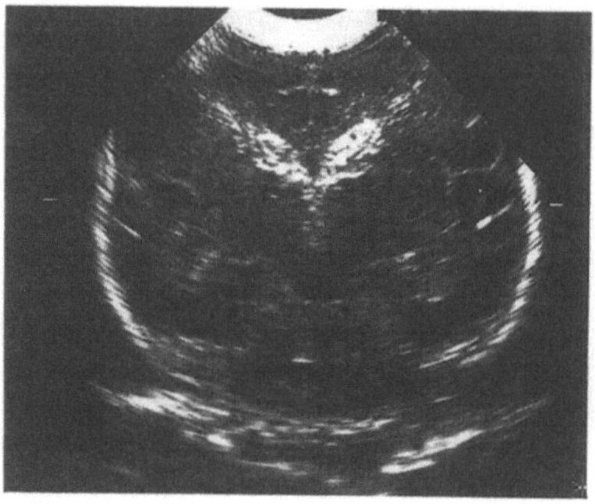

Fig. 9. Coronal scan showing moderate bilateral intraventricular hemorrhage

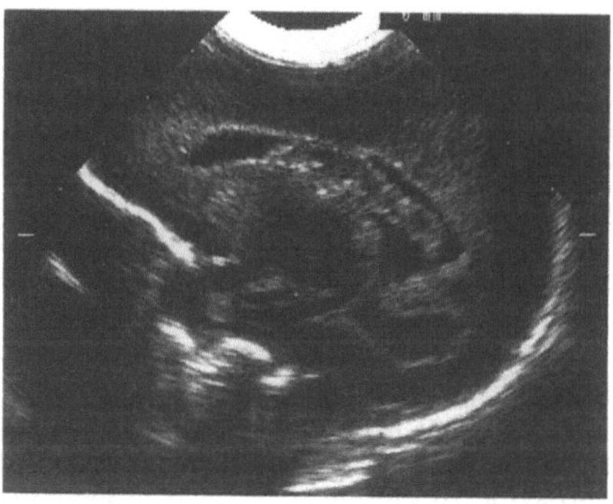

Fig. 10. Parasagittal scan demonstrating an intraventricular hemorrhage forming an echogenic clot adherent to the choroid plexus

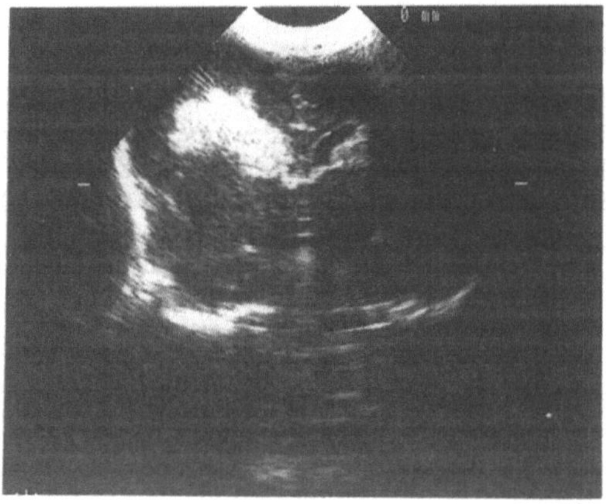

Fig. 11. Coronal scan of an intraparenchymal hemorrhage associated with an intraventricular hemorrhage

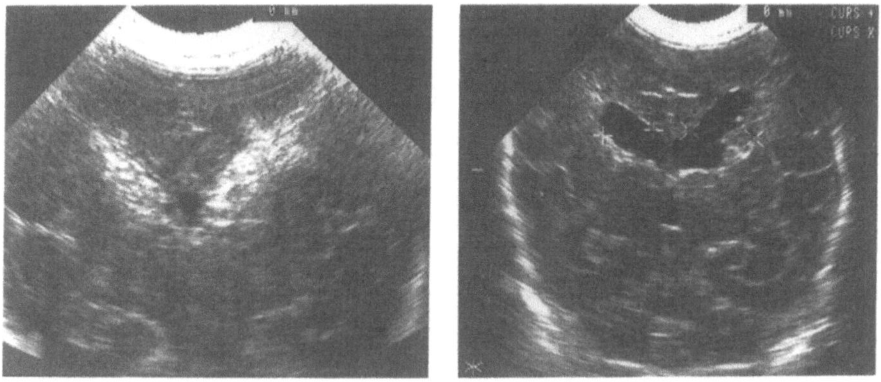

Fig. 12. *Left panel* hemorrhagic intracerebral involvment with evolution to porencephalic cyst. *Right panel*, at 3 weeks of age the cystic cavity becomes apparent as necrotic tissue and clot retract

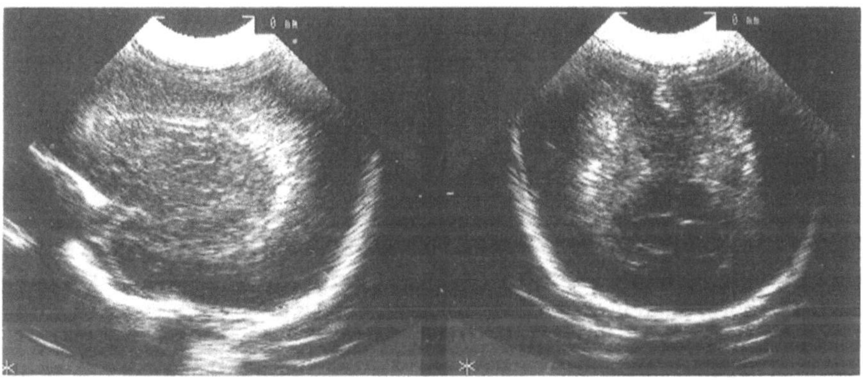

Fig. 13. Early stage of periventricular leukomalacia: bilateral, increased echogenicities visible on both parasagittal (*left panel*) and coronal views (*right panel*)

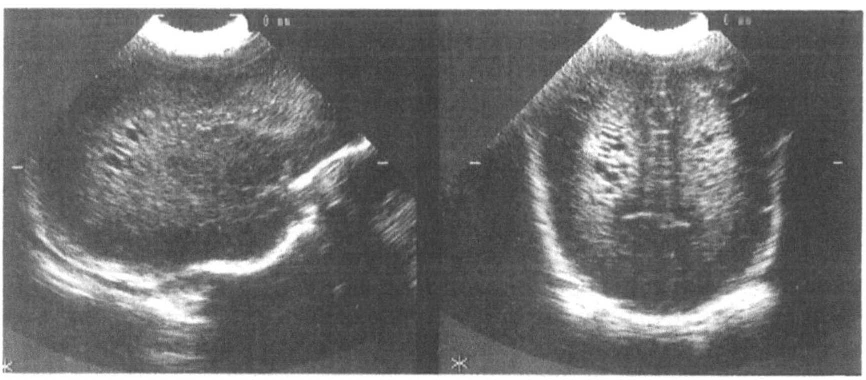

Fig. 14. Parasagittal and coronal scans showing late cystic stage of periventricular leukomalacia with multiple echo-free areas in the periventricular white matter

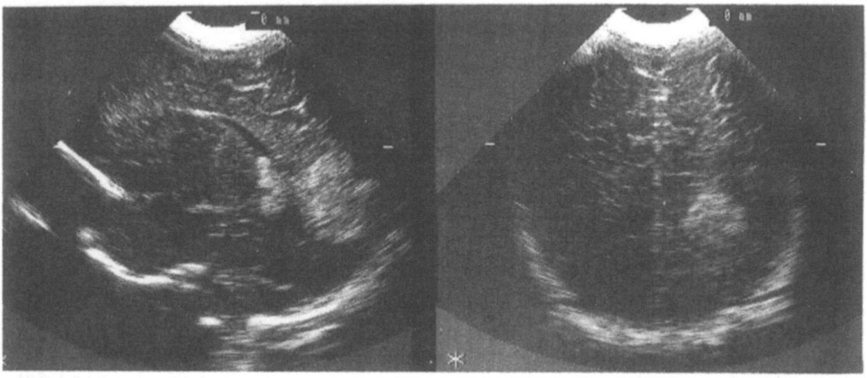

Fig. 15. Prolonged flare. Occipital echogenetic area in the periventricular white matter lasting for more than 2 weeks

on both coronal and parasagittal views and usually disappear within the first week of life. These patterns may represent transient and reversible ischemic injury and (b) prolonged flares refer to increased echogenicity in the periventricular region persisting for at least 2 weeks but not leading to cystic degeneration (Fig. 15). On microscopy, persistent echogenicity correspond to microcalcifications and glial scars.

Optimal Timing of Ultrasonography in Preterm Infants. The optimal timing of ultrasonography relates to the time of occurrence of IVH and ischemic lesions. Approximately 90% of IVH can be detected by a scan performed by the end of the third day of life. A second scan near the end of the first postnatal week is necessary to assess the maximal extent of IVH or to confirm PVL.

Hypoxic–Ischemic Lesions in the Term Infant

As described previously, various patterns of cerebral damage occur in the mature neonate. A variety of corresponding abnormalities seen on US examination have been reported. These may be focal or generalized.

Subcortical Leukomalacia. These lesions refer to a necrosis of white matter in the subcortical area at the watersheds between the major cerebral arteries and at watershed zones at the depth of the sulci in term infants. With advanced gestational age and the development of convolutions and gyri, a relatively avascular triangle forms in the white matter at the depth of the sulcus. This could explain the different sites of ischemic lesions among term infants. The US patterns are identical to those in premature infants with increased echogenicity in the early stage and cyst formation later (Fig. 16).

More generalized echodensities throughout the brain are also described after severe perinatal asphyxia. In these instances there is a partial or

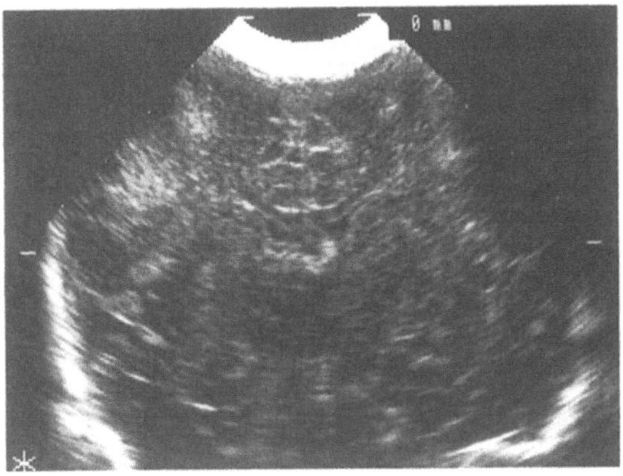

Fig. 16. Subcortical leukomalacia occurring after severe birth asphyxia

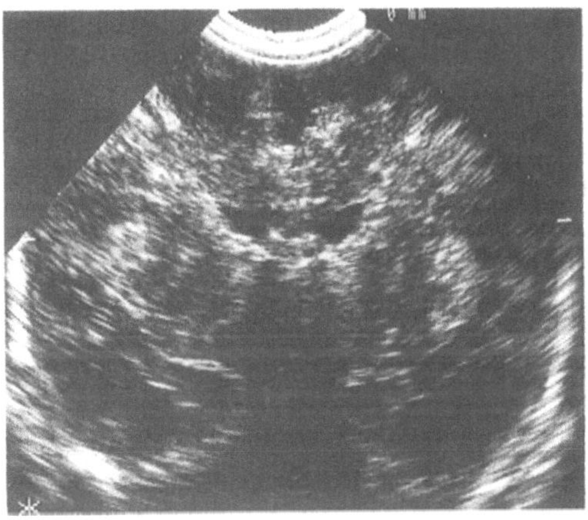

Fig. 17. Coronal scan showing diffuse increased echogenicities following severe perinatal asphyxia

complete obliteration of anatomical structures which are normally seen clearly. This condition is probably due to cerebral edema (Fig. 17).

Focal and multifocal ischemia are related to necrosis that occurs within the distribution of the major vessels. The involvement of the middle cerebral artery occurs in approximately 50% of cases. Necrosis of these lesions is associated with dissolution of the tissue and cavity formation (Fig. 18).

Lesions of the Basal Ganglia and Thalamus. Bilateral echodensity of the thalamus and putamen has been demonstrated by real-time US in infants who suffered severe perinatal asphyxia [6]. These US patterns were correlated with autopsy findings of hemorrhagic necrosis of the thalamus and

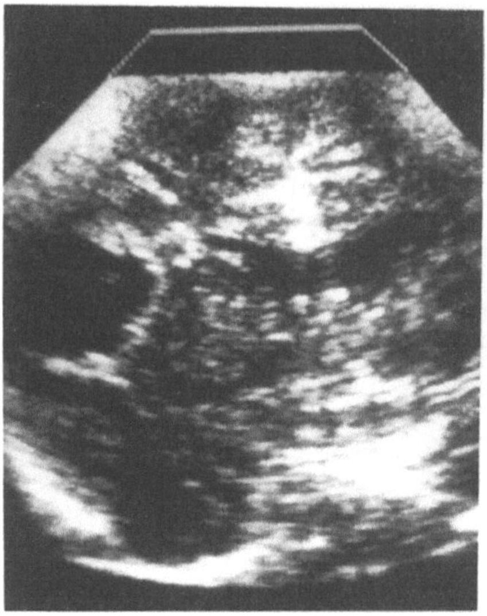

Fig. 18. Middle cerebral artery infarction in term infant. Coronal view showing a well-circumscribed echo-free area in the region supplied by the left middle cerebral artery

basal ganglia. These lesions have also been observed in infants who survived with significant neurodevelopmental sequelae.

Brain-stem nuclei and other sites involved in hypoxic–ischemic insults in term newborns are incompletely visualized by US, and damage to these areas cannot be detected. Thus, an apparently normal brain at ultrasonography does not preclude a neurodevelopmental impairment. In our experience, computed tomography (CT) scan seems to be a much better method than US for assessing deep cerebral structures in asphyxiated full-term neonates.

Although there are constant technological advances in exploring the neonatal brain, US imaging is still the first choice for the routine assessment of the brain in the premature newborn. It is of major value for the diagnosis of IVH. It also may be of major value for the diagnosis of PVL. However, in the first weeks of life there are limitations in diagnosing ischemic injury. This relates to the inability of US to accurately identify mild injury. Another factor is the poor specificity of increased echogenicity in distinguishing between hemorrhagic and ischemic injury in cerebral parenchyma.

In the term newborn, ultrasonography is of limited value in assessing hypoxic–ischemic parenchymal injury. This is due to the limited technology available (i.e., deep structure) and the subjectivity involved in interpreting the images. Furthermore, these lesions are unpredictable, may appear several days after the insult, and are not always directly related to the severity of the clinical course. Disturbances in brain metabolism and circulation are most commonly present well before the occurrence of morphological changes. CT scanning or US appear to have a better predictive value only after the first week of life. The prognosis of severely asphyxiated term newborns should be therefore based on the global appreciation of the following modalities: clinical examinations, brain imaging, cerebral hemodynamics (i.e., Doppler, single photon emission computed tomography), and EEG.

References

1. Anton G (1893) Über die Betheiligung der basalen Gehirnganglien bei Bewegungsstörungen und insbesondere bei der Chorea; mit Demonstrationen von Gehirnschnitten. Wien Klin Wochenschr 6:859–861
2. Armstrong DL, Sauls CD, Goddard-Finegold J (1987) Neuropathologic findings in short-term survivors of intraventricular hemorrhage. Am J Dis Child 141:617–621
3. Banker BQ, Larroche JC (1962) Periventricular leukomalacia of infancy. Arch Neurol 7:32–50
4. Brann AW, Myers RE (1975) Central nervous system findings in the newborn monkey following severe in utero partial asphyxia. Neurology (Minneap) 25:327
5. Gould SJ, Howard S, Hope PL, Reynolds EOR (1987) Periventricular intraparenchymal cerebral haemorrhage in preterm infants: the role of venous infarction. J Pathol 151:197–202
6. Kreusser KL, Schmidt RE, Shackelford GD, Volpe JJ (1984) Value of ultrasound for identification of acute hemorrhagic necrosis of thalamus and basal ganglia in an asphyxiated term infant. Ann Neurol 16:361–363
7. Larroche JC (1984) Perinatal brain damage. In: Adams JH, Corsellis JAN, Duchen LW (eds) Greenfield's neuropathology, 4th edn. Arnold, London
8. McMenamin JB, Shackelford GD, Volpe JJ (1984) Outcome of neonatal intraventricular hemorrhage with periventricular echodense lesions. Ann Neurol 15:285–290
9. Volpe JJ, Herscovitch P, Perlman JM, Kreusser KL, Raichle ME (1985) Positron emission tomography in the asphyxiated term newborn: parasagittal impairment of cerebral blood flow. Ann Neurol 17:287–296

Magnetic Resonance Imaging and Hypoxic – Ischemic Brain Injuries

J. Haddad and J. Messer

Introduction

Hypoxic–ischemic brain injuries are a major problem in perinatology and are responsible for later disabilities in children. Two kinds of lesions are encountered and could be interrelated: ischemia and hemorrhage. The initiating stimulus for hypoxic–ischemic injury is impaired cerebral blood flow (CBF) autoregulation (Volpe 1987). It is important to carry out the imaging of central nervous system (CNS) injuries early in the neonatal period. Indeed, nowadays it is essential to know the nature, the location, and the extent of the lesions since the decision to continue treatment or not depends on this information.

In the past two decades computed tomography (CT) scan followed by ultrasound (US) was used to identify CNS injuries. Much information regarding diagnosis and prognosis has been drawn from those techniques. However, there are limitations using these methods relating to: (a) the lack of specificity of certain radiological signs such as hypodensity on CT scan (ischemia or normal feature of immature brain) or hyperechogenicity on US (ischemia, hemorrhage, gliosis, tumor, or lipoma) and (b) CNS injuries may lie outside the range of the US probe. Difficulties in diagnosing posterior fossa lesions are also a limitation. Magnetic resonance imaging (MRI) is a new noninvasive method of imaging the CNS which uses the magnetism of the hydrogen nucleus (proton) present in virtually all biological materials. This technique has gained interest in the neonatal period because of the lack of ionization and high-quality images. However, even with the powerful properties of MRI, one must not neglect the fact that the neonate is removed from the neonatal ward and the benefit of such an examination must be studied with regard to the timing of the examination in the neonatal period. In this chapter, we will see to what extent MRI is useful in diagnosing and providing a prognosis of ischemic and hemorrhagic lesions in the neonate.

Methods

Sedation and Patient Handling

Satisfactory sedation is essential for a proper examination and for improving the quality of the images. The sedation protocol we use is dependent on the age of the infant: In premature babies, the examination was performed after oral feeding, in full-term babies, chloral hydrate was given via a nasogastric tube (50 mg/kg) 30 min before the exam, and in early infancy (<2 years), 50–100 mg/kg chloral hydrate was required.

Neonates were wrapped on a special nonmagnetic bed and were moved to and from the magnetic resonance (MR) facility. Older infants did not require this type of handling. In addition to direct monitoring by touching, seeing, and hearing the patient, we monitored vital parameters (electrocardiogram monitor, pulse oximeter "CRITIKON", and blood-pressure monitoring). Infants needing ventilation were ventilated by hand with a breathing circuit devoid of ferromagnetic material or by a "Monal ventilator." Sources of oxygen and suction were available and a neonatologist was always present in the magnet room. After careful monitoring of the patients, no adverse effects were noted. A new sophisticated material for the caring of neonates and children who need resuscitation is now available. This material which assembles incubator, ventilator, blood-gas monitoring as well as arterial-tension monitoring has been investigated by Boesch and Martin (1988) in Zurich and Bern, Switzerland.

Magnetic Resonance Parameters

In order to be familiar with MRI, one must be aware of some of the terms used. The strength of magnetic fields can range from 0.1 to 2.35 T (1 T = 10 000 G). For instance, differences regarding hemorrhagic signals are observed between low and high fields (see Hemorrhage, this chapter; Zuerrer et al. 1991).

The sequences used in our institution in neonates are mainly spin echo (SE). Relaxation times (T1 and T2) are very important determinants of image contrast. T1-weighted images (5–7 min) are characterized by a repetition time (TR) of 400–600 ms with an echo time (TE) of 20–32 ms. Those sequences show good anatomical details. T2-weighted (10–15 min) images are characterized by a TR of 2000–3200 ms with a TE of 40–120 ms. They allow good recognition of different parenchymal components of the CNS.

Gradient echo images use a TR of 400–600 ms, a TE of 15 ms with a flip angle of 90°. It exhibits images similar to SET1 but with a lesser acquisition time. With a flip angle of 25°, images are similar to SET2.

Interpretation of MRI needs a perfect knowledge of CNS anatomy and injuries in neonates, as well as knowledge of the sequences used. An abnormal signal exhibited on MRI is always related to the signal of the white matter in the sequence used. It could be isosignal or isointense, hyposignal or hypointense, or hypersignal or hyperintense. Whether T1 or T2 is used, the signal displayed is different.

Ischemia

Pathophysiology and Neuropathology

Ischemia is one of the major neurologic problems of the perinatal period in both premature and full-term babies. It also constitutes a major determinant of neurologic disabilities in infancy. Impaired autoregulation of CBF seems to be the most important factor in the genesis of ischemia and hemorrhage. This impairment is mainly the result of perinatal asphyxia (hypoxia, hypercapnia). Other factors can also lead to a disturbance of CBF, especially in the premature baby. These include hypocapnia, patient handling and suctioning, apnea and bradycardia. In this section, we shall review several patterns of ischemic injury in neonates with respect to Volpe's classification of neuropathologic features (Volpe 1987):

- Selective neuronal necrosis refers to necrosis of neurons of the cerebral cortex, diencephalon, basal ganglia, midbrain, pons, medulla, or cerebellum.
- Status marmoratus of basal ganglia and thalami is characterized by hemorrhagic necrosis and the occurrence of hypermyelination in early infancy.
- Parasagittal cerebral injury refers to a lesion of the cerebral cortex and subcortical white matter. The distribution of this injury is intimately related to the border zones and end fields of the anterior, posterior, and middle cerebral arteries (parasagittal cerebral convexity).
- Cerebral leukomalacia refers to a necrosis of white matter either adjacent to the external angle of the lateral ventricles (periventricular leukomalacia, PVL) or in the subcortical area at the depth of the sulcus (subcortical leukomalacia, SL). Abnormal vascular development appears to be the main predisposing factor leading to ischemia in these two particular regions. Indeed, in the premature baby, distal fields in the periventricular arterial border and end zones determine a watershed area vulnerable to systemic hypotension, hypoxemia, hypercapnia, etc. Lesions occurring within these areas are classified as PVL (Dereuck et al. 1972). With advanced gestational age and the development of convolutions and gyri, the depth of the sulcus reveals a relatively avascular triangular area vulnerable to the same factors mentioned above and predisposing to SL (Takashima and Tanaka 1978). However, recent studies by Kuban and Gilles (1985) have provided new features on the histogenesis and regional development of telencephalic microvasculature. The result does not confirm either the end arterial bed (watershed area) or the subsulcal hypovascular region. Although, the periventricular area and subsulcal area are supplied by arteries, the latter seems to be devoid of apparent muscularis – an important feature of vasoregulation until the final weeks of gestation. Thus, assumptions about the genesis of neonatal white matter lesions may rest on uncertain grounds.
- Focal or multifocal ischemia is related to necrosis that occurs within the distribution of the major cerebral vessels. Necrosis is associated with

dissolution of the tissue and cavity formation. The terms of porencephaly (single cavity), hydranencephaly (hemispheres reduced to cerebrospinal fluid-filled sacs), or multicystic encephalomalacia (multiple cavities of necrosis) are used to describe those patterns of ischemia.

Even though the neuropathologic features are well classified, as mentioned above, distinction between these ischemic lesions is sometimes difficult and in some instances they seem to be interrelated. Other factors, such as gestational age, must be taken into consideration. With either advanced gestational age or in the full-term baby, selective neuronal necrosis, parasagittal cerebral injury, status marmoratus, and focal or multifocal ischemia seem to be more frequent; in premature babies, cerebral leukomalacia is the most common lesion. Thus, to illustrate ischemia in neonates by MRI, we have adopted the following classification: (1) MRI of cerebral ischemia following perinatal asphyxia (abnormal cardiotocography, meconium-stained aspiration, Apgar <5 at 5 min, abnormal neurological signs, seizures, coma, etc.) and (2) MRI of cerebral leukomalacia.

MRI of Cerebral Ischemia

Ischemia exhibits different signals whether SET1 or SET2 is used. Generally, on SET1, ischemia appears as hypointense with a later cystic appearance. On SET2, ischemia shows a hyperintensity which persists. However, in premature infants, deep white matter lesions seem to be difficult to interpret since immature white matter appears similar to ischemia on SET1 or SET2, Despite this, we will attempt to define in this chapter MR findings in acute, subacute, and chronic phases of perinatal asphyxia. Our experience is based on 14 neonates with a gestational age (GA) of 33–42 weeks who suffered birth asphyxia and in whom MRI was performed. Eight of them showed ischemic lesions on MRI (Table 1) and a delay in their neurologic development. The remainder developed normally.

Acute Phase (0–24 h)

Early detection of ischemia following the asphyxial event is rather difficult in the neonate. In our experience, neither SET1 or SET2 have allowed us to depict abnormalities on MRI with in the first 24 h following asphyxia, whereas in adults MR changes have been shown to occur as early as 1–2 h after the event (Yuh et al. 1990). In fact, in adults an abnormal vascular finding is seen, the so-called absence of flow void phenomenon. This latter phenomenon is due to normal arterial structures being devoid of intraluminal signals due to rapid or turbulent flow. In ischemia, possible causes of absent flow are slow flow or occlusion. Another early finding in adults is arterial enhancement after the injection of an MR contrast agent (gadopentetic acid) (Yuh et al. 1990). This might be due to the accentuation of flow derangement. SET1 morphological changes, SET2 signal changes, and SET1 signal changes occur after. Those modifications were related to cytotoxic edema and blood-brain barrier breakdown.

Table 1. Perinatal data and patient outcome

Patient	GA (weeks)	Signs	MRI[a]	Outcome
1	33	Coma seizures	D2 (basal ganglia injury)	Died at D3
2	35	Coma seizures	D5 (multifocal ischemia)	Died at D5
3	35	Hypotonia seizures	D28 (parasagittal cerebral lesion)	Severely handicapped epilepsy
4	40	Hypotonia seizures	D10 (multifocal ischemia)	Severely handicapped epilepsy
5	40	Hypotonia seizures	D16 (multifocal ischemia)	Severely handicapped epilepsy
6	43	Seizures	D15 (multifocal ischemia)	Severely handicapped epilepsy
7	40	Seizures	D20 (focal ischemia)	
8	40	Coma seizures	D16 (parasagittal cerebral lesions)	Delayed development

GA, gestational age at birth; D, day.
[a] Day after birth on which MRI was performed and results.

Discrepancies between adults and neonates in detecting an early feature of ischemia may lie in the pathophysiological event leading to ischemic lesions. In adults, the initiating phenomenon is arterial occlusion whereas in perinatal asphyxia, cerebral hypoxia and hypoperfusion are implicated (Volpe 1987). Other explanations could account for these discrepancies. First, ischemic lesions in the immature brain seem to be difficult to assess; confusion arises on the signal exhibited by ischemia in the immature brain. Second, the use of nonspecific MR sequences; in an adult rat experimental model of ischemia, early detection of insults have been realized by diffusion-weighted MRI (Baker et al. 1991). As early as 45 min after arterial occlusion, a hyperintensity was observed on diffusion-weighted images over T2, whereas on SET2 hyperintensity was delayed for 1.5–3.0 h. Diffusion images are sensitive to the microscopic motion of water protons. A reduction in CBF would attenuate brain pulsatile motion, thereby leading to a decrease in apparent proton diffusion-weighted hyperintensity in the ischemic region. Hyperintensity could also be related to postischemic hypothermia, increased tissue osmolarity, or cytotoxic edema. Even though diffusion-weighted MRI enables us to detect earlier ischemic lesions, no similar evidence is yet available on neonatal brains. A single study has reported anisotropically restricted diffusion images in neonates, but not in the acute stage of asphyxia (Rutherford et al. 1991).

Unlike in adults, perfusion imaging of acute cerebral ischemia by using a paramagnetic contrast agent is of limited value in neonates, since the enhancement is dependent on blood-brain barrier breakdown which is delayed after the ischemic event and is not uniformly present.

Combined MRI and spectroscopy seems to be of great interest and may be particularly important in assessing the potential for the reversibility of ischemic brain damage. Indeed, image-guide 31 Pand H1 MR may help to determine whether there are threshold phosphocreatine/inorganic phosphorus (PCR/Pi) or lactate/*N*-acetylaspartate ratios above which injury is irreversible and will open new horizons on a possible treatable condition in acute stages of perinatal asphyxia.

In conclusion, in acute phase ischemia, there is no benefit in removing a newborn undergoing intensive care. Imaging the brain in early ischemic stage (if needed) could be performed by an US bedside examination, unless the detection of hemorrhage is necessary and requires a CT scan or MRI. Prognostic factors could be drawn from other investigations such as EEG and hemodynamic assessment. However, the future seems promising with the development of new MR sequences and combined MRI and spectroscopy.

Subacute Phase of Ischemia (24 h–5 days)

Two features are detected in this phase by MRI: cerebral edema and ischemic lesions.

Cerebral Edema. Brain swelling is not a prominent feature of perinatal asphyxia or of a dominant lesion. It occurs mainly in neonates after 36 weeks' GA and its severity is related to concomittant tissue necrosis (Volpe 1987). It is considered to be a cytotoxic edema (abnormal accumulation of intracellular water as a consequence of the decreased availibility of adenosine triphosphate – ATP – and the resultant dysfunction of Na–K–ATP pump).

Cerebral edema shows several features on MRI (Fig. 1): slit-like ventricles – this radiological sign could also be detected by US or CT scan. Furthermore it does not constitute a specific sign of cerebral edema since normal term neonates could exhibit a slit-like ventricle appearance (Siegel et al. 1984); morphological changes caused by tissue swelling were found on SET2, These consist of an abnormal appearance of the white matter with high-signal intensities. This could be related to the accumulation of intracellular water (free water). Convolutions and gyri are not easy to identify. We do think that it may be caused by structural distortions.

Ischemic Lesions. These could be identified as early as 2 days following the asphyxial event. To illustrate the pattern of lesions encountered in the subacute phase, we have chosen two cases in which MRI was performed at 2- and 5-days old, respectively (cases 1 and 2, Table 1).

In the first patient (33 weeks' GA, Fig, 2a,b), ischemia was localized to the right occipital lobe with cortical necrosis. Ischemia exhibited a low-intensity signal more pronounced than that of white matter on SET1. By contrast, a high-intensity signal was noted in the basal ganglia. On SET1, the high signal intensity is related to hemorrhage. In this patient, we deter-

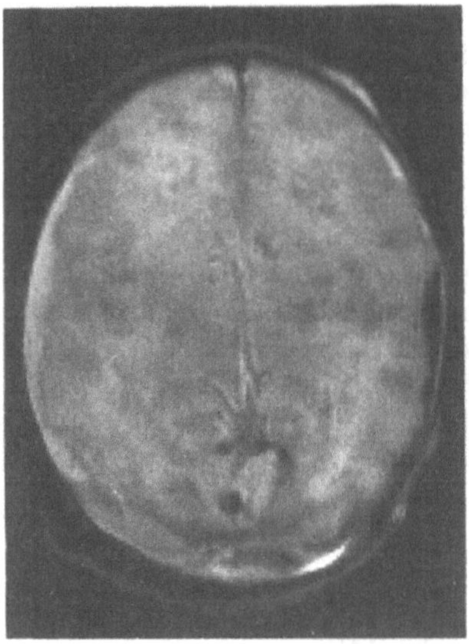

Fig. 1. MRI at day 3 in a newborn of 40 weeks' GA, SET2, 1800/120 ms, axial view. Note slit-like ventricles and morphological changes: unrecognizable convolution, distortion of structures

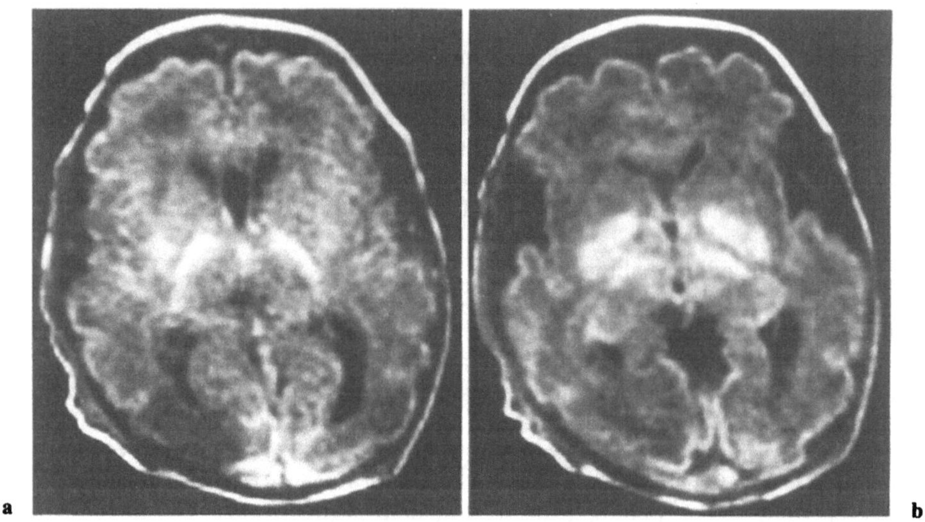

Fig. 2a,b. MRI in a newborn of 33 weeks' GA at **a** day 2 and **b** day 5. SET1, 500/28 ms, axial view. Note cerebral ischemia, hyposignal on the right occipital lobe with cortical necrosis associated with hypersignal on lenticular nuclei (basal ganglia injury)

mined that ischemia was associated with basal ganglia injury (hemorrhagic necrosis).

In the second patient (35 weeks' GA), several brain disorders could be noted and right temporal infarction was considered (Fig. 3a,b). The infarct exhibited a low signal intensity more pronounced than that of the white

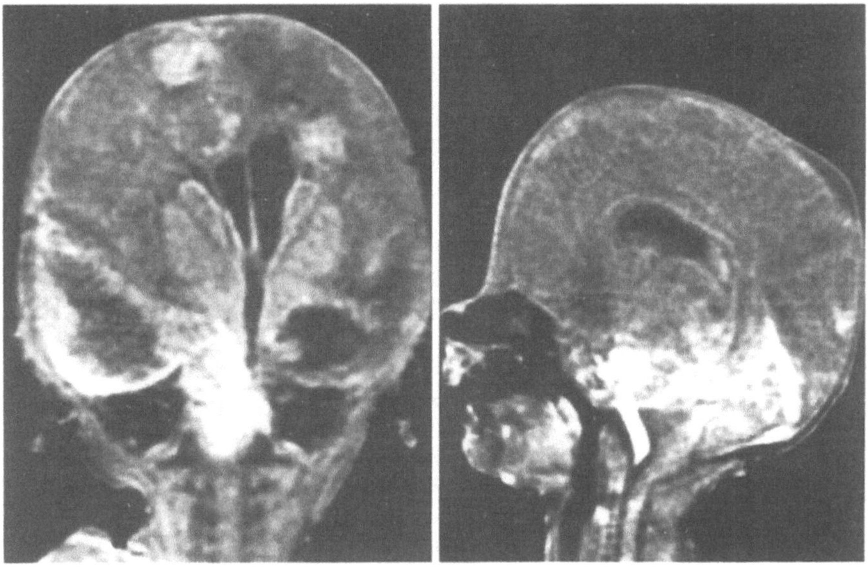

a b

Fig. 3a,b. MRI in a newborn of 35 weeks' GA. GE 450/12 ms 90°. **a** Sagittal view: hemorrhage in basal cisterna and posterior fossa (hypersignal). **b** Coronal view: multiple parenchymal involvement. Note midline shift and temporal infarction

matter on gradient echo images. We noted also a midline shift and compression of the lateral ventricle. This seems to occur in the early phase of cerebral infarction (Haddad et al. 1991). It was also obvious that the white matter signal was abnormal based on the postconceptional age. Thus, the cerebral parenchyma had undergone dissolution (multifocal ischemia?). Besides necrosis, hemorrhagic parenchymal involvement was identified in the left periventricular area and in the right subcortical parietal parenchyma. Hemorrhage was also seen in both the magna cisterna and the right subdural space. We determined that focal ischemia preceded multifocal ischemia and was associated with hemorrhage. In the two patients, US and CT scan showed abnormalities. However, the injuries diagnosed were less obvious regarding their nature and topography.

In conclusion, several ischemic patterns could be identified in the subacute phase of perinatal asphyxia. The hemorrhagic component is easily identified. Thus, MRI could constitute a useful tool for diagnosis and prognosis. Identifying such important lesions must lead to the decision to discontinue intensive care. In the subacute phase of asphyxia, this might be possible since neonates are ventilator dependent.

Chronic Phase (Less or more than 1 Week)

At this phase, it seems difficult to raise the problem of withdrawal since neonates are generally ventilator independent. Thus MRI is useful in

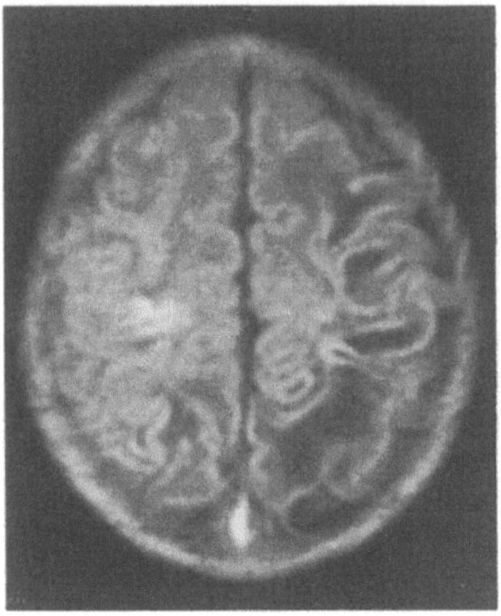

Fig. 4. MRI in a near-term infant. GE 450/12 ms 90°, axial view showing parasagittal cerebral lesions. Note the hyposignal on the right hemisphere

elaborating the cartography of ischemic lesions. Patterns of ischemia encountered include:

1. *Parasagittal cerebral lesions* (Fig. 4; cases 3 and 8, Table 1). This lesion is observed in the near-term infant. The areas affected are in the border zones between the end fields of the major cerebral arteries and thus a necrosis of the cortical and adjacent parenchyma occurs which is most marked in the posterior cerebrum. The reason why premature brains are spared is related to the presence of meningeal interarterial anastomoses between the three major cerebral arteries. Parasagittal cerebral lesions seem to be difficult to assess with either US or CT scan. MRI has proved to be useful in diagnosing such lesions especially with multiplan images (Fig. 5).

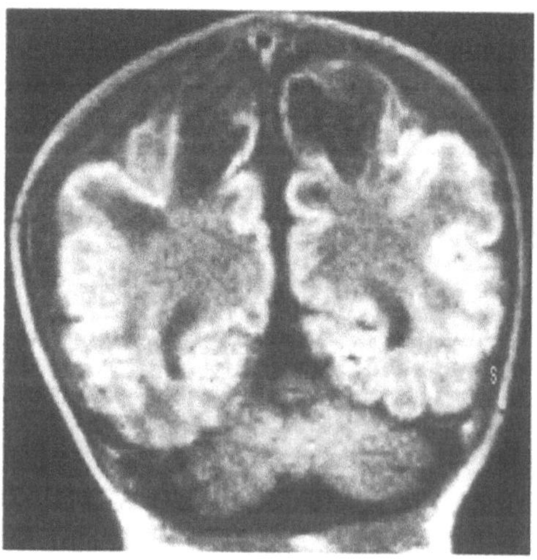

Fig. 5. MRI in a near-term infant. GE 450/12 ms 90°, coronal view showing parasagittal cerebral lesions. Note the hyposignal in right and left hemispheres

2. *Multicystic encephalomalacia* refers to multiple cavitations within the distribution of major cerebral vessels. It is mainly caused by generalized systemic circulatory insufficiency. In our experience this lesion has been observed as early as 10 days following the asphyxial event (Fig. 6; case 4, Table 1). It can also be detected by US and CT scan. However, delimitation of the lesions is better defined by MRI.
3. *Basal ganglia injury* (Figs. 7, 8). This type of lesion refers to status marmoratus, the characteristic feature being hypermyelination. How-

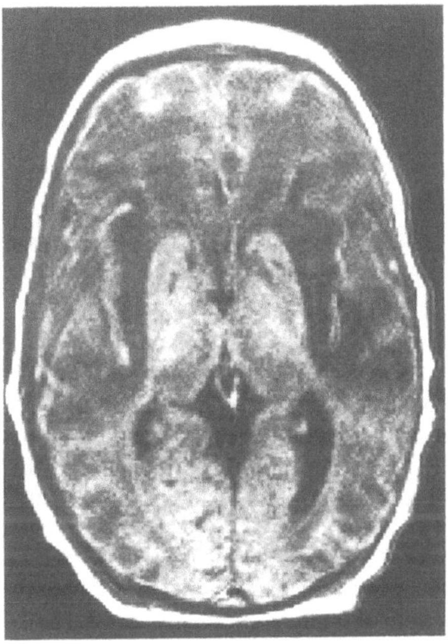

Fig. 6. SET1, 600/32 ms, axial view showing multicystic encephalomalacia. Note diffuse hyposignal within cerebral parenchyma

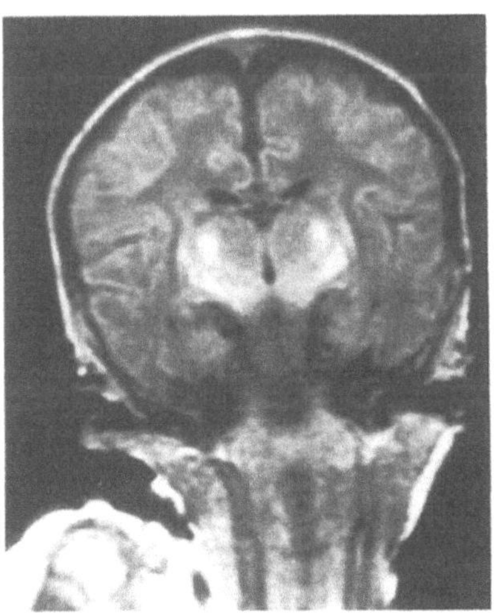

Fig. 7. SET1, 600/32 ms, coronal view showing basal ganglia injury. Note their hypersignal (hemorrhagic necrosis)

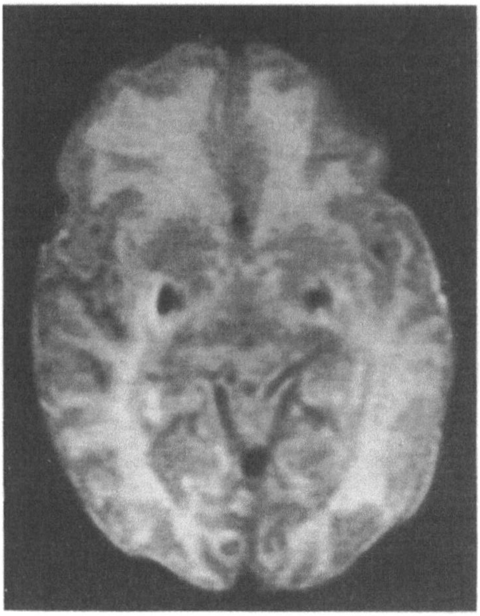

Fig. 8. SET1, 2200/120 ms, axial view showing basal ganglia injury. Note their hyposignal relative to hemorrhagic necrosis

ever, hypermyelination is not seen until the end of the first year of life and gives a marbled appearance to basal ganglia (*état marbré*). In the neonatal period, MRI features of basal ganglia injury seems to be hemorrhagic necrosis (Fig. 8) rather than capillary proliferation. Furthermore, this lesion, which is apparently more common in term infants, is also encountered in preterm neonates (Fig. 2a,b).

4. *Focal ischemia*. This refers to necrosis within an area supplied by one of the major cerebral arteries. Besides perinatal asphyxia, focal ischemia could be caused by cerebrovascular insufficiency (vascular maldevelopment, vasculopathy, embolus, thrombus) or inflammation. Proximal or distal branches of cerebral arteries could be involved (e.g., middle cerebral antery). Distal infarction of the middle cerebral antery has a better prognosis, whereas proximal infarction is responsible for contralateral hemiparesis (Haddad et al. 1991). MRI is useful in delimitating proximal from distal territories of main cerebral arteries (Figs. 9, 10).

Conclusion (Fig. 11)

MRI seems to be the imaging modality of choice for infants with severe perinatal asphyxia, although it must be performed at a safe time for the neonates. Thus, we believe that it constitutes a useful tool in assessing subacute and chronic phases of perinatal asphyxia leading to a better delimitation and topography of CNS injuries. Furthermore, it provides much information regarding the prognosis. Indeed, all neonates who exhibited severe cerebral edema and/or ischemic injuries such as parasagittal cerebral lesions, multicystic encephalomalacia, or basal ganglia injuries died

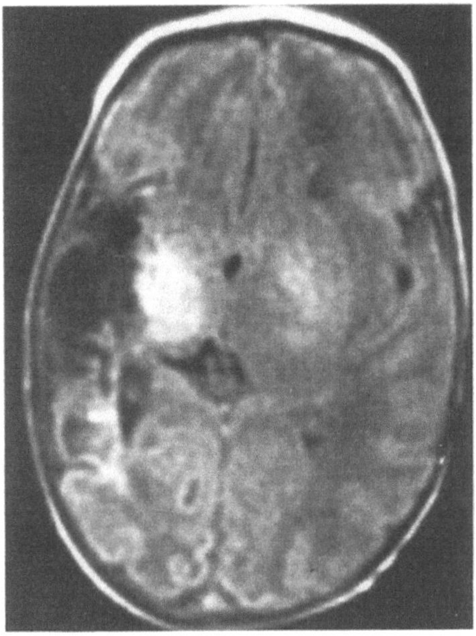

Fig. 9. GE 450/12 ms 90°, axial view showing cerebral infarction focal ischemia of proximal territory of MCA. Note the hyposignal on the left hemisphere and the associated hemorrhage (hypersignal)

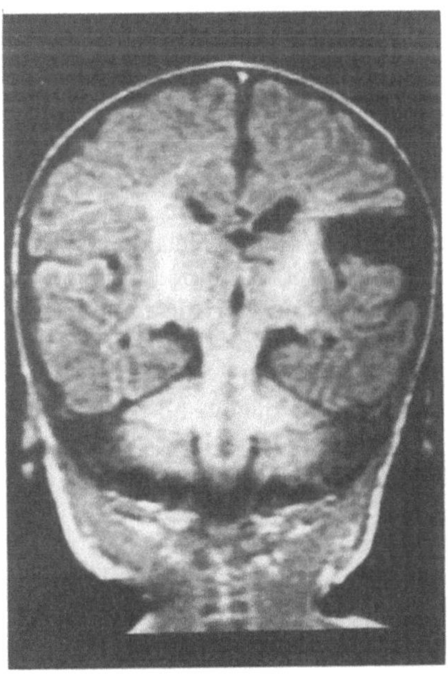

Fig. 10. SET1 600/30 ms, coronal view. Note the cerebral infarction of the distal territory of MCA

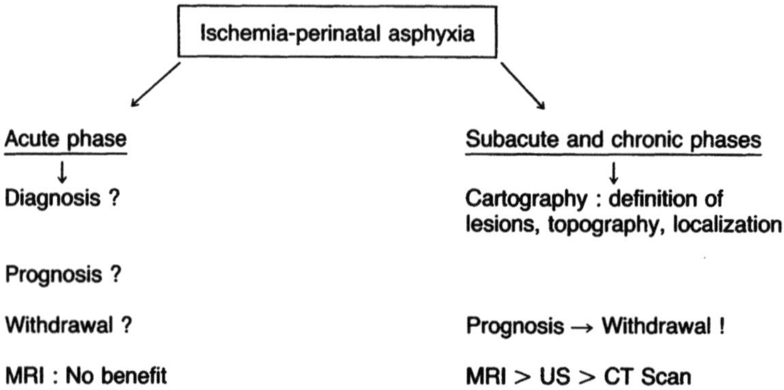

Fig. 11. Place of MRI in the assessment of ischemia and perinatal asphyxia

in the neonatal period or developed severe disabilities (Table 1). Unlike severe ischemic injury, focal ischemia showed a better neurodevelopmental outcome.

MRI of Cerebral Leukomalacia

Cerebral leukomalacia refers to necrosis of the white matter. It can be found in the periventricular area or the subcortical white matter at the depth of the sulcus (subcortical leukomalacia).

Ischemia may involve either one or two hemisphere lobes and may be located either in the frontal, parietal, occipital, or temporal lobes. The centrum semiovale is also involved. Corticospinal tracts, from their origin in the motor cortex, optic (occipital horn) and acoustic (temporal horn) radiations, and the visual area may also be affected by cerebral leukomalacia and are responsible for the abnormal neuromotor outcome and neurosensorial loss. A serious complication that can occur and which seems to be deleterious is hemorrhage into the lesion (Haddad et al. 1991). An attempt to identify the injuries mentioned above and their characteristics, and to link them to prognostic information is required in the neonatal period. Recently, this goal has been achieved using US (Fawer and Calame 1991). There is a close relationship between cerebral leukomalacia, as diagnosed by US, and cerebral palsy later in infancy (Table 2). Nonetheless, it is particularly difficult to predict minor impairments which may occur after the age of 5 years.

In our experience, MRI performed in the neonatal period has provided the same information as that obtained from US, with two additional criteria that seem to contribute to the ability to predict the neurological outcome: hemorrhagic leukomalacia (Fig. 12) and centrum semiovale injury (Fig. 13). Hemorrhagic leukomalacia was associated with a very poor outcome. This is not surprising since hemorrhaging into the ischemic lesion occurs when the latter is most critical. The process is referred to as hemorrhagic infarction

Table 2. Neonatal brain imaging classification of cerebral leukomalacia and its outcome (Haddad et al. 1991)

Classification	Outcome
Ultrasound	
– Unilateral leukomalacia	Contralateral motor impairment
– Bilateral leukomalacia	Bilateral motor impairment
– Topography	
• PVL	Bilateral motor impairment
• SL	Bilateral motor impairment with visual loss
• Mixed	Severe
Magnetic resonance imaging	
– Localization	
• Frontal	Uneventful short-term motor impairment
• Parietal	
• Occipital	Visual impairment
• Temporal	Auditory impairment
– Number and diameter of cysts	Mild to severe
– Pure leukomalacia	Moderate to severe
– Hemorrhagic leukomalacia	Severe
– Destruction of centrum semiovale	Moderate to severe

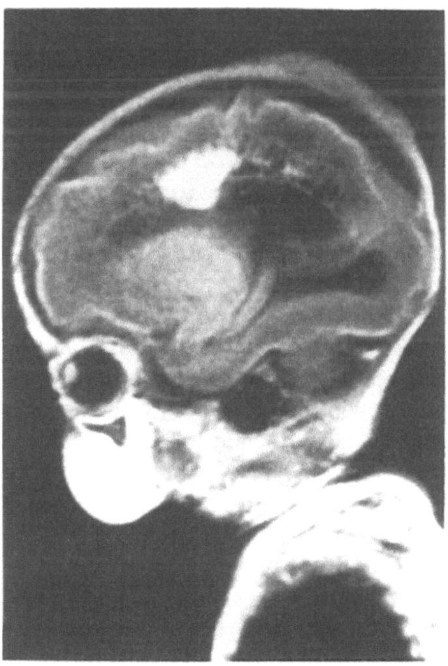

Fig. 12. GE 450/12 ms 90°, axial view showing hemorrhagic periventricular leukomalacia. Note cystic leukomalacia and hypersignal related to hemorrhage

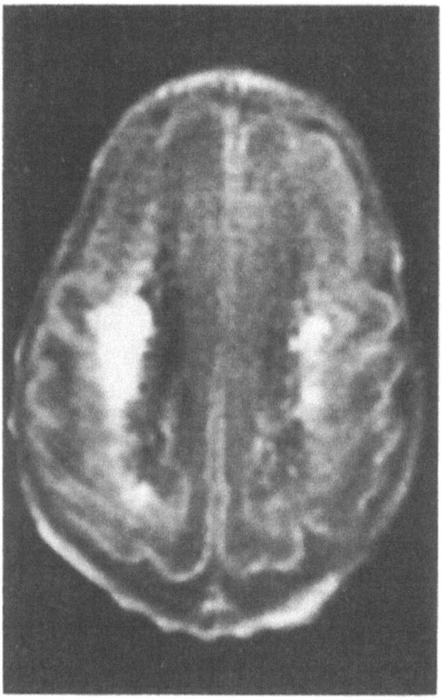

Fig. 13. GE 450/12 ms 90°, axial view showing hemorrhagic periventricular leukomalacia and destruction of the centrum semiovale

(hemorrhage into reperfused ischemic areas). Some 25% of cerebral leukomalacias seem to be hemorrhagic. Hemorrhagic leukomalacia is suspected on US when a significant and very high echogenicity is exhibited. However, a similar echogenicity could also be related to other brain pathologies. On MRI, hemorrhage shows a signal different to that of ischemia and can last for several weeks. Thus, hemorrhagic leukomalacia can be easily diagnosed (Figs. 12, 13). The other diagnostic criterion obtained from MRI is the destruction of the centrum semiovale. This area is of particular importance since all motor fibers descending from their origin cross it. MRI, by providing multiplanar images, contributes to the evaluation of the injury of the centrum semiovale. Its destruction is associated with severe motor impairment. There are other benefits that can be obtained from MRI in the neonatal period: (1) better location of leukomalacia, especially in the frontal and occipital lobes, (2) determination of whether a periventricular area exhibiting increased echogenicity is due to ischemic injury or considered a normal variant (periventricular flares on US scan seem to exhibit an abnormal signal on MRI), and (3) better and easier identification of small ischemic lesions.

We can conclude that MRI is a useful tool for studying the pathology of the neonatal period since it provides information regarding diagnosis and prognosis. However, one must not neglect the fact that it requires the removal of premature infants from their neonatal units and can thus be, in some instances, deleterious. MRI could be applied at a safe time in the neonatal period and in any instances in which cerebral leukomalacia is diagnosed or suspected (abnormal neuromotor examination at discharge and limited US data, (Figs. 14, 15).

In early infancy, US is limited in the follow-up of such lesions with respect to the size of the anterior fontanelle; MRI is more helpful. We have measured the effect of leukomalacia on brain development and have found the following features:

- No abnormal white to gray matter differentiation.
- Dilated lateral ventricles with irregular ventricular walls due to the resorption of cysts (Fig. 16). However, in some patients cysts do persist and are still obvious.

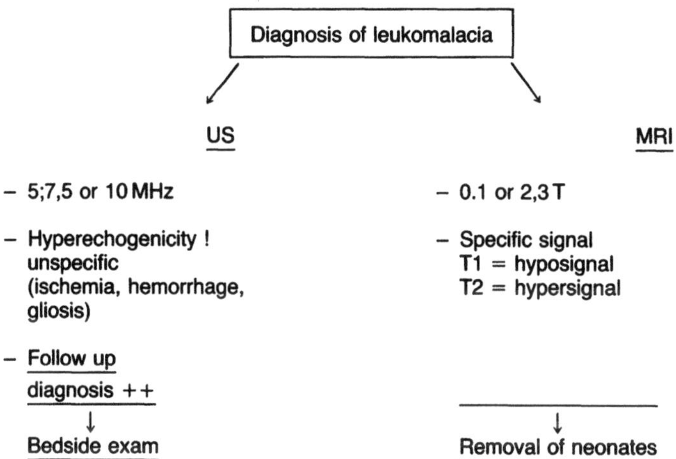

Fig. 14. Cerebral leukomalacia US vs MRI: diagnosis, benefit and limitations

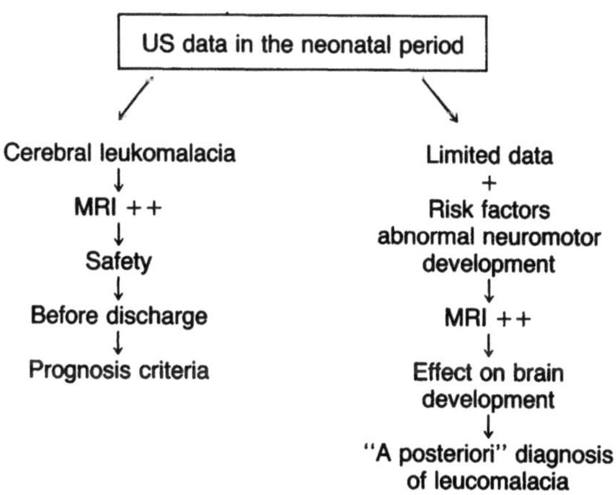

Fig. 15. Place of MRI in the assessment of cerebral leukomalacia

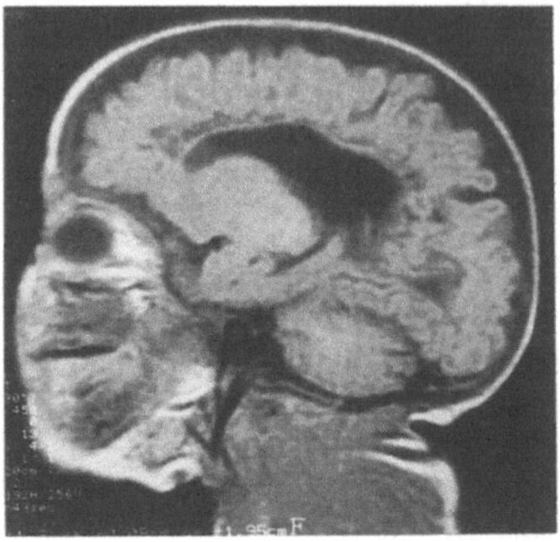

Fig. 16. MRI at 3 months old, GE 450/12 ms 90°, sagittal view. Note the irregular ventricular walls in a neonate who showed periventricular leukomalacia in the neonatal period

- Gliosis (hyperintensity on SET2) of corticospinal tracts (centrum semiovale, internal capsula especially posterior limb and lenticular nuclei; Figs. 17, 18).
- Basal ganglia abnormalities caused by iron accumulation due to the abnormal axonal transfer of iron towards subcortical white matter (Fig. 19).
- Reduction in thickness of the corpus callosum either on its anterior, medium, or posterior part (Fig. 20).

The features described above could be seen in the same infant; however, their occurrence depends on the severity of cerebral leukomalacia.

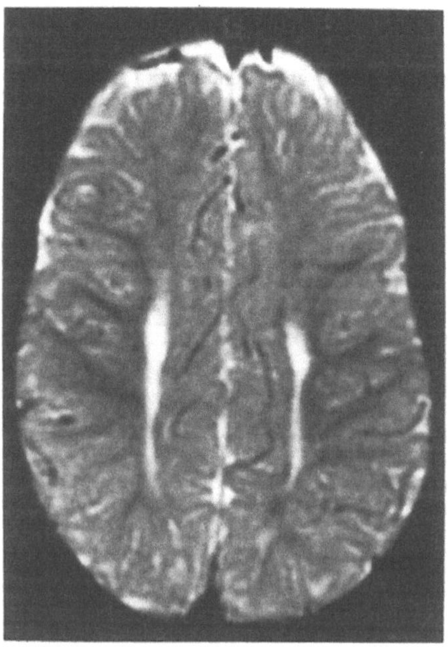

Fig. 17. MRI at 12 months old, SET2 2200/120 ms, axial view. Note the hypersignal within the centrum semiovale (gliosis)

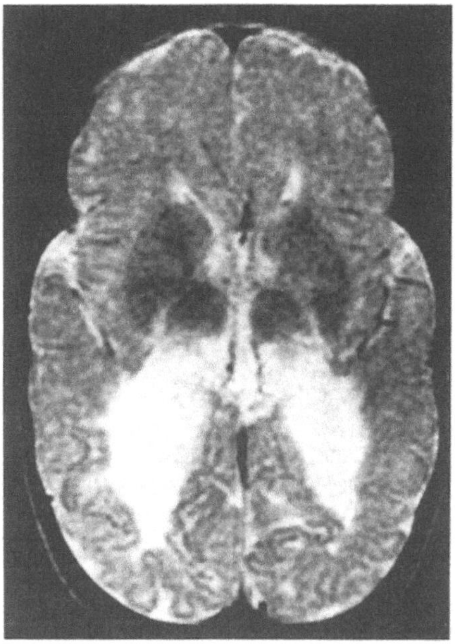

Fig. 18. MRI at 1 year of age, axial view, SET2 2200/120 ms. Note dilated ventricles with irregular walls and hypersignal within the capsula interna at the posterior limb

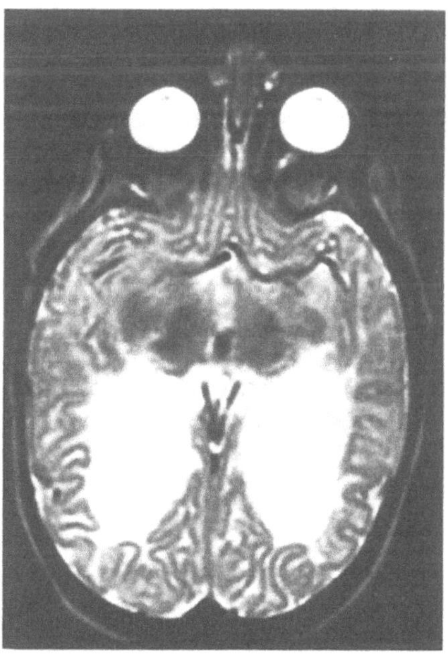

Fig. 19. MRI at 1 year of age, axial view, SET2 2200/120 ms. Note dilated ventricles with irregular walls and the hyposignal of posterior basal ganglia due to iron accumulation

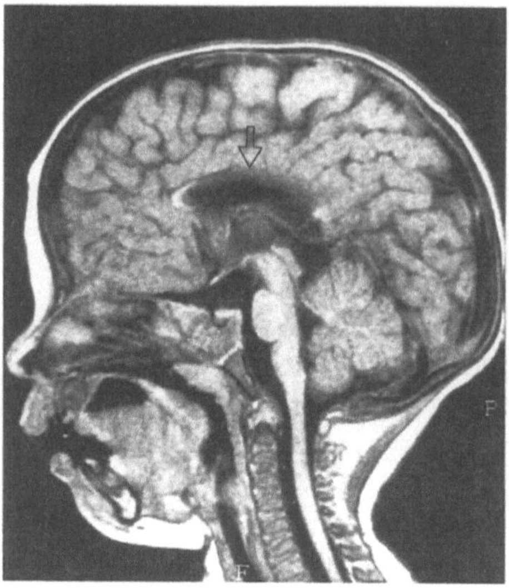

Fig. 20. MRI at 12 months old, GE 450/12 ms 90°, sagittal view. Note the reduction thickness of corpus callosum (*arrowhead*) following PVL

Thus, MRI allows "a posteriori diagnosis" of cerebral leukomalacia in infants who exhibit abnormal neuromotor development without complete neonatal data by identifying the effects of ischemia on brain development (Fig. 15).

Hemorrhage

Perinatal asphyxia is one of the major factors implied in intracranial hemorrhage (ICH) in neonates. Several types of ICH are encountered in the neonatal period (Table 3). US and CT scan are very helpful in diagnosing hemorrhage. However, MRI carries powerful criterion to identify hemorrhage mainly due to the hemorrhagic signal. The latter must be studied in different stages using SET1 and SET2.

Table 3. Types of ICH in the neonatal period

	Frequency	Full-term	Prematures
Pericerebral hemorrhage			
– Subarachnoidal	Common	±	+
– Subdural	Uncommon	+	–
Intracerebral hemorrhage	Uncommon	+	–
Parenchymal involvement	Common	–	+
Thalamic hemorrhage	Uncommon	+	–
Periventricular intraventricular hemorrhage	Common	–	+
Intracerebellar	Uncommon	±	+

+, Specific; ±, less specific; –, not specific.

Acute Phase of Hemorrhage. On SET1, hemorrhage exhibits an isosignal in the first 24–48 h, followed by the appearance of a high-intensity signal. On SET2, acute hemorrhage is marked by the appearance of both an isosignal and a low-intensity signal in the first 24–48 h (Fig. 21).

Subacute Phase of Hemorrhage (Fig. 22). In this phase, the hemorrhage begins to exhibit a high-intensity signal on both SET1 and SET2.

Chronic Phase of Hemorrhage. After 2 weeks, the hemorrhage exhibits a high-intensity signal on both SET1 and SET2 which can last for several weeks, even in the ventricular system (Fig. 23).

Development of Hematoma (Figs. 24, 25). The characteristic MRI patterns during the development of a hematoma are as follows: There is a central hypointensity on SET2 during the acute stage and an isosignal on SET1. The change to hyperintensity, initially on SET1 and then on SET2, extends from the hematoma periphery inwards in the subacute phase. The rim of parenchymal hypointensity on SET2 abutting the hematoma is found during both the subacute and chronic stages, especially with higher magnetic fields and is related to hemosiderin deposits.

Pericerebral Hemorrhage. This is well displayed on MRI; however, differentiation between subdural and subarachnoidal hemorrhage is still a

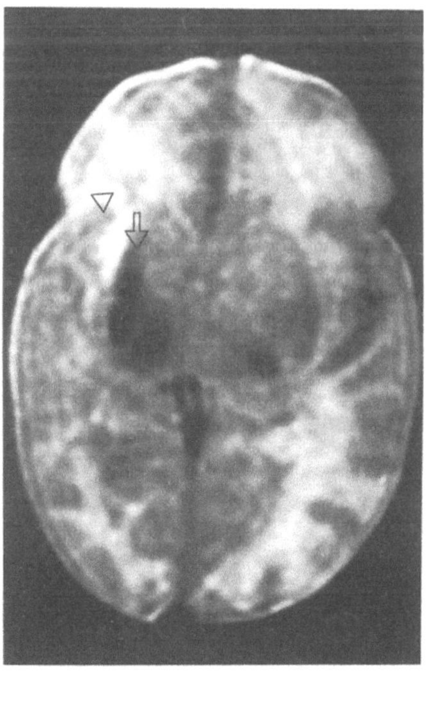

Fig. 21. SET2, 2200/12 ms 90°, axial view. Note the hyposignal (*arrowhead*, fresh hemorrhage) beside a hypersignal (*arrowhead*, old hemorrhage)

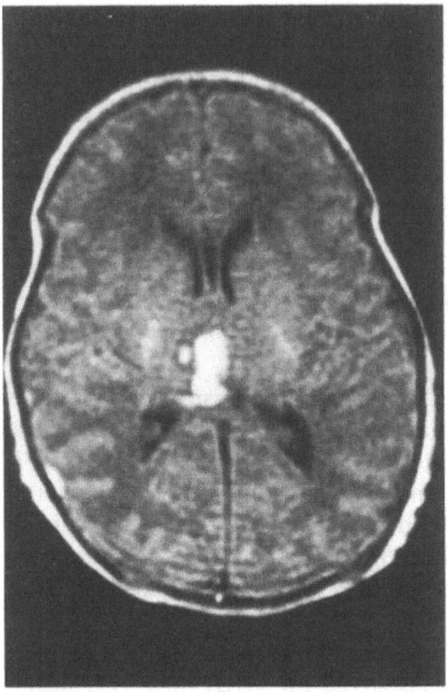

Fig. 22. MRI at 12 days old, GE 450/12 ms 90°, axial view showing hemorrhagic signal (hypersignal) at right basal thalamus

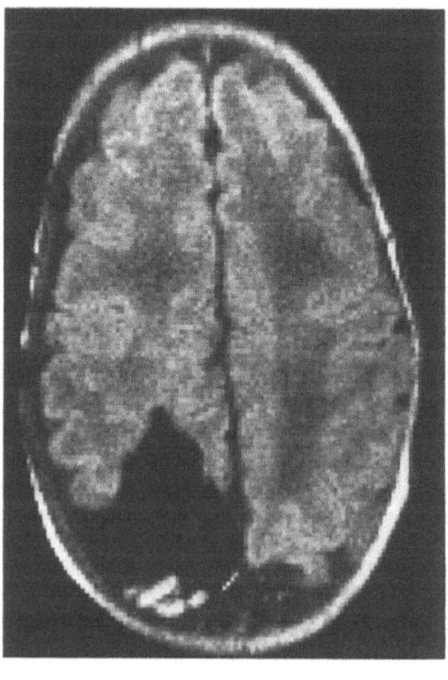

Fig. 23. MRI at 14 weeks old, SET1, 600/32 ms, axial view. Note the hypersignal within the right parietal porencephalic cavity

challenge since they are often interrelated in the newborn. Generally, subdural hemorrhage leads to compression of the cerebral parenchyma (Fig. 26). Subarachnoidal hemorrhage fills the convolutions and gyri and is often present in the sylvian fissure and the subarachnoidal space of the posterior fossa. It seems that SET1 is more sensitive than SET2 in detecting pericerebral hemorrhage, especially in differentiating subdural from subarachnoidal bleeding (Figs. 24, 25).

The appearance of hemorrhage on MRI is complex and mainly related to the breakdown of oxyhemoglobin to deoxyhemoglobin and methemoglobin. Other factors such as the magnetic field could interfere with the hemorrhagic signal. Higher fields of 1.5 T exhibit better hemosiderin signals than lower fields in the neonates (Zuerrer et al. 1991). Fresh hemorrhage (acute phase), made up essentially of deoxyhemoglobin, is not detected by SET1; however, on SET2 it exhibits a low-intensity signal easy to detect, since the high-intensity signals of the unmyelinated cerebral parenchyma and the cerebral spinal fluid (CSF) provide an ideal background to the low intensity of fresh hemorrhage.

After the first few days (subacute phase), the hemorrhage is more easily identified by MRI. Visibility of the hemorrhage depends on the formation of methemoglobin, which begins 24–48 h after the hemorrhage occurs and increases during the first week. An intermediate to the high-intensity signal is seen on both SET1 and SET2. At this stage SET1 provides better recog-

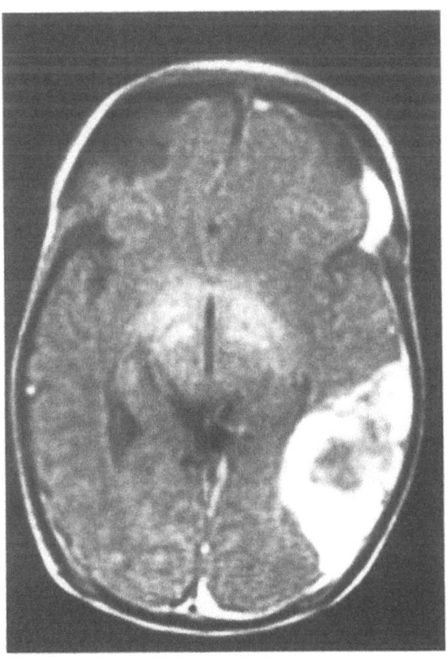

Fig. 24. MRI at 8 days old, GE 450/12 ms 90°, axial view showing left parenchymal hematoma. Note two different signals from the hemorrhage: a central hyposignal surrounded by a peripheral hypersignal. Note also the subdural hemorrhage

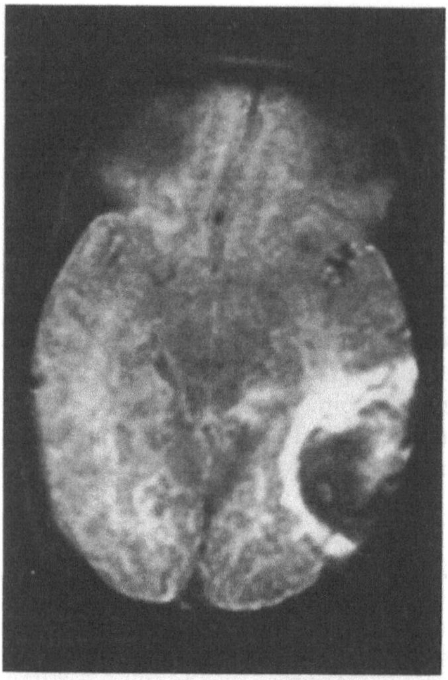

Fig. 25. MRI at 8 days old, SET2, 2000/120 ms, axial view showing the same hematoma as in Fig. 21. Two different signals are noted: a very large central hyposignal surrounded by a thin peripheral hypersignal. Extracerebral hemorrhage is rather difficult to identify

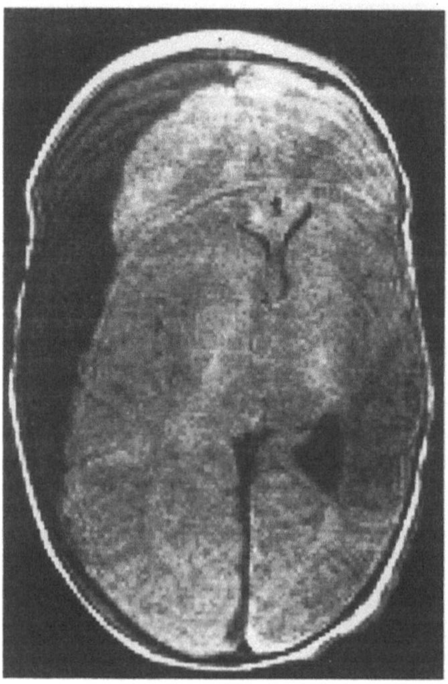

Fig. 26. SET1, 600/32 ms, axial view showing subdural hemorrhage at a later stage (3 weeks old) with parenchymal compression

nition of the hemorrhage, since it exhibits a hyperintensity signal different from that of cerebral parenchyma. On SET2, confusion can result due to the hypersignal exhibited by unmyelinated cerebral parenchyma or CSF.

The increased intensity of hemorrhage during the chronic phase on SET1 and SET2 is most likely due to the continued presence of methemoglobin. At this stage, SET1 and SET2 are equally sensitive in detecting hemorrhage. With respect to the appearance of hematoma, the signal related to hemorrhage is also related to oxyhemoglobin, deoxyhemoglobin, and methemoglobin as described above. The hypointense rim found in the subacute and chronic stages is due to hemosiderin and can be depicted several months after the onset of hemorrhage.

Regarding pericerebral hemorrhage, it seems that the signals exhibited do not last as long as those for intracerebral or periventricular intraventricular hemorrhages (PV-IVH; Figs. 22, 25; Haddad et al. 1991).

PV-IVH are of particular importance in the premature baby. They occur early in the neonatal period and can lead to cerebral impairment. The current classification of PV-IVH is as follows: grade 1, subependymal hemorrhage (germinal matrix hemorrhage) from which the blood enters the ventricles; grade 2, IVH with blood covering 10%–50% of the ventricular area; and grade 3, IVH with blood covering more than 50% of the ventricular area with ventricular dilatation.

US is highly effective in diagnosing grades 1, 2, or 3. Grades 1 and 2 are generally associated with a good outcome, whereas grade 3 is often associated with a poor outcome. However, the outcome with PV-IVH depends essentially on the absence or presence of complications which can alter the course of the lesion, e.g., hydrocephalus and parenchymal involvement. Regarding hydrocephalus, there is no doubt that early diagnosis leads to early treatment and thus to a better outcome. The most reliable technique for diagnosing hydrocephalus is US, which identifies ventricular dilatation that occurs before the appearance of clinical signs. Parenchymal involvement refers to hemorrhagic infarction or hemorrhagic leukomalacia. Here again, US is highly effective in diagnosing such lesions. MRI adds useful information by depicting ischemia more accurately. However, MRI requires the removal of sick premature babies from their neonatal care units. Thus, US seems the most reliable technique in diagnosing PV-IVH in prematures due to its safety and the possibility of performing serial examinations. Moreover, decisions concerning continuous intensive care (prognosis) in patients with PV-IVH are mainly based on US; here, CT and MRI are neither beneficial nor desirable.

Finally, comparing the sensitivity and sensibility of US, CT, and MRI in the detection of hemorrhage, we can conclude that MRI is a useful tool in detecting hemorrhage in the subacute and chronic stages. In allows one to date the occurrence of the hemorrhage, in those in whom it occurred antenatally. Indeed, MRI applied early in the neonatal period enabled us to detect a hemorrhagic signal within the ventricles which was compatible with an old hemorrhage (Fig. 27). Findings of ventricular dilatation and irregular walls also led us to conclude that IVH occurred antenatally with a

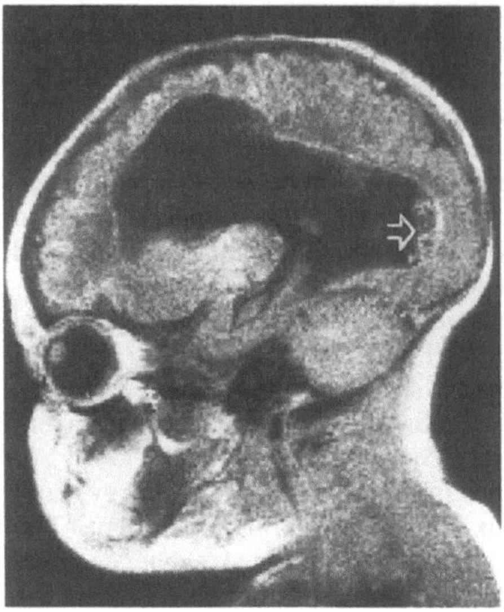

Fig. 27. MRI at 5 days old, SET1, 600/32 ms, sagittal view. Note intraventricular clot (hypersignal) and ventricular dilatation due to resorption of parenchymal involvement which occurred antenatally

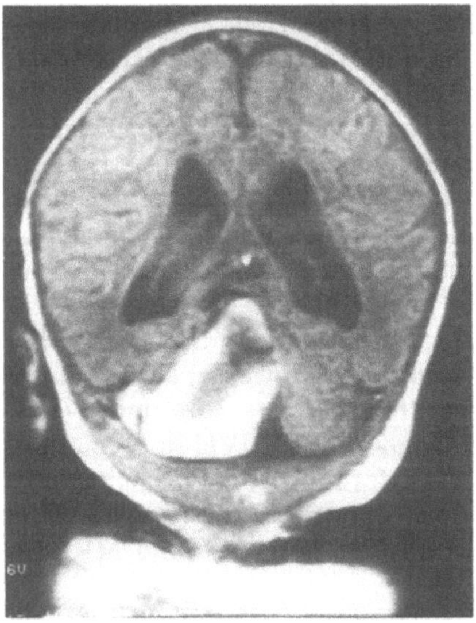

Fig. 28. MRI at 3 days old, GE 450/12 ms 90°, coronal view. Note hemorrhagic signal of the right cerebellar hemisphere. The signal exhibited is compatible with the subacute phase of hemorrhage

parenchymal involvement. Furthermore, MRI performed at day 3 after birth showed a hemorrhagic signal in a neonate (Fig. 28). The latter is compatible with the subacute phase of hemorrhage. US performed at the same time showed the appearance of hypoechogenicity within the posterior fossa which

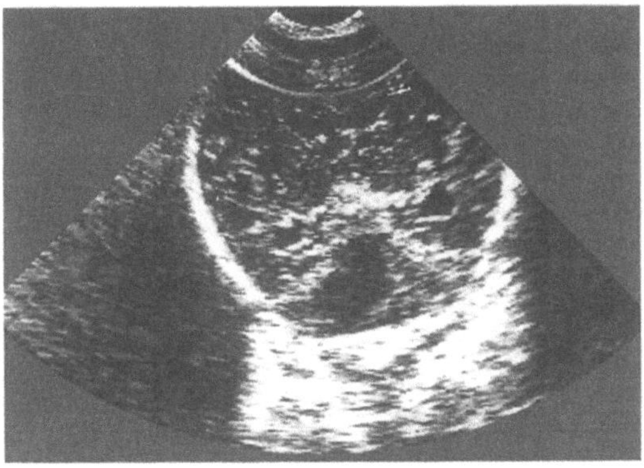

Fig. 29. Ultrasound, coronal view of the same as in Fig. 28. Only hypoechogenicity of posterior fossa was noted

is difficult to allocate to hemorrhage (Fig. 29). Again MRI was highly effective in detecting the antenatal occurrence of hemorrhage.

Conclusion

MRI is a powerful imaging method for infants suffering from perinatal asphyxia. Ischemia and hemorrhage are diagnosed well with an emphasize on the prognosis. From our experience, it is the current practice to screen neonates at the bedside using US and to perform MRI rather than CT scan at a safe time. The future is promising with the development of newer and faster MR sequences, the use of angio MRI and tridimensionnal images.

References

1. Baker L, Kucharczyk J, Sevick R, Mintorovitch J, Moseley M (1991) Recent advances in MR imaging. Spectroscopy of cerebral ischemia. AJR 156:1133–1143
2. Boesh C, Martin E (1988) Combined application of MR imaging and spectroscopy in neonates and children: installation and operation of 2.35 T system in a clinical setting. Radiology 168:481–488
3. Dereuck J, Chattha AS, Richardson EP (1972) Pathogenesis and evolution of periventricular leucomalacia in infancy. Arch Neurol 27:229–236
4. Fawer CL, Calame A (1991) Ultrasound. In: Haddad J, Christmann D, Messer J (eds) Imaging techniques of the CNS of the neonate. Springer, Berlin Heidelberg New York, pp 79–106
5. Haddad J, Christmann D, Messer J (eds) (1991) Imaging techniques of the CNS of the neonate. Springer, Berlin, Heidelberg New York
6. Kuban KC, Gilles FH (1985) Human teleencephalic angiogenesis. Am Neurol 39:709–712
7. Rutherford MA, Cowan FM, Mansur A, Pennock JM, Hajnal JU, Young IR, Bydder GM (1991) MR imaging of anisotropically restricted diffusion in infants. J Comput Assist Tomog 15:188–198
8. Siegel MJ, Schackeford GD, Perelman J, Fulling K (1984) Hypoxic-ischemic encephalopathy in term infants: diagnosis and prognosis evaluated by ultrasound. Radiology 152:395–399

9. Takashima S, Tanaka K (1978) Development of cerebrovascular architecture and its relationship to periventricular leucomalacia. Arch Neurol 35:11–16
10. Volpe J (1987) Neurology of the newborn. Saunders, Philadelphia, pp 209–233
11. Yuh WTC, Crain MR, Loes DJ, Greene GM, Ryals TJ, Sato Y (1990) MR imaging of cerebral ischemia. Findings in the first 24 hours. AJR 156:1133–1143
12. Zuerrer M, Martin E, Bolsthauser E (1991) MR imaging of intracranial hemorrhage in neonates and infant at 2.35 T. Neuroradiology 33:223–229

Doppler Assessment of the Cerebral Hemodynamics in Perinatal Asphyxia

E. Saliba and J. Laugier

Doppler ultrasound offers the neonatologist the opportunity to monitor cerebral hemodynamics noninvasively. Since the pioneering work of Bada et al. in 1979 [3], Doppler ultrasound has been used for research into many physiological conditions in the neonate.

A brief review of the principles and some practical aspects of the technique is followed by a discussion of the clinical application of Doppler in perinatal asphyxia.

Cerebral Hemodynamics

The principles of hemodynamics are the basis for understanding Doppler findings. Cerebral hemodynamics are determined by many factors. The viscosity of blood, cerebral vascular resistance, cerebral perfusion pressure, and the action of cerebral autoregulation.

Blood Viscosity. Viscosity is the property of a fluid whereby it offers resistance to shear. Oil is an example of a highly viscous fluid, whereas water is an example of a liquid with relatively low viscosity. Blood is not a simple fluid. It is essentially a suspension of red blood cells in plasma, and its viscosity is mainly dependent on hematocrit.

Cerebral Vascular Resistance. Blood can only flow where a pressure gradient (P) exists. Cerebral blood flow (CBF) is tightly controlled under normal physiologic circumstances by an alteration in cerebral vascular resistance (CVR). CVR changes in response to changes in cerebral metabolic rate oxygen consumption ($CMRO_2$), arterial oxygen pressure (PaO_2), arterial carbon dioxide pressure ($PaCO_2$), and cerebral perfusion pressure (CPP). CVR is determined primarily within the arterioles, including the precapillary sphincters.

Cerebral Perfusion Pressure and Blood Flow. The net driving force for the cerebral circulation is usually defined as the perfusion pressure, which is normally the arterial blood pressure (ABP) minus the intracranial pressure (ICP). When ICP is raised, the pontine veins connecting subdural veins to the sinus collapse because of their special construction of soft walls. As a result of this, the intracranial venous pressure (IVP) rises until it equals ICP

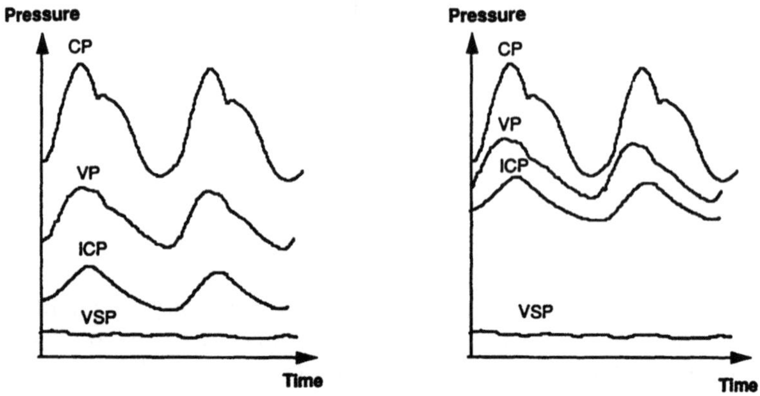

Fig. 1. Waveforms of capillary pressure (*CP*), cerebral venous pressure (*VP*), intracranial pressure (*ICP*), and venous sinus pressure (*VSP*), evaluated in the case of normal ICP (*left*) and raised ICP (*right*)

(Fig. 1). This venous collapse increases outflow resistance; because of this special property of the outflow from the subdural space, the cerebral venous pressure can be defined as being equal to the ICP. The CBF is then equal to: CBF = CPP/CVR [23].

Cerebral Autoregulation. Autoregulation is the term used to describe the maintenance of a constant CBF across a broad range of CPP.

Doppler Principle

The ultrasonic Doppler velocimeter works by sensing the change in frequency imposed on an ultrasonic wave as it is reflected or scattered by moving red blood cells. That change in frequency is given by the Doppler equation $\Delta F = \frac{2\ Fo\ V + COS\ \theta}{C}$, where ΔF indicates the Doppler frequency shift (Hz), Fo is the transmitted frequency, V the velocity of the red blood cells (cm/s), θ the angle between the ultrasound beam and the direction of flow, and C the velocity of ultrasound in tissue (=1550 m/s). For practical purposes, Fo and C are constant from one determination to another. Thus, the Doppler shift is proportional to the blood velocity when the angle of the probe (θ) remains constant. The Doppler equation also demonstrates the influence of the angle θ on the Doppler shift. A maximum value is found at $\theta = 0°$; at $\theta = 90°$ the Doppler shift is zero. The measuring error caused by the angle of insonation is 6% or less when θ is 20° or less.

Doppler Techniques

Continuous Wave and Pulsed Doppler Instruments

There are two distinct types of Doppler velocimeter – continuous wave (CW) and pulsed wave (PW).

Continuous Wave. CW Doppler velocimeters employ a transducer containing two piezoelectric crystals. One of the crystals is used to continuously transmit a beam of ultrasound into the tissue, while the other is used for receiving. These devices have no depth resolution and signals are measured from the entire penetration depth of the ultrasonic waves.

Pulsed Wave. PW Doppler velocimeters usually contain only a single crystal that serves both as a transmitter and a receiver. With this technique ultrasonic pulses of only a millisecond in duration are transmitted intermittently. After each pulse has been transmitted, there is a delay before a gate in the receiving circuit is opened for a short period of time to admit returning signals from a small volume of tissue. The delay between transmission and the gate opening may be altered by the operator to determine the depth from which signals are gathered. PW Doppler has the obvious advantage of making it possible to select signals from a particular depth.

Doppler Signal Processing

Once the returning ultrasound has been received by the transducer it is necessary to process it in order to obtain a Doppler signal. The preferred method of processing Doppler signals is to perform a full spectral analysis using a real-time spectrum analyzer, and to display the results in the form of a sonogram. In this type of display, time is represented along the abscissa (x-axis), Doppler shift frequency along the ordinate (y-axis), and the power of the signal at a given frequency and time by the intensity of the corresponding pixel. A maximum frequency envelope may be derived by following the outline of the sonogram either automatically or manually using a digitizer.

Interpretation of Doppler Signals

There are several options available to interpret the Doppler signal, including waveform analysis, velocity measurement, and flow measurement.

Waveform Analysis. In the application of Doppler techniques to pediatrics, waveform analysis techniques are the most widely used. Several of these techniques are based on finding the ratio of the height of one feature of a waveform to that of another. The advantage of taking such a ratio is that both the numerator and the denominator include the cosine of the angle between the Doppler probe and the blood vessel. The cosine term therefore cancels, and the index is independent of angle.

Velocity Measurement. Measurements of mean blood velocity are relatively easy to make with duplex scanners which combine the two-dimensional imaging technique with PW Doppler. They can be used to measure proportional changes in blood flow, provided that the diameter of the vessel under study does not alter significantly and the Doppler angle is kept constant. There is some evidence, as yet unconfirmed, that it may not always be possible to assume a constant vessel cross section. The mean velocity may be obtained directly from the temporal average of the mean frequency envelope or, if the flow is unidirectional and the velocity profiles fully established (which is the case in the cerebral arteries), may be estimated from half the temporal average of the maximum frequency envelope.

Volumetric Flow. Volumetric flow is given by the product of mean velocity and cross-sectional area. Unfortunately, it is presently impossible to measure the size of cerebral vessels accurately in vivo.

Clinical Application

Intracranial Hemorrhage in the Preterm Infant

Among the factors involved in the pathogenesis of germinal-matrix intraventricular hemorrhage (IVH), those related to the regulation of CBF and pressure in the microvascular bed of the germinal matrix are probably the most important.

In 1976, Hambleton and Wigglesworth [6] suggested that increases in CBF play an important pathogenetic role in IVH. The possibility of a pressure-passive cerebral circulation in some premature infants was suggested initially by studies of CBF with the Xenon technique [12]. These observations were carried out in the first hours after birth in a series of 19 infants, ten of whom had Apgar scores less than seven at 1 min. Thus, it appears likely that under certain circumstances (i.e., hypoxic–ischemic insult) the cerebral circulation of the premature infant in the first hours or days of life is pressure passive. When pressure flow autoregulation is abolished, arterial blood pressure peaks may reach the capillary bed in the germinal matrix. It is reasonable to postulate that due to hypoxic–ischemic insult, the already fragile matrix capillaries are more likely to rupture, especially in the presence of a pressure-passive circulation.

Many investigators hoped that using Doppler technique to assess cerebral hemodynamics in preterm infants would help in finding a reliable marker for an increased risk of IVH.

Fluctuating Cerebral Blood Flow. Using a CW Doppler method, Perlman et al. [17] showed that velocity waveforms had a fluctuating pattern (beat-to-beat variation in velocity greates than 10%) in premature babies who were at a greater risk of developing intracranial hemorrhage. The velocity fluctuations decreased with the use of muscle paralysis. In a later study, these authors demonstrated that infants paralyzed with pancuronium bromide had a lower incidence of intracranial hemorrhage than controls [18]. These fluctuations appear to be the transmission of fluctuations in systemic arterial pressure and may be a cardiovascular manifestation of the respiratory lack of coordination [14]. More recently, Kuban et al. [9] failed to establish an association between velocity fluctuations and hemorrhage.

Increases in Cerebral Blood Flow. As previously mentioned in this chapter, all factors that cause an abrupt elevation in CBF are important to detect. Among them hypertension and hypercarbia are likely to play an important role. The causes of abrupt elevations in ABP in the premature infant are numerous and include tracheal suctioning, rapid exchange transfusion, rapid infusion of colloid, pneumothorax, or seizures [5, 8, 15, 16]. In almost all these conditions, the Doppler technique was able to detect a simultaneous increase in ABP and cerebral blood flow velocity (CBFV). Hypercarbia, a common accompaniment of respiratory distress syndrome, has been demonstrated conclusively to be a potent means for increasing CBF in experimental studies [26]. In human infants, Pryds et al. [19] using the ^{133}Xe clearance technique reported an increase in CBF by 7%–8% for each torr increase in PCO_2. Using Doppler methods, Archer et al. [2] reported a 5.9%–7% change in velocity for each torr of $PaCO_2$ change in preterm infant.

The role of decreased hematocrit in causing increases in CBF has been reported by Younkin et al. [27] in 15 premature infants by the ^{133}Xe clearance technique. A striking inverse relationship between hematocrit (range 24%–48%) and CBF was defined, such that every 5% decrease in hematocrit was associated with an increase in CBF of 11 ml/100 g per minute. We examined the effect of hemodilution on CBF, CBFV, and $CMRO_2$ in newborn piglets rendered anemic by isovolumic exchange. With decreasing hematocrit, CBF and CBFV increased significantly and $CMRO_2$ remained unchanged. These data indicate that CBF presumably increased to maintain $CMRO_2$ constant [25]. Low hematocrit may also modulate the CO_2 reactivity of the cerebral hemodynamics. In a group of healthy newborn baboons, when the baseline hematocrit was low, all the velocity components of the Doppler signal increased with increasing $PaCO_2$ [21].

Increases in Cerebral Venous Pressure. The particular importance of increased venous pressure in the pathogenesis of IVH relates in part to the venous anatomy in the region of the germinal matrix [10]. Thus, a marked curve in the direction of venous flow occurs at this site as the internal cerebral veins are formed from the confluence of the terminal, thalamostriate, and choroidal veins. This situation would be expected to predispose to venous stasis and, perhaps, increased intravascular pressure leading to rupture of capillaries.

During asphyxia, increased cerebral venous pressure may result from hypoxic–ischemic cardiac failure. In preterm fetal sheep, Reynolds et al. [22] reported the intricate relationship between asphyxia, increased venous pressure, and IVH. Thus, it seems likely that increased venous pressure may contribute to the occurrence of IVH observed after severe perinatal asphyxia. Doppler measurements of blood flow velocity in the superior sagittal sinus may be of particular help in demonstrating changes in venous circulation in relation to changes in venous pressure [4].

Posthemorrhagic Hydrocephalus. The effect of an abnormally high resistance index has been described in posthemorrhagic hydrocephalus [7, 24]. This may reflect an increased resistance secondary to increased ICP. Following ventricular peritoneal shunt and drainage of cerebrospinal fluid, the velocity indices were normalized. These Doppler abnormalities may help in monitoring the progress of hydrocephalus and in deciding the time for intervention.

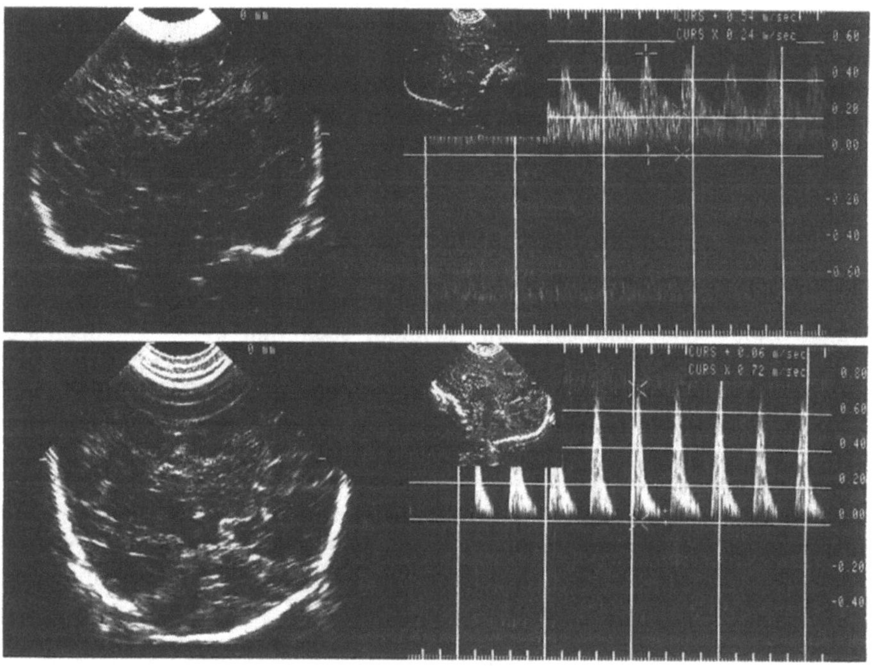

Fig. 2a,b. Perinatal asphyxia in a term newborn. **a** *Left panel*, coronal scan at 2 days of life showing diffuse increased echogenicities; *right panel*, Doppler sonogram in the same baby showing high velocities and a low resistance index (RI = 0.55). **b** Evolution at 5 days of life. *Left panel*, coronal scan showing diffuse echogenicities with loss of the main structural landmarks; *right panel*, the flow velocities decreasing with increasing resistance index (RI = 0.92)

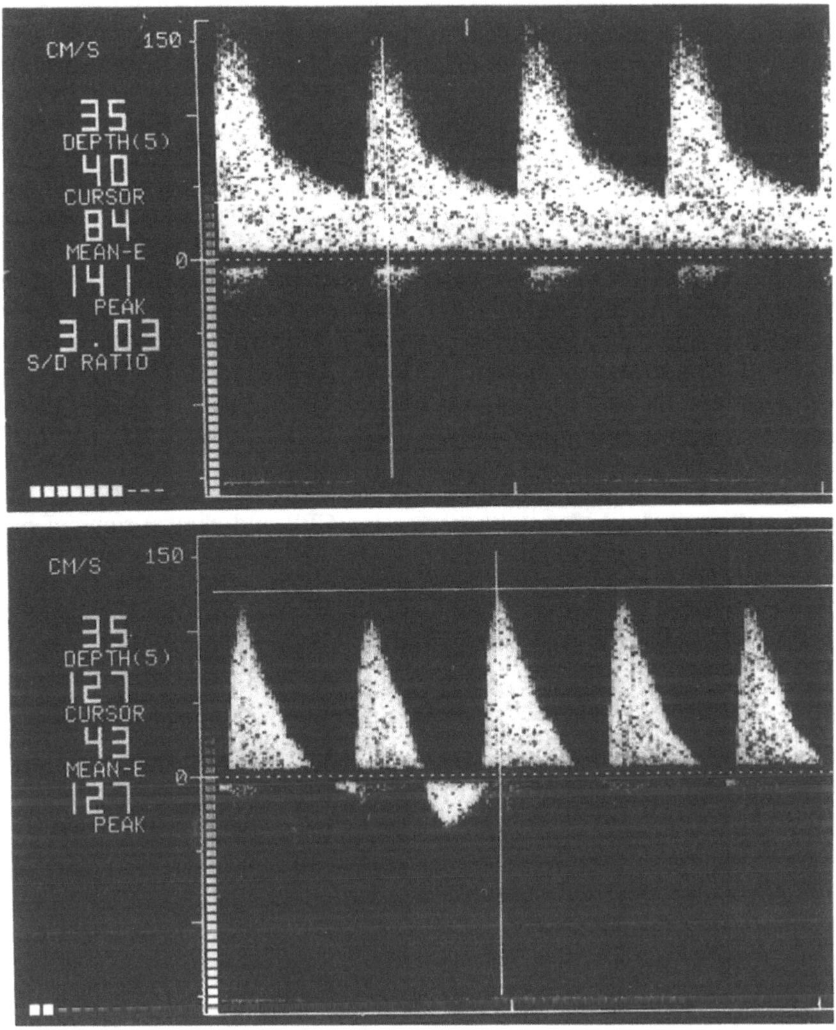

a

b

Fig. 3a,b. Occurrence of diastolic back flow with decreasing perfusion pressure and isoelectric EEG following severe perinatal asphyxia. **a** Reduced diastolic velocities during increased ICP; **b** diastolic retrograde flow and isoelectric EEG

Vasoparalysis in Asphyxiated Term Infants

Bada et al. [3] were first to report abnormally low Pourcelot's resistance index in neonates with perinatal asphyxia. Reduced resistance was considered to reflect luxury reperfusion secondary to asphyxia, and reductions in vascular resistance distal to the site of recording. Luxury perfusion has been reported in comatose adult patients following blunt head injury [13].

Recently, Pryds et al. [20] reported high CBF, measured by the Xenon technique, after severe asphyxia in term infants. Luxury perfusion refers to the dissociation between cerebral metabolic rate and CBF when flow exceeds metabolic requirements. During asphyxia, lactate and CO_2 are accumulated in the brain, resulting in a decrease in the extracellular cerebrospinal fluid pH with subsequent arteriolar vasodilatation. The cerebral metabolism is often depressed after such insults. Along the same lines, Archer et al. [1] and more recently Levene et al. [11], demonstrated that in a group of infants suffering from birth asphyxia, abnormalities in the resistance index (low-resistance index) and velocities (high CBFV) predicted poor outcome. An abnormally low Pourcelot's index (of less than 0.55) was recorded during the neonatal period in 16 of 27 infants who later developed moderate and severe neurologic sequelae. Thus, the combination of low resistance and high CBFV supports the concept of cerebral vasoparalysis, a marker of severe perinatal asphyxia. (Fig. 2a). In some severe cases, it is possible to follow the changing pattern of Doppler sonogram towards a progressive reduction in velocities with increasing ICP (Fig. 2b). In a few cases a diastolic back flow with decreasing CPP and clinical signs of brain death and isoelectric EEG are observed (Fig. 3). The mechanisms by which

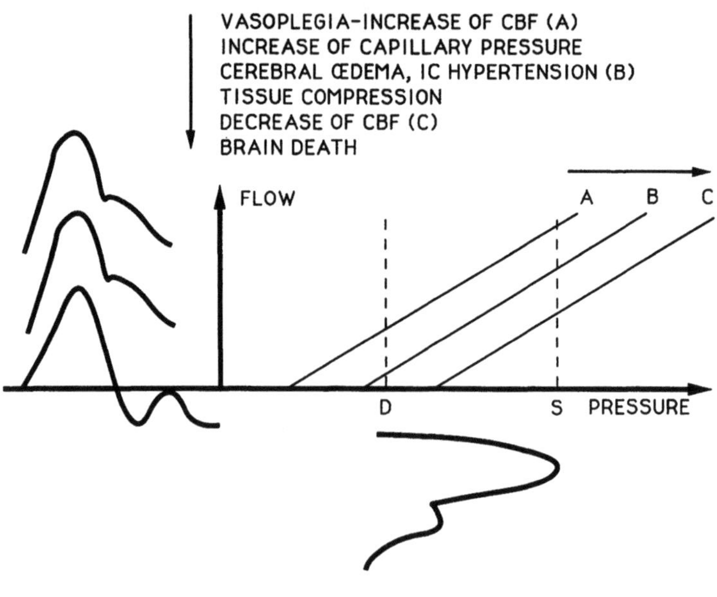

Fig. 4. Following a severe perinatal asphyxia the initial increase in flow velocity, mainly the diastolic (*D*) component, is considered as a vosoparalysis (*A*). When intracranial pressure increases and approaches the level of the diastolic systemic artery blood pressure, the diastolic part of the flow velocity spectrum starts to decrease (*B*). When the intracranial pressure is above the value of the diastolic pressure a reversed diastolic flow velocity appears (*C*). *S*, systolic

ICP leads to a decrease in CBFV and to a retrograde flow during the diastole, are shown in Fig. 4.

The CO_2 reactivity of the cerebral hemodynamics is of special interest. Reduced CO_2 reactivity was found in adults in the presence of severe head trauma and has been used as a prognostic criterion. Recently Pryds et al. [20] showed that the disappearance of vasoreactivity to $PaCO_2$ changes in term newborn after severe perinatal asphyxia was a marker of poor prognosis. Although these are interesting findings because these different types of pathologic patterns (i.e., hyperperfusion) may occur at different time intervals following hypoxic–ischemic injury, the clinical value of Doppler monitoring in cases of perinatal asphyxia needs to be determined.

References

1. Archer LNJ, Levene MI, Evans DH (1982) Cerebral artery Doppler ultrasonography for prediction of outcome after perinatal asphyxia. Lancet 2:1116–1118
2. Archer LNJ, Levene MI, Evans DH (1986) Controlled hypercapnia and neonatal cerebral artery Doppler ultrasound waveforms Pediatr Res 20:218–221
3. Bada HS, Hajjar W, Chua C, Summer DS (1979) Non-invasive diagnosis of neonatal asphyxia and intraventricular hemorrhage by Doppler ultrasound. J Pediatr 95:775–779
4. Cowan F, Thorensen M (1985) Changes in superior sagittal sinus blood velocities due to postural alterations and pressure on the head of the newborn infant. Pediatrics 75:1038–1047
5. Goldberg RN, Chung D, Goldman SL et al. (1980) The association of rapid volume expansion and intraventricular hemorrhage in the preterm infant. J Pediatr 96:1060–1063
6. Hambleton G, Wigglesworth JS (1976) Origin of intraventricular hemorrhage in the preterm infant. Arch Dis Child 51:651–659
7. Hill A, Volpe JJ (1982) Decrease in pulsatile flow in the anterior cerebral arteries in infantile hydrocephalus. Pediatrics 69:4–7
8. Hill A, Perlman JM, Volpe JJ (1982) Relationship of pneumothorax to occurrence of intraventricular hemorrhage in the premature newborn. Pediatrics 69:144–149
9. Kuban KCK, Skouteli H, Cherer A et al. (1988) Hemorrhage, phenobarbital, and fluctuating cerebral blood flow velocity in the neonate. Pediatrics 82:548–553
10. Larroche JC (1977) Developmental pathology of the neonate. Excerpta Medica, Amsterdam
11. Levene MI, Fenton AC, Evans DH, Archer LNJ, Shortland DB, Gibson NA (1989) Severe birth asphyxia and abnormal cerebral blood flow velocity. Dev Med Child Neurol 31:427–434
12. Lou HC, Lassen NA, Friis-Hansen B (1979) Impaired autoregulation of cerebral blood flow in the distressed newborn infant. J Pediatr 94:118–121
13. Obrist WD, Langfitt TW, Yaggi JL (1984) Cerebral blood flow and metabolism in comatose patients with acute head injury. Relationship to intracranial hypertension. J Neurosurg 61:241–253
14. Perlman JM, Thach BB (1988) Respiratory origin of fluctuations in arterial blood pressure in premature infants with respiratory distress syndrome. Pediatrics 81:339
15. Perlman JM, Volpe JJ (1983) Seizures in the pre-term infant: effects on cerebral blood flow velocity, intracranial pressure and arterial blood pressure. J Pediatr 102:288–293
16. Perlman JM, Volpe JJ (1983) Suctioning in the preterm infant: effects on cerebral blood flow velocity, intracranial pressure and arterial blood pressure. Pediatrics 72:329–334
17. Perlman JM, McMenamin JB, Volpe JJ (1983) Fluctuating cerebral blood flow velocity in respiratory distress syndrome. N Engl J Med 309:204–209

18. Perlman JM, Goodman S, Kreusser KL, Volpe JJ (1985) Reduction in intraventricular hemorrhage by elimination of fluctuating cerebral blood flow velocity in preterm infants with respiratory distress syndrome. N Engl J Med 312:1353–1357
19. Pryds O, Greisen G, Lou H, Friis-Hansen B (1989) Heterogeneity of cerebral vasoreactivity in preterm infants supported by mechanical ventilation. J Pediatr 115:638–645
20. Pryds O, Greisen G, Lou H, Friis-Hansen B (1990) Vasoparalysis is associated with brain damage in asphyxiated term infants. J Pediatr 66:685–689
21. Raju TNK, Kim SY (1991) The effect of hematocrit alterations on cerebral vascular PCO_2 reactivity in newborn baboons. Pediatr Res 29:385
22. Reynolds ML, Evans CAN, Reynolds EOR, Saunders NR, Durbin GM, Wigglesworth JS (1979) Intracranial hemorrhage in the preterm sheep fetus. Early Hum Dev 3(2):163–186
23. Saliba E, Laugier J (1992) Doppler assessment of the cerebral circulation in pediatric intensive care. Crit Care Clin 8:79–91
24. Saliba E, Santini JJ, Arbeille PH, Chergui A, Gold F, Pourcelot L, Laugier J (1985) Mesure non invasive du flux sanguin cérébral chez le nourrisson hydrocéphale. Arch Fr Pediatr 42:97–102
25. Saliba E, Dharmapuri V, Shankararao R, Philobos M (1991) Effects of hemodilution on cardiac ouptut, cerebral blood flow and cerebral oxygen consumption. Pediatr Res 29:1378
26. Saliba E, Raju TNK, John EG, Fornell LC et al. (1991) Acute effect of hypo- and hypercapnia on systemic, cerebral and renal hemodynamics. Pediatr Res 29:2161
27. Younkin DP, Reivich M, Jaggi JL, Obrist WD, Delivorio-Papadopoulos M (1987) The effect of hematocrit and systolic blood pressure on cerebral blood flow in newborn infants. J Cereb Blood Flow Metab 7:295–299

The Electroencephalogram in Birth Asphyxia

M.-F. Radvanyi-Bouvet and J.-P. Relier

In the first days of life following birth asphyxia, the assessment of cerebral function is very important. At this critical stage, clinical examination, neuro-radiological imaging, and EEG should be included to establish the neurological prognosis. In full-term babies (FTB) and in premature babies (PB), EEG is of help in evaluating the cerebral repercussion of hypoxic episodes.

The developmental changes of cerebral rhythms and their changes related to sleep states must be considered in interpreting the tracings and this constitutes one of the difficulties of this method.

Numerous authors [8–10, 16, 18, 20, 21, 41], in particular Dreyfus-Brisac and Monod in France, have described these changes in normal babies, related to the central nervous system (CNS) maturational level and to sleep states. They have also described EEG patterns observed in sick babies. The EEG was shown to be helpful in the recognition of seizures and in monitoring treatment [1, 5, 10, 28, 31, 33, 39, 41]. The prognostic value of the neonatal tracing was assessed, taking into account the quality of the background EEG activity [8, 18, 22, 25, 32, 38], the presence of discharges (with or without clinical seizures), and the presence of a particular anomaly, positive rolandic sharp waves (PRSs), mainly observed in PB [6].

Without going into too much detail on the EEG recording technique [see 20], we would emphasize some points: it is important to observe newborns continuously, to note their behavior, and also to record their respiration and ECG. These data help to identify the babies' state and are indispensable for interpreting the EEG, detecting subtle seizures, and identifying and excluding artifacts [14–41].

In this chapter, analyzing data from the literature and from our own experience, we summarize information on normal and abnormal neonatal EEGs. During the last 12 years we have analyzed the EEGs of 73 FTB and of 239 PB who suffered birth asphyxia (outcome: Table 1).

Normal Neonatal EEG

Correlations with Gestational Age or Conceptional Age. Cerebral rhythms are related to CNS maturation [see 8, 9, 11, 16, 18, 21, 41]. Recognizable EEG patterns (criteria of maturation) representing maturation levels with a precision of 1 week (wk) either way, have been described, thus allowing the assessement of the gestational age (GA) at birth and later of the

Table 1. Outcome of 73 full-term babies and 239 premature babies who suffered birth asphyxia

Outcome	Full-term babies	Premature babies
Death	33	144
Major sequelae[a]	10	16
Moderate sequelae[b]	1	16
Minor sequelae	3	7
Favorable outcome	19	56
Unknown outcome (age <3 years)	7	0

[a] Major sequelae are defined as motor sequelae (however severe) and mental retardation, and spastic quadriplegia with or without mental retardation.
[b] Moderate sequelae, spastic diplegia.
Follow-up ranged from 3 to 10 years.

conceptional age (CA; CA = GA + legal age). In PB, in the absence of neurological complications, the cerebral rhythms continue to evolve in the same way as they did in utero and are related to the CA. For instance a 12-week-old PB, born at 28 wks has a CA of 40 wks and a tracing similar to that of the FTB.

Correlations with Wakefulness and Sleep State. In a normal FTB one can discern, besides wakefulness, two main sleep states characterized by behavioral [26] and EEG criteria (Anders et al. in [8]).

In active sleep (AS) there are many rapid eye movements (REM), movements of the extremities, facial movements, and irregular breathing. There is continuous EEG activity with slow waves prevailing on posterior areas. AS is also named irregular sleep or state 2 by Prechtl [26].

In quiet sleep (QS), there are no REM, body movements are rare (except some startles), nonnutritive sucking can be present, and breathing is regular. The *tracé alternant* described by Samson-Dolfus [in 8] as typical of QS, is characterized by a period of slow waves synchronous on the two posterior areas, lasting 3–5 s and alternating with equally long sequences of low-voltage theta rhythms. QS is also named regular sleep or state 1 by Prechtl [26].

In wakefulness and in the period of AS following QS, low-voltage theta rhythms are recorded called *tracé d'activité moyenne* [21].

In PB, criteria vary according to the GA. The discontinuous EEG characterizes the bioelectrical activity of the immature brain.

In early prematurity (GA < 32 wks) AS and QS can be defined with only two criteria: the presence of REM and continuous EEG in AS and the absence of REM and discontinuous EEG in QS; body movements are also less frequent in QS. High-voltage bursts of slow waves and theta rhythms (up to 300 μV) are separated by intervals of "silent" periods (activity < 5 μV), these interburst intervals lasting up to 40 s at 26 wks of GA (Fig. 1). The discontinuity decreases as GA increases. After 32 wks (Fig. 2), the

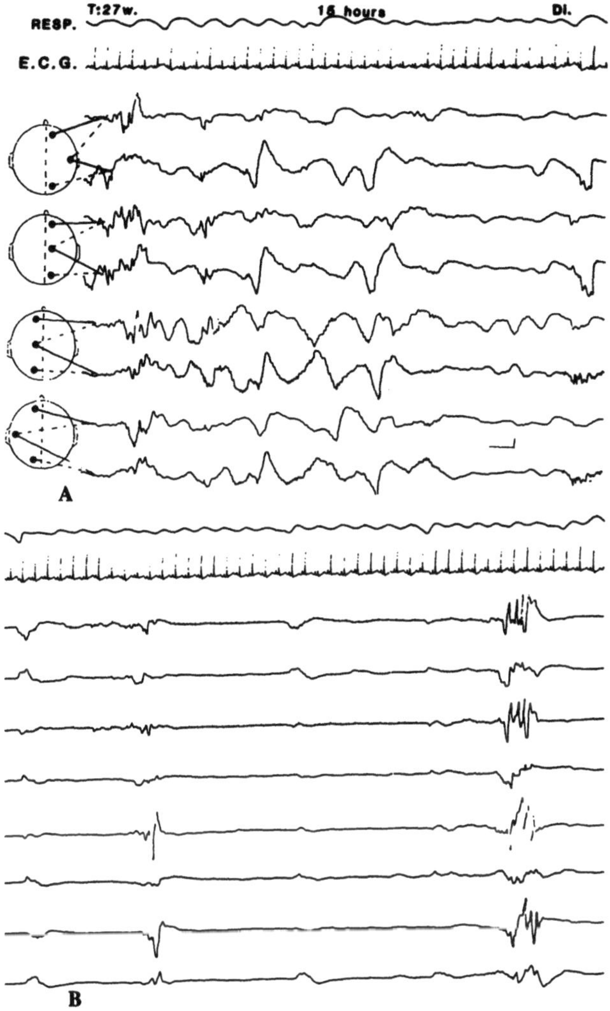

Fig. 1. GA 27 wks. First day of life, two types of trace: **A**, continuous (with some REM): high-voltage (up to 300 μV) slow waves prevailing on the occipital areas, some on the rolandic regions sometimes superimposed with occasional theta and rapid rhythms; **B**, discontinuous: some slow waves on temporal or posterior areas superimposed with some theta rhythms, high-voltage bursts of theta rhythms prevailing on fronto-temporal areas. Interburst intervals lasting 4–5 s. Calibration in all the figures is: 1 s, 50 μV, time constant 0.3

amplitude of slow waves decreases and theta rhythms are replaced by rapid rhythms (delta brushes) superimposed upon the slow waves, these decrease after 39 wks. At about 34 wks, some frontal sharp transients (*encoches frontales*) with various amplitudes and frequencies can often be observed during sleep. At about 36–37 wks, slow anterior dysrythmia, also called *dysrythmie lente antérieure*, can appear on the frontal regions (1–4 cycles/s,

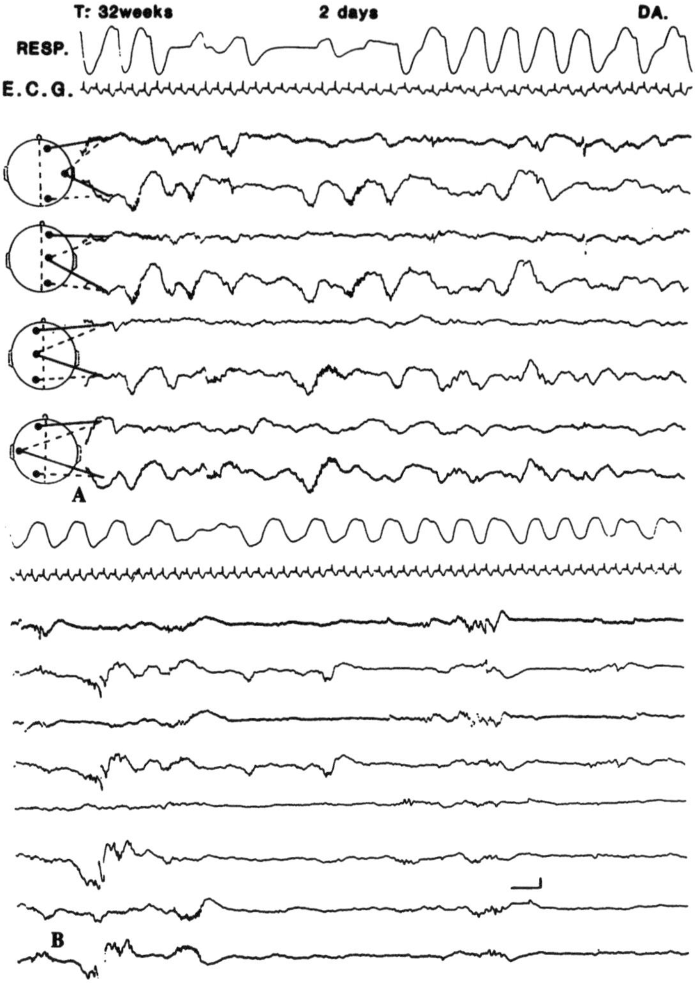

Fig. 2. GA 32 wks. Second day of life, two types of trace: **A**, continuous (with REM): the amplitude of the slow waves has decreased; numerous rapid rhythms and rare theta rhythms; **B**, discontinuous: slow waves on temporal or occipital areas, some theta rhythms, and short interburst intervals. Note a gross interhemispheric asynchrony often observed in "normal" premature babies

$50-150\,\mu V$). The discontinuous trace is replaced by the *tracé alternant* during QS.

To conclude, any valid interpretation of neonatal EEG must take the maturation and sleep states into account. The GA or the CA of the babies must be known (a very discontinuous EEG, for instance, does not have the same significance at 28 wks as it does at 34 wks of GA) and recordings should be of a sufficient duration to obtain the different types of EEG: some anomalies appear mainly during QS and the presence of both AS and QS (sleep cycling) denotes the maintenance of some CNS regulation.

EEG After Birth Asphyxia

Neonatal asphyxia may have different consequences depending on GA. In (full-term) FT and near-term (NT) newborns (GA ⩾ 35 wks), diffuse neuronal necrosis of the cortical, basal ganglia, and brain-stem regions is the main lesion; leukomalacia can also occur. In PB lesions are mainly intraventricular or periventricular hemorrhages and leukomalacia which may sometimes evolve later into porencephalic cysts. Lesions of neuronal necrosis are less frequent than in the FT newborn. As a consequence, after severe asphyxia or hypoxia EEG rhythms are more depressed in FT and NT babies than in early PB.

The EEG can be analyzed by considering (a) patterns and significances of background EEG activity and (b) anomalies, either isolated or associated, superimposed on this background: EEG discharges with or without clinical seizures and PRSs.

Patterns of Background EEG Activity

Inactive and Paroxysmal Records in FTB

In FTB, two patterns usually have a bad prognostic significance as they are related to severe neuronal necrosis [8]: inactive records, i.e., activity <5 μV (nearly isoelectric) and paroxysmal records, i.e., activity with bursts lasting <10 s, interburst intervals with "silent" periods (lasting 10 s to >1 min), morphology of burst rhythms different from any normal pattern and varying from one burst to another, varying amplitude and frequency of the EEG activity within a single burst, and interhemispheric asynchrony and antero-posterior asynchrony of burst activity being seen frequently. A paroxysmal EEG (also called a suppression-burst pattern) must be distinguished from the discontinuous EEG normally recorded in PB.

In these very abnormal EEGs, recording is necessary for at least 2 h, with periods of tactile or auditory stimulation showing no modification of the EEG. An inactive EEG should be confirmed by recordings of long duration, high-gain and long-distance montage. An inactive tracing may be transformed into paroxysmal tracing after 1–2 days of life. A paroxysmal tracing may then last for 1–2 wks and is followed by a very abnormal recording. These patterns are now seldom observed. The EEGs of two cases of severe birth asphyxia and unfavorable outcome are shown in Figs. 3 and 4.

The Time of Recording

The EEG pattern is important, but in FTB, its significance depends also upon the time of recording. Pezzani et al. [25] have described the good prognostic value of the EEG pattern in the first 24 h of life. At this time, the prognosis appears favorable with a normal EEG (two types of trace and sleep cycles), whatever the neurological state (Fig. 5.). The prognosis is

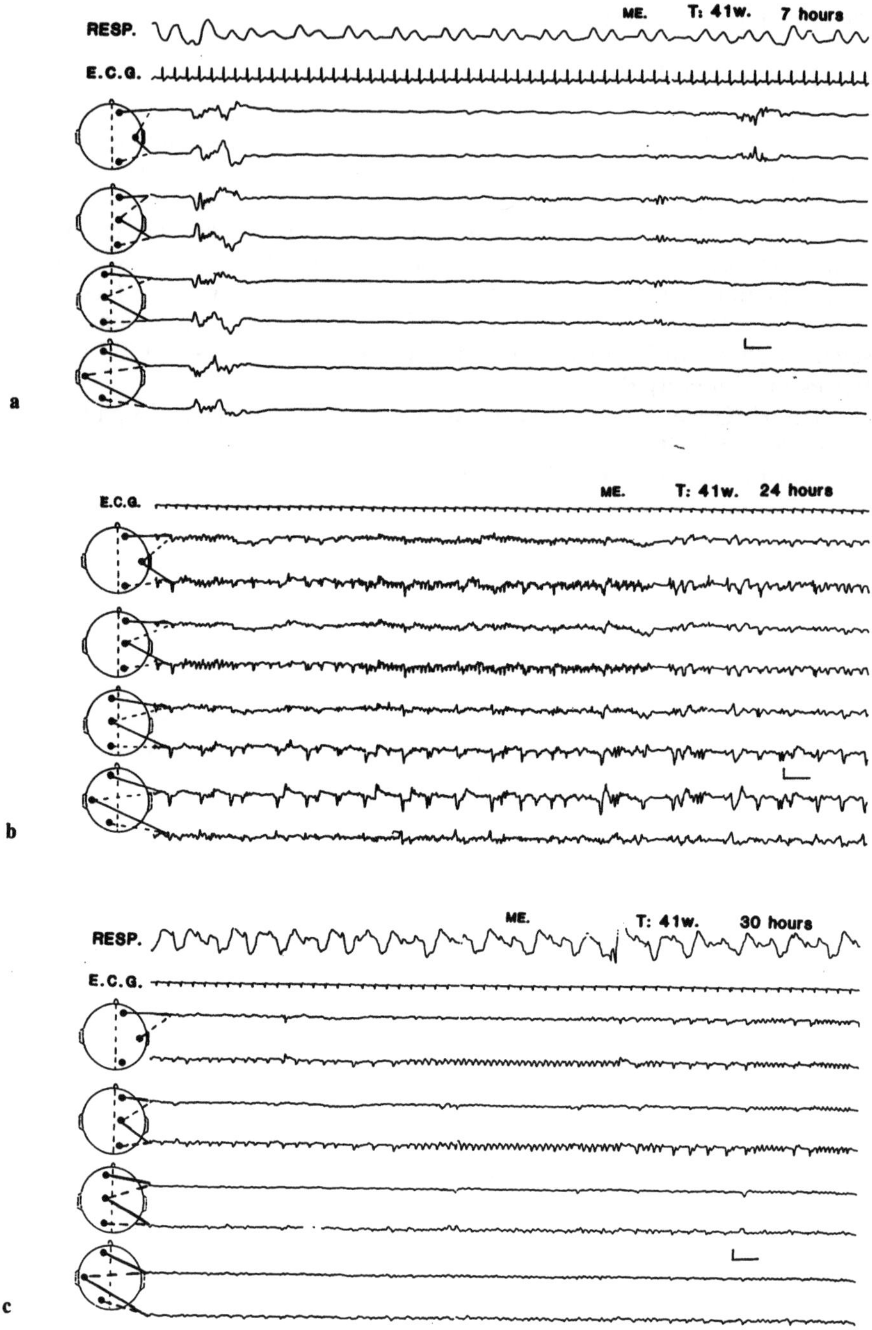

Fig. 3a–c. GA 41 wks. Severe birth asphyxia leading to death. Neuropathological examination revealed diffuse neuronal necrosis. **a** 7 h (coma): very discontinuous EEG; occasional slow waves on the temporal, rolandic, and occipital regions; theta rhythm bursts prevailing on the rolandic areas. No physiological rhythms related to the GA. (This pattern persisted for several hours and discharges appeared after 13 h – Fig. 3b and c). **b** The same baby at 24 h (coma): multifocal discharges with various frequencies: bursts of rapid rhythms followed by rhythmic slow waves on the right hemisphere, slower periodic spikes on the left hemisphere. **c** The same baby at 30 h (coma): discharges of lower voltage prevailing on the right occipital areas

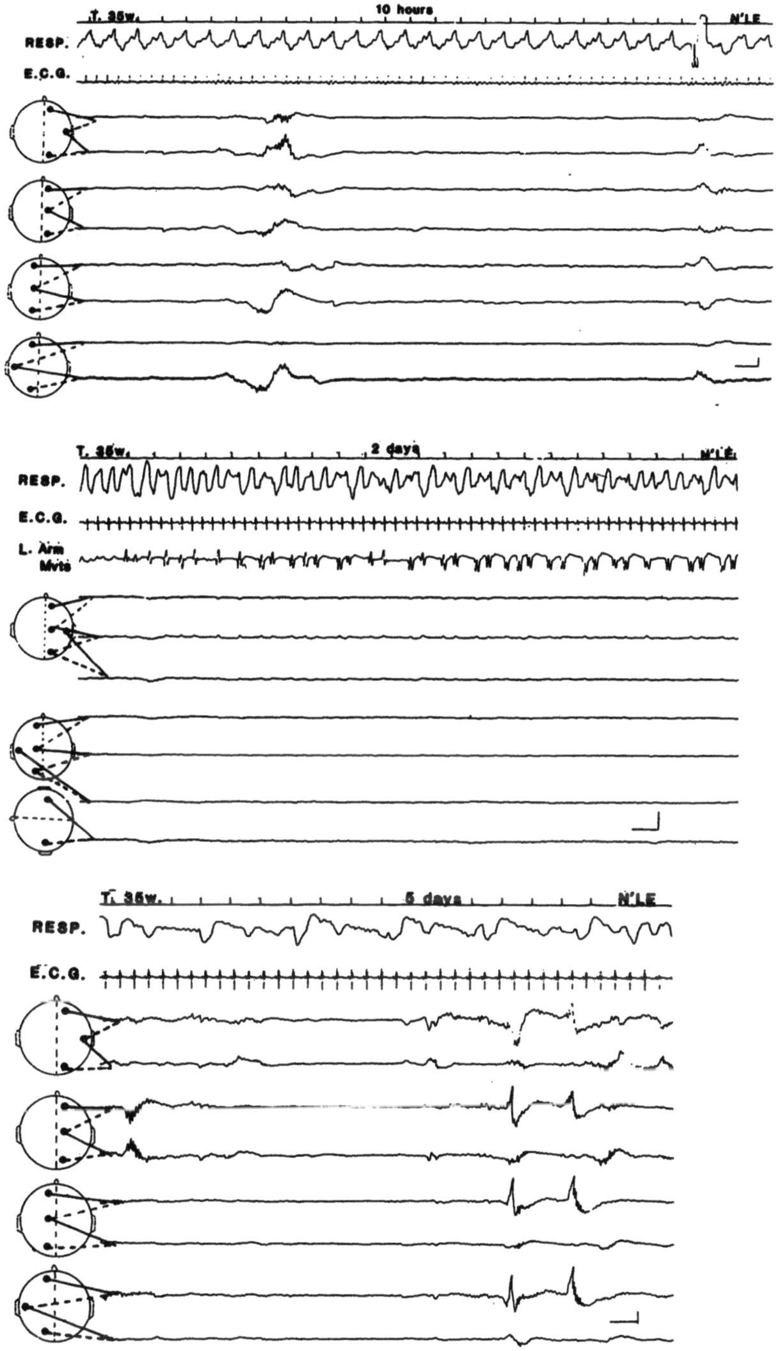

a

b

c

Fig. 4a–c. GA 35 wks. Severe birth asphyxia leading to death. **a** 10 h: EEG too discontinuous for the GA. Presence of some physiological rhythms with slow waves and rapid rhythms prevailing on the occipital regions. **b** The same baby on the second day: Clonic movements of the left arm recorded (3rd channel). Inactive background EEG with some periodic spikes of low voltage on the right temporal areas. **c** The same baby on the fifth day: Very abnormal EEG with some slow waves, spikes and rapid rhythms prevailing on the frontal areas, occasional slow waves and rapid rhythms on the right rolandic and occipital regions

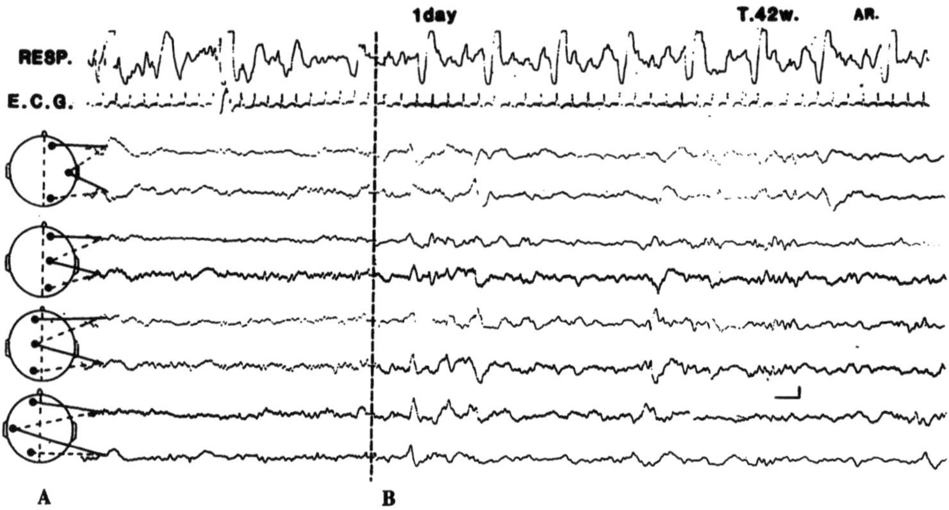

Fig. 5a,b. GA 42 wks. After severe birth asphyxia this baby had a slightly abnormal EEG on the first day of life. Seizures occurred during the second day of life but the outcome was favorable. At 17 h, two types of trace: **A**, slow waves superimposed with theta rhythms prevailing on the occipital areas, and **B**, High-voltage activity, with some slow waves, and occasional spikes on temporal and rolandic areas, rolandic theta bursts. (This initial EEG allowed a good prognosis, in spite of the depressed interictal EEG observed later, probably due to anticonvulsant drugs)

poor for inactive or paroxysmal traces and also for traces with bursts lasting < 6 s and with silent periods lasting > 40 s and never < 3 s.

In an EEG study of 20 FTB in a comatose state lasting for 4–15 days after birth asphyxia, Scavone et al. [32] have shown that the initial tracing on the first day of life has the most significant prognostic value. A transitory EEG improvement on the second or third day was followed by a worsening and an unfavorable outcome in three of these cases. We have also recently observed the same outcome in three more cases.

An inactive EEG during the first 12 h of life does not always imply a poor prognosis: six newborns had a favorable outcome although they had an inactive EEG during this period [25, 34 plus three personal cases]. It is therefore necessary in these cases to record long-lasting and serial EEGs. In the three personal cases we observed the rapid recovery to a normal background EEG within the first 24 h.

Takeuchi and Watanabe [35] have found in a study of 176 babies in the first 3 weeks of life that all babies with a normal EEG before the seventh day developed normally, but those with an EEG which was only normal after 7 days of life occasionally developed handicaps. A marked or maximal depression of EEG rhythms at any time, a moderate depression after the seventh day, or a mild depression after the 12th day all precede neurological handicaps.

Depressed EEGs are defined as follows:

- Minimal depression: various patterns are recognizable, the low-amplitude phase of *tracé alternant* is abnormally attenuated (5–20 μV)
- Mild depression: high-voltage slow wave pattern is decreased, the low-amplitude phase of *tracé alternant* is greatly attenuated
- Moderate depression: two EEG patterns made up of discontinuous or abnormal *tracé alternant* and a very low voltage irregular pattern
- Marked depression: a single pattern consisting of bursts of high- to medium-amplitude activity of various kinds interrupted by near isoelectric activity of long duration
- Maximal depression: nearly isoelectric

Inactive and Paroxysmal EEG in PB

In PB inactive and paroxysmal EEG can be observed with the same poor prognostic significance [36]. However, some physiological rhythms may persist in the days following even a severe insult in early prematurity. This is due probably to a reduced sensitivity of immature neurons to hypoxia. In 100 babies of 26–32 wks' GA with early death and major cerebral lesions found at autopsy, we observed inactive tracing in only 6%. Inactive tracings were only recorded in cases of very large parenchymal hemorrhages or hydranencephaly. The percentage of inactive tracing was 17% (4 out of 23 cases) in babies with a GA of 33–36 wks whose death was due to neonatal asphyxia. Three of the four cases with inactive EEGs on the first day of life also had discharges and later they had very discontinuous EEGs with absent physiological rhythms.

Ellison et al. [11] attributed a bad prognostic significance to an excessively discontinuous EEG (interval interbursts lasting > 60 s or intervals lasting 15–30 s plus seizure activity > 10 s). We studied the evolution of 18 very discontinuous EEGs (interburst intervals up to 250 s), recorded on the first day of life. In the following days, the discontinuity decreased and numerous Positive Rolandic Sharp Waves (PRSs) appeared in 16 cases. In two cases, the EEG appeared normal on the third day and the outcome was favorable (the depressant effect of anticonvulsant drugs could be suspected in these two cases).

The Absence of an EEG Pattern Related to GA or CA

In FTB and PB the absence of an EEG pattern related to GA or CA is also markedly abnormal and has a poor prognostic significance, even without the characteristics of a paroxysmal trace. This type of abnormality can be observed at birth or appears progressively on successive EEGs [29]. Discrepancies between the maturational level can be observed during the two sleep states. Usually the maturational level seems to be related to a younger GA during QS than during AS. A delay in maturation >4 wks is considered a marked anomaly at 30–35 wks and a mild anomaly after 35 wks by Ellison et al. [11].

The Effects of Depressant Drugs in FTB and in PB

Inactive, paroxysmal, and excessively discontinuous traces in FTB and PB cannot be interpreted satisfactorily if the presence or absence of CNS depressant drugs (mainly anticonvulsants) is not known: these drugs can affect EEG background activity [41]. It is important to know the dose, time-schedule, and blood level of these drugs and also whether the mother received such treatment at the end of the pregnancy or during labor. Couto-Sales et al. [in 30] did not find EEG depression on the first day of treatment with diazepam or phenobarbital, but Plouin et al. [in 25] have observed two FTB with paroxysmal recordings and a favorable outcome after anticonvulsant therapy. In FTB and PB with high levels of benzodiazepine or phenobarbital, we observed low-voltage hypoactive or high-voltage very discontinuous EEGs (Fig. 6), lasting several hours (interburst intervals lasting up to 240 s in PB). In cases with a favorable outcome, the excessive discontinuity disappeared and continuous activity was observed, sleep cycles occurring in spite of persistent high blood levels of anticonvulsant. With

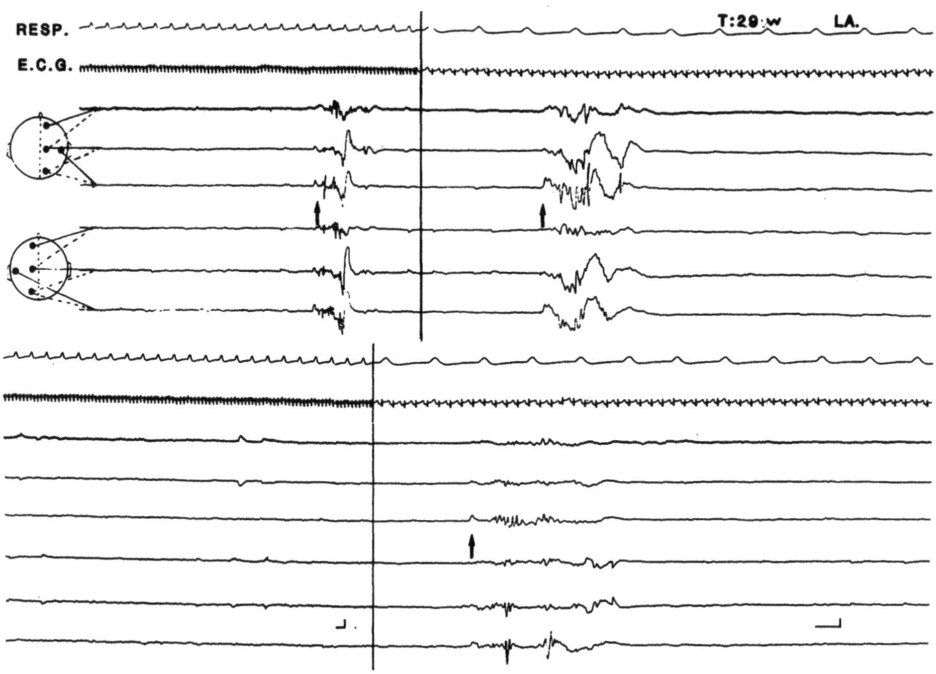

Fig. 6. GA 29 wks. At 15 days abnormal movements were observed and phenobarbital was administered IV 1 h before the tracing (blood level: 80 μg/ml.). Eventual favorable outcome. Very discontinuous EEG (coma): high-voltage bursts with slow waves and theta rhythms prevailing on occipital areas spontaneously or after stimulation (*arrow*, noise: clap). (Probably related to this high-anticonvulsant blood level, the discontinuity decreased within 4 h, normalization rapidly occurred, and periods of continuous activity were then recorded)

lidocaine infusion used as anticonvulsant treatment, we have observed (in about two out of three cases in FTB and PB) a very discontinuous EEG with high-voltage theta rhythms and sharp waves prevailing on the frontal areas. This pattern can occur either at the beginning of infusion or after a delay (up to 24 h in FTB). In cases with a favorable outcome, the discontinuity disappears while lidocaine is still being infused [30].

Abnormalities Superimposed on the Background EEG

Seizures and Electrical Discharges

Various types of seizures can be observed: electroclinical seizures with clinical manifestations occurring simultaneously with all or part of the EEG discharges, EEG discharges without any clinical phenomena (motor, ocular or vegetative), and clinical manifestations without associated EEG discharges.

Since 1960 (Harris and Tizard in [25]), EEG recordings have been shown to be useful in FTB as well as in PB for recognizing seizures in the neonatal period which are often subtle and atypical [1, 7, 8, 10, 16, 17, 18, 22, 27, 28, 33, 39]. EEGs help in identifying seizures, monitoring the effectiveness of treatment, and deciding when anticonvulsant therapy can be stopped [31, 39]. An EEG discharge is usually defined as an abnormal rhythmic activity different from the background EEG, lasting for at least 10 s and sometimes several hours. Various patterns of EEG discharges are seen in Fig. 3b,c and Fig. 4b. The morphology, amplitude, frequency, polarity, and localization of the abnormal rhythms may vary during the same discharge or during successive discharges and are not related to etiology or to GA. Patterns are rarely polymorphic and are often synchronous on various regions in PB. FTB exhibit more polymorphic pattern discharges, with multifocal seizures often discharging independently on the various regions (as seen in Fig. 3b); electroclinical seizures are often observed with discharges lasting more than 11 min [27]. Unilateral status epilepticus or discharges are rarely observed in the neonatal period; Bour et al. [4] described strictly lateralized seizures or discharges always beginning on the same hemisphere in 21 cases out of 250; five cases had suffered birth asphyxia, one of them had a cerebral infarct.

The percentage of babies presenting seizures after birth asphyxia is difficult to assess, but birth asphyxia is the most frequent etiology of seizures in neonates. In 150 cases of clinical seizures (in FTB and PB) analyzed by Calciolari et al. [5] two out of three were related to hypoxic–ischemic encephalopathy. In Port-Royal, out of 90 cases with seizures confirmed or discovered by EEG in FTB, 75% were related to birth asphyxia. We registered 73 FTB who suffered severe birth asphyxia and found EEG discharges in 67 cases, four had abnormal movements without EEG discharges. Of 43 babies who had a tracing on the first day of life, 15 out of 26 with an unfavorable outcome and eight out of 17 with a favorable outcome presented seizures. Thus, the presence of seizures on the first day of life, classically considered as having a poor prognostic significance [8, 22, 25], does not always imply a bad prognosis. In ten cases which had an inactive EEG on the first day of life (unfavorable outcome), discharges (with or without clinical manifestations) appeared during the following days on a very abnormal background. Three babies who had a slightly abnormal EEG in the same period (as in the case of Fig. 5) and later presented seizures, developed favorably. The tracing on the first day of life allowed us to hope for a good prognosis in spite of seizures, with an interictal EEG the second day probably depressed by anticonvulsant drugs.

We found seizures occurring more frequently in PB with a "greater" GA than in early premature babies. In 100 cases with cerebral lesions confirmed after early death, 40 had seizures confirmed by EEG and their GA was significantly greater than in 60 cases without seizures (mean ± SD, 30.8 wks ± 3.1 versus 29.6 ± 2.93; *t* test 0.03).

Occasional seizures are not considered as grave as status epilepticus. However, the definition of status epilepticus is not easy during this period [7].

In all cases, the interictal EEG, irrespective of the pattern and duration of discharges, has the most important prognostic value. The presence of physiological rhythms and two types of EEG imply a good prognosis even if seizures are frequent and long-lasting, whereas a markedly abnormal background EEG (inactive or paroxysmal traces or traces without any physiological pattern) precedes an unfavorable outcome (Fig. 3b,c and Fig. 4b). In FTB and PB we have found the worst outcome to be associated with short discharges in newborns with an abnormal background EEG, a badly damaged brain probably being unable to sustain long-lasting discharges [27]. Clinical phenomena without electrical discharges, but with very abnormal EEGs also precede an unfavorable outcome.

In order to assess the background EEG quality it is useful to try and obtain tracings before initiating anticonvulsant treatment. Discharges without clinical manifestations occur frequently after anticonvulsant therapy which may also depress the background EEG. Anticonvulsant therapy often suppresses the clinical part of seizures without stopping discharges which are only shortened. Curarization also suppresses the clinical events. In both these cases, the EEG detects persisting discharges.

In FT newborns who needed admission into our intensive care unit, the prognosis of EEG-confirmed seizures has remained poor in recent years: only 43% (out of 21 FTB between 1985 and 1989) had a favorable outcome. In PB, the prognosis was even worse, with only 8% out of 50 cases of seizures having a favorable outcome [28]. However, Andre et al. [1] described a favorable outcome after seizures in about 50% of PB with an older GA (43% < 33 wks compared with 60% in the previous study).

Considering the seriousness of the prognosis of seizures in the neonatal period, it is important to assess this diagnosis. Some authors do not record the EEG, thus missing seizures not associated with clear clinical phenomena or considering abnormal movements, which are sometimes only reflections of excessive motility during active sleep, as seizures.

Positive Rolandic Sharp Waves (PRSs)

Sharp waves with positive polarity on the rolandic areas are observed mainly in PB, rarely in FTB, and never in older infants or adults. PRSs were described in 1972 by Cukier et al. [6] in 23 out of 36 PB with extensive intracerebral hemorrhages (two out of three confirmed at autopsy). Blume and Dreyfus-Brisac [3] differentiate between two types of PRSs, isolated (type A) and grouped (type B), with only type A PRSs having a bad prognostic significance. Type A PRSs are isolated events in which voltage and morphology clearly differ from the background EEG. Type B PRSs occur in bursts with the voltage differing less from the background activity than those of type A. Characteristic PRSs isolated or associated with some slow waves and rapid rhythms are sometimes better seen on a discontinuous EEG during QS (Figs. 7 and 8). To evaluate the occurrence and frequency of PRSs on successive EEGs, it is necessary to compare recordings taken during the same sleep states, preferably during QS with discontinuous EEG.

When numerous (>1/min), whatever the background EEG activity, type A PRSs precede severe sequelae (Murat in [29]). With the use of neuroradiological techniques, they are now considered as markers of periventricular lesions (ischemic and/or hemorrhagic) rather than of intraventricular hemorrhages only [24]. PRSs are seldom observed in FTB, sometimes related to a softening of the white substance [8].

We have analyzed 75 cases (during the last 10 years) who had numerous PRSs in order to evaluate the time they occurred, their course, and their

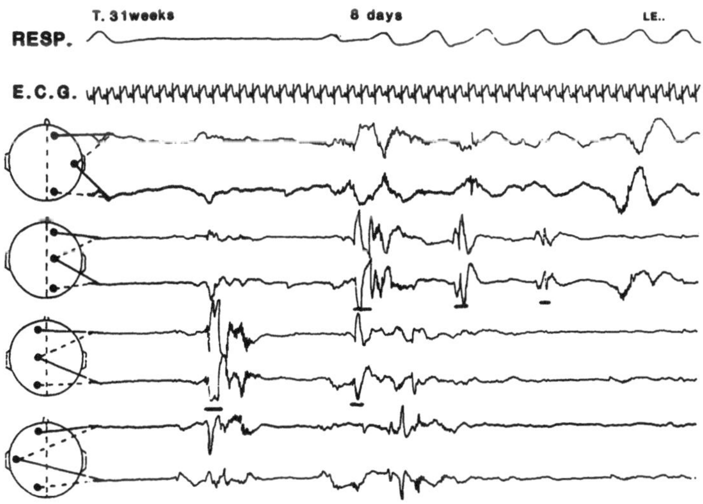

Fig. 7. GA 31 wks, major sequelae. Discontinuous EEG during quiet sleep. *Underlining* indicates numerous PRSs, bilateral or prevailing either on the left or on the right, associated with some slow waves and rapid rhythm. Some physiological rhythms only on the right occipital area

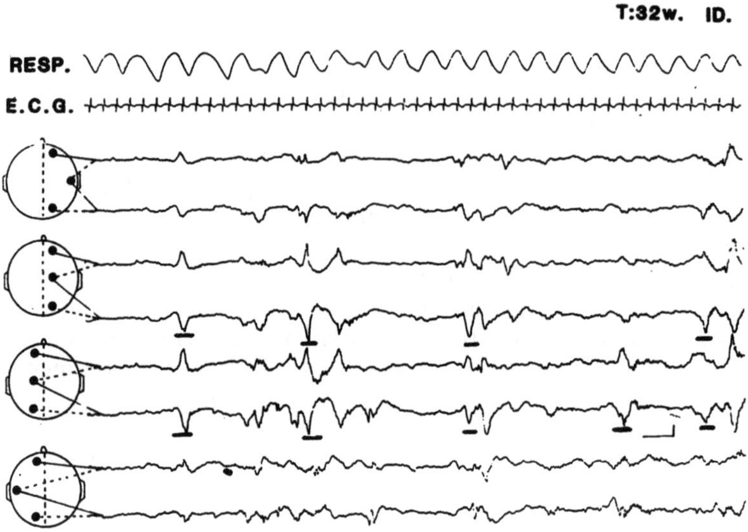

Fig. 8. GA 32 wks, major sequelae. Slightly discontinuous EEG during quiet sleep. *Underlining* indicates PRSs, bilateral or prevailing on the left. Presence of some physiological rhythms with slow waves and rapid rhythms on the occipital areas

pathological and prognostic significance (264 EEGs have been recorded in these cases during the first 2 weeks of life).

Time of Occurrence of PRSs. EEGs were recorded on the first day of life in 24 cases. PRSs were already present in five cases; an abnormally discontinuous EEG was recorded in 16 cases (two also having numerous PRSs) and a normal EEG in five cases. PRSs were present on the following 2–4 days (EEGs subsequently recorded between the second and the fourth day), in all these cases. Out of the five cases with a normal initial EEG, numerous PRSs appeared between the third and ninth day of life.

The GA of these babies with numerous PRSs varied from 27 to 35 wks (mean ± SD = 30.8 ± 2).

Pathological Significance. Early death occurred in 52 cases (GA 25–36 wks). Autopsy findings from 38 cases showed large leukomalacia (associated with hemorrhage in nine cases, with neuronal necrosis in seven and cerebral infarctus in one).

We compared the GA of these 38 cases with PRSs and major cerebral lesions to the GA of 57 other cases with comparable cerebral lesions but without PRSs and found no significant differences (mean ± SD = 30.5 wks ± 2.1 versus 29.3 ± 3.08).

Prognostic Significance. Twenty-three babies with numerous PRSs survived – 14 developed into severe sequelae and five into moderate sequelae. In these 19 cases, PRSs persisted for many days (3–12 wks) and were bilateral.

The maturation of background EEG was ongoing during this period, even in the cases of major sequelae. One case had minor sequelae. In this case, which had a normal EEG on the first day of life, numerous PRSs appeared on the third day and persisted on four subsequent EEGs (up to 32 days), but remained strictly unilateral. Three cases had a favorable outcome. In these cases, PRSs were bilateral and numerous but they were only present once (on one recording respectively the first, second, and fifth day of life and disappearing the following days).

To summarize PRSs may (1) already be present on the first day of life at various GA in PB, (2) rapidly appear when cerebral lesions occur or follow an abnormally discontinuous EEG, and (3) persist for several weeks, having an important poor prognostic value related not only to presence but to frequency (a poor prognosis is implied when they are numerous – >1/min, bilateral, and persisting on successive EEGs), irrespective of the background EEG activity.

Prognostic Value of a Normal or Mildly Abnormal EEG After Birth Asphyxia

The prognostic value of a normal or a slightly abnormal EEG depends on the GA of the baby and on the time of recording. In FT newborns these patterns imply a good prognosis in the first 24 h of life, but their prognostic value decreases with time and does not have any value after 12 days of life.

In premature newborns, a normal EEG or only slightly abnormal EEG in the first days of life may precede moderate sequelae: in 16 cases with major sequelae, we observed only one case with normal EEGs (at 1, 2, and 18 days) but 11 with moderate sequelae (eight out of ten with a GA < 30 wks, three out of six with a GA > 30 wks). Thus a normal EEG in early prematurity is of less prognostic value. It is sometimes later in development that abnormalities appear; for example, in one of our published cases born with a GA of 28 wks, cerebral rhythms did not evolve on successive EEGs (29). The EEGs of five babies who died with hydrocephaly (death between 14 and 90 days) did not show any significant abnormalities.

Conclusion

In Full-Term Babies. The EEG gives more information when recorded in the first 24 h of life: a normal EEG implies a good prognosis, a very abnormal background EEG after 12 h of life precedes severe sequelae. Serial EEGs may sometimes be helpful, when the EEG on the first day does not provide a definite prognosis.

In Premature Babies. A very abnormal background EEG and inactive EEGs have a bad prognostic significance, but are rarely observed. More often abnormalities consist of numerous abnormal PRSs superimposed on a more or less normal background EEG. Numerous and persisting PRSs imply a high risk of severe sequelae. These abnormalities can be present on the first day of life, more often, however, they appear after a delay and one EEG on the first or second day of life is insufficient, especially when a very discontinuous EEG is recorded. The poor prognostic value of numerous persisting and bilateral PRSs does exist even with the persistence of physiological rhythms. With these abnormalities it is important to monitor these babies carefully as they are at risk of severe sequelae. The early institution of care can be helpful for the babies and their families.

A normal EEG or an EEG without significant anomalies has various significant values depending on the GA. It has less prognostic value in early PB born at a term of ≤30 wks than later. Other abnormalities may rarely appear after some delay and physiological rhythms do not develop on successive EEGs.

In FT and Premature Newborns. In the presence of a very discontinuous, paroxysmal or inactive EEG, caution must be observed as regards prognosis as a good recovery can occur with this pattern in the first hours following birth, or with various depressant drugs. In babies with seizures, prognosis depends mainly on the interictal EEG background, sometimes depressed after anticonvulsant therapy.

To conclude, combined clinical examination, neuroradiological imaging, and EEG [40] are helpful in evaluating the risks of severe handicap following birth asphyxia. Some uncertainty still remains with these methods but it is necessary to avoid imprecise decisions.

Newer techniques of integration and spectral analysis of cerebral rhythms, recordings on magnetic tapes are not yet commonly used [2, 12, 13, 37]. They may visualize discharges, recognize discontinuous or hypoactive EEG, and follow the course of these patterns, but they are not used to identify PRSs which are a valuable EEG indicator in premature babies.

Acknowledgements. We are indebted to C. Fallet, J.C. Larroche, and F. Narcy who performed neuropathological examinations, to C. Bordarier for helpful discussions, to the EEG technicians for skillful assistance, and specially to F. Morel-Kahn, L. Rapon, and L. Salignac.

References

1. Andre M, Matisse N, Vert P, Debruille CH (1988) Neonatal seizures. Recent aspects. Neuropediatrics 19:201–207
2. Bell AH, McClure BG, Hicks EM (1990) Power spectral analysis of the EEG of term infants following birth asphyxia. Dev Med Child Neurol 32:990–998
3. Blume WT, Dreyfus-Brisac C (1982) Positive rolandic sharp waves in neonatal EEG: types and significance. Electroencephalogr Clin Neurophysiol 53:277–282
4. Bour F, Plouin P, Jalin C et al. (1983) Les états de mal unilatéraux an cours de la période néonatale. Rev Electroencephalogr Neurophysiol Clin 13:162–167
5. Calciolari G, Perlman JM, Volpe JJ (1988) Seizures in the neonatal intensive care unit of the 1980s. Clin Pediatr 27:119–123

6. Cukier F, Andre M, Dreyfus-Brisac C (1972) Apport de l'E.E.G. au diagnostic des hémorragies intra-ventriculaires des prématurés. Rev Electroencephalogr Neurophysiol Clin 2:318–322
7. D'Allest AM, Navelet Y (1983) Electroencéphalogramme et crises convulsives néonatales. Rev de Ped 19:189–196
8. Dreyfus-Brisac C (1979) Neonatal encephalography. In: Scarpelli EM, Cosmi EV (eds) Review in perinatal medicine, vol 3. Raven, New York, pp 397–472
9. Dreyfus-Brisac C (1981) Veille et sommeil au cours des premières semaines de la vie. In: Minkowski A (ed) Biologie du développement. Flammarion, Paris, pp 152–176
10. Dreyfus-Brisac C, Peschanski N, Radvanyi M-F et al. (1981) Convulsions du nouveau-né. Aspects clinique, électrographique, étiopathogénique et pronostique. Rev Electroencephalogr Neurophysiol Clin 11:367–378
11. Ellison P, Franklin S, Brown P et al. (1989) The evolution of a simplified method for interpretation of EEG in the preterm neonate. Acta Paediatr Scand 78:210–216
12. Eyre JA, Oozer RC, Wilkinson AR (1983) Diagnosis of neonatal seizures by continuous recording and rapid analysis of the electroencephalogram. Arch Dis Child 53:785–790.
13. Greisen G, Pryds O, Rosen I et al. (1988) Poor reversibility of EEG abnormality in hypotensive preterm neonates. Acta Paediatr Scand 77:785–790
14. Guidasci S, Pajot N, Leautier N (1977) Artefacts recueillis sur des enregistrements de nouveau-nés et de prématurés. J Electrophysiol Technol 3:45–55
15. Holmes G, Rowe J, Hafford J et al. (1982) Prognostic value of the electroencephalogram in neonatal asphyxias. Electroencephalogr Clin Neurophysiol 53:60–72
16. Hrachovy RA, Mizrahi EM, Kellaway P (1990) Electroencephalography of the newborn. In: Daily DD, Pedley TA (eds) Current practice in electroencephalography. Raven, New-York, pp 201–242
17. Lombroso CT (1983) Prognosis in neonatal seizures. Adv Neurol 34:101–114
18. Lombroso CT (1985) Neonatal polygraphy in full-term and premature infants. A review of normal and abnormal findings. J Clin Neurophysiol 2:89–103
19. Marret S, Parain D, Samson-Dollfus D et al. (1986) Positive rolandic sharp waves and periventricular leukomalacia in the newborn. Neuropediatrics 17:199–202
20. Monod N (1981) L'électroencéphalogramme au cours de la première enfance. Rev Int Pediatr 112:21–36
21. Monod N, Tharp B (1977) Activité électrographique normale du nouveau-né et du prématuré au cours des états de veille et de sommeil. Rev Electroencephalogr Neurophysiol Clin 7:302–315
22. Monod N, Pajot N, Guidasci S (1972) The neonatal EEG: statistical studies and prognostic value in full-term and preterm babies. Electroencephalogr Clin Neurophysiol 32:529–544
23. Moussali-Salefranque F, Mises J, Plouin P (1983) Significations du tracé discontinu chez le nouveau-né à terme. Rev Electroencephalogr Neurophysiol Clin 13:240–244
24. Novotny EJ Jr, Tharp BR, Coen RW et al. (1987) Positive rolandic sharp waves in the EEG of the premature infant. Neurology 37:1481–1486
25. Pezzani C, Radvanyi M-F, Relier J-P et al. (1986) Neonatal electroencephalography during the first 24 hours of life in full-term newborn infants. Neuropediatrics 47:7–11
26. Prechtl HFR (1974) The behavioural states of the newborn infants (a review). Brain Res 76:185–212
27. Radvanyi-Bouvet M-F, Cukier-Hemeury F, Morel-Kahn F (1981) Décharges critiques chez les prématurés et les nouveau-nés à terme. Rev Electroencephalogr Neurophysiol Clin 11:404–411
28. Radvanyi-Bouvet M-F, Vallecalle M-H, Morel-Kahn F et al. (1985) Seizures and electrical discharges in premature infants. Neuropediatrics 16:143–148
29. Radvanyi-Bouvet M-F, de Bethmann O, Monset-Couchard M et al. (1987) Cerebral lesions in early prematurity. EEG prognostic value in the neonatal period. Brain Dev 9:399–405
30. Radvanyi-Bouvet M-F, Torricelli A, Rey E et al. (1990) Effects of lidocaine on seizures in the neonatal period: some electroclinical aspects. In: Wasterlain CG, Vert P (eds) Neonatal seizures. Raven, New York, pp 275–283

31. Scarpa P, Chierici R, Tamisari L et al. (1983) Criteria for discontinuing neonatal seizure therapy: a long-term appraisal. Brain Dev 5:541–548
32. Scavone C, Radvanyi-Bouvet M-F, Morel-Kahn F et al. (1985) Coma après souffrance foetale aiguë chez le nouveau-né à terme. Evolution électroclinique. Rev Electroencephalogr Neurophysiol Clin 15:279–288
33. Scher MS, Painter MJ (1990) Electroencephalographic diagnosis of neonatal seizures: issues of diagnostic accuracy, clinical correlation, and survival. In: Wasterlain CG, Vert P (eds) Neonatal seizures. Raven, New York, pp 15–25
34. Sternberg B, Frenkel AL, Plouin P et al. (1983) Diagnostic à évoquer devant un tracé inactif de nouveau-né. Rev Electroencephalogr Neurophysiol Clin 13:153–156
35. Takeuchi T, Watanabe K (1989) The EEG evolution and neurological prognosis of perinatal hypoxia neonates. Brain Dev 11:115–120
36. Tharp BR, Cukier F, Monod N (1981) The prognostic value of the electroencephalogram in premature infants. Electroencephalogr Clin Neurophysiol 51:219–236
37. Thornberg E, Thiringer K (1990) Normal pattern of the cerebral function monitor trace in term and preterm neonates. Acta Paediatr Scand 79:20–25
38. Volpe JJ (1987) Brain death determination in the newborn. Pediatrics 80:293–297
39. Volpe JJ (1989) Neonatal seizures: current concepts and revised classification. Pediatrics 84:422–428
40. Vries LS (de), Dubowitz LMS, Dubowitz V, Pennock T (1990) A colour atlas of brain disorders in the newborn. Wolf, Cowell, Ipswich
41. Werner SS, Stockard JE, Bickford RG (1977) Atlas of neonatal electroencephalography. Raven, New York

Multimodality-Evoked Potentials in Perinatal Asphyxia

E. Scalais

Introduction

Asphyxia implies a reduction in the supply of oxygen and substrates to brain tissue. Electrophysiologic measures have been recognized as being coupled to brain function, metabolism, and cerebral blood flow (CBF; Shapiro 1986). Therefore neurophysiologic methods have been applied to assess hypoxic and ischemic neonates and have been used as a prognostic tool. Recent studies of evoked potentials (EPs) gave good results as a means of prognostic evaluations (Willis et al. 1987, 1989; Majnemer et al. 1988; Klimach and Cooke 1989; Muttitt et al. 1991; Taylor et al. 1991, 1992). In this article we discuss the techniques of multimodality-evoked potentials (MEPs) that allow an evaluation of cerebral function and prognostic outcomes in asphyxia.

Neurophysiologic Techniques for Cerebral System Evaluation

Technical Aspects of Evoked Potential Recording

The electrical activity recorded from electrodes placed on the scalp includes the EEG and the EPs. The EEG represents the spontaneous, ongoing electrical activity of the brain. In contrast the EPs represent electrical activity induced in the brain by external stimuli (visual, auditory, or somatosensory stimuli). Because of the low amplitude of the EPs (μV) and their admixture with normal background EEG (mV), signal averaging is accomplished by separating the buried EPs (the signal) from the background EEG (the noise; Regan 1989; Duffy et al. 1989). The number of presentations of the stimulus depends on the signal-to-noise ratio or the size of the EPs relative to the background EEG. To extract visual-evoked potentials (VEPs; 20–30 μV) from the background EEG (60–100 μV), 100 repetitions of the visual stimulus are normally used. But because of the lower amplitude, brain-evoked auditory responses (BEARs; $<1\,\mu$V) required more stimulus repetitions (2000–3000).

Visual-Evoked Potentials

Visual evoked potentials represent electrical activity induced in the visual system by light stimuli that reach the retina. At the level of the retina different regions are specialized for different functions. The macula (rich in cones) is specialized for fine visual discrimination and good visual acuity (photopic vision), while the periphery of the retina (poor in cones and rich in rods) has a high sensitivity for the detection of light but poor discrimination (scotopic vision) (Celesia 1985).

The retinocortical projections of the photopic and scotopic system are different: the fibers originating from the cone-dominated retina project mainly to the primary visual cortex (area 17 of Brodmann) via the lateral geniculate body and the optic radiations. The fibers originating from the rod-dominated retina project mainly to the pretectal and tectal area and to the cortex (area 19 of Brodmann, frontal and parietal area) (Regan 1989; Mayer 1990; Guerit 1991). Few studies, although not confirmed by animal studies, have suggested that in early life VEPs could be mediated at least in part by the subcortical pathway (Dubowitz et al. 1986). VEPs can be elicited by flashing lights (light-emitting diodes – LEDs – incorporated into goggles) or by pattern-reversal stimuli. Pattern VEPs (pVEPs) are more specific to assess the visual discrimination (photopic vision). In contrast the flash VEPs (fVEPs) are more specific to assess the scotopic system and give better information about the hemispheric structures because the retinocortical projections of the scotopic system are more diffuse.

In preterm infants the initial wave emerging at 24 weeks is a large negativity at 300 ms, the N300 (N2). After about 27 weeks of gestation a positivity emerges, the P450 (P3), and after 32–35 weeks an earlier positivity, the P200 (P2; Taylor et al. 1987). In term infants the major components of fVEPs are P200 (P2), N300 (N2), and P450 (P3; Fig. 1). The N70 and P100 components are often present as well. The morphological development of the VEPs has been correlated in one study with the neuronal maturation of the visual cortex in neonates (Purpura 1977). In preterm infants, there appears to be accelerated maturation of the visual system as reflected by the earlier emergence of the VEPs components (Taylor et al. 1987).

Somatosensory-Evoked Potentials

Somatosensory-evoked potentials (SSEPs) represent brain response to electrical stimuli. We usually test the median and the posterior tibial nerves. We stimulate the median nerve in front of the forearm, close to the wrist. For the tibial nerve it is at the medial aspect of the ankle. The signals travel from the lower or upper extremities in the dorsal columns of the spinal cord and reach the lower end of the medulla. At this point the fibers synapse, cross to the opposite side, and form the medial lemniscus. The medial lemniscus carries the impulses to the thalamus. From the thalamus the

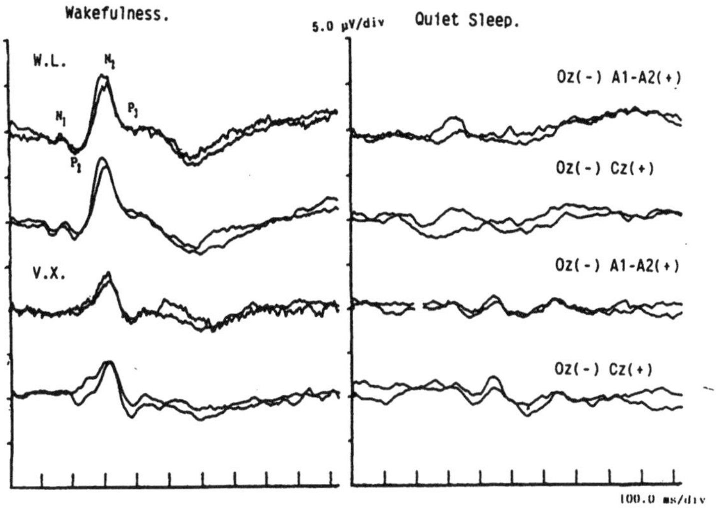

Fig. 1. fVEPs recorded from normal term newborns in the waking state and in quiet sleep. fVEPs were recorded on two channels Oz^{-}, referred to either Cz^{+} or linked ears. In quiet sleep there was a change in amplitude of the N2 (N300) and the P2 (P200) disappeared

sensory signals reach the sensory cortex through the thalamocortical fibers. Potentials can be recorded over the plexus, the spinal cord, and the sensory cortex. Figure 2 shows normal SSEPs recorded in neonates. Peripheral and spinal activities are recorded, followed by subcortical (lemniscal) and cortical (frontal and parietal) activities. The test examines the afferent sensory-pathway exclusively. The SSEPs are likely to be affected by an insult to the motor pathway which lies in close proximity to the thalamocortical sensory pathway in the periventricular region. But other areas not involving the tested structure may be impaired. Our recordings were made usually when the child was in quiet or active sleep without medication. During the course of development the EPs change drastically with age. The relationship between N20 cortical wave latency (y-axis) and postconceptional age in weeks (x-axis) is illustrated in Fig. 3. N20 latency decreases with increasing gestational age; the relationship is linear. The same profile exists between spinal cord latency and postconceptional age (Fig. 4). It is crucial, therefore, to compare an individual result against age-matched normative data.

Auditory-Evoked Potentials

Brain-stem auditory-evoked potentials are generated in the auditory pathway in response to auditory stimuli. The electrical signals coming from the cochlea are carried through the auditory nerve and the brain stem. The auditory nerve makes a connection with cochlear and superior olivary

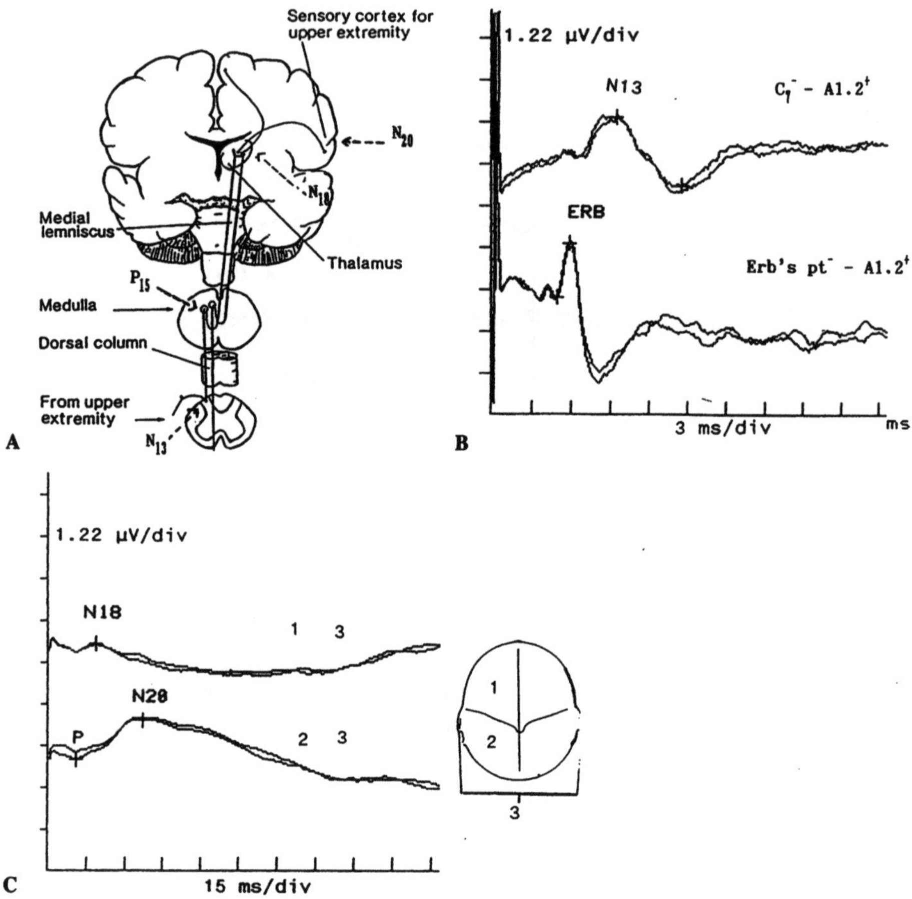

Fig. 2. A Somatosensory pathway. Early components of the SSEPs to electrical stimulation of the right median nerve at the wrist. **B** Peripheral (Erb's point, *ERB*) and spinal (*N13*). **C** Spinal (*P*), subcortical (*N18*) and cortical components (*N20*). Electrodes were placed 4.5 cm to the left before (*1*) and behind (*2*) the vertex, referred to linked ears (*3*). Negative-upward convention. (Adapted from Duffy 1989)

nuclei. Fibers cross to the opposite side forming trapezoid bodies. Axons arising from the superior olivary nucleus travel dorsally in lateral lemnisci. The BEARs consist of several components (waves I, II, III, IV, and V). In neonates the response consists primarily of waves I, III, and V (Fig. 5). Waves II and IV are often absent; this has not been demonstrated to be clinically significant (Mayer 1990). Wave I is the afferent volley in the proximal part of the auditory nerve. Wave III is probably generated by the trapezoid bodies wave V is not generated in the colliculus inferior but in a more caudal structure in the region of the lateral lemniscus (Guerit 1991). The amplitudes of waves I and V are smaller in preterm than in term

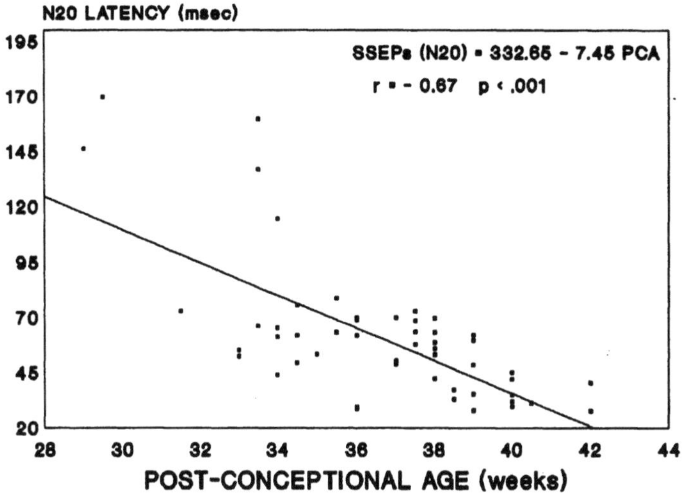

Fig. 3. Relationship between postconceptional age and absolute latency of N13 ranging from 31 to 40 weeks

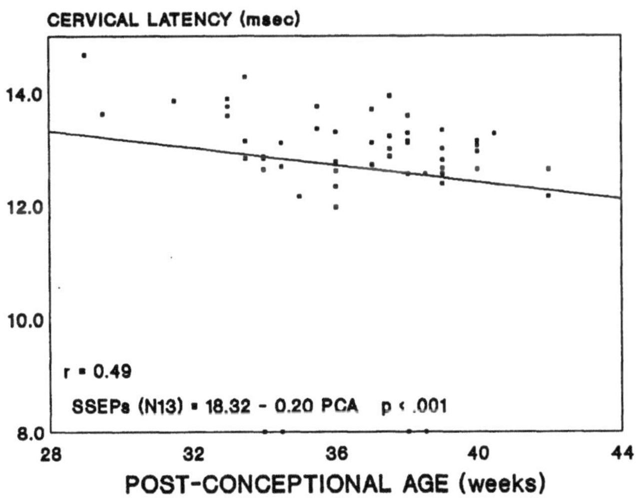

Fig. 4. Relationship between postconceptional age and absolute latency of N20 ranging from 29 to 42 weeks

neonates, and absolute latencies are longer in preterm than in term infants. Changes with maturation occur quite rapidly in the preterm period with the absolute and interpeak latencies shortening and the amplitudes increasing. Changes in latency and amplitude of the early BEARs' peaks seen with maturation are more rapid than those seen for the later BEARs com-

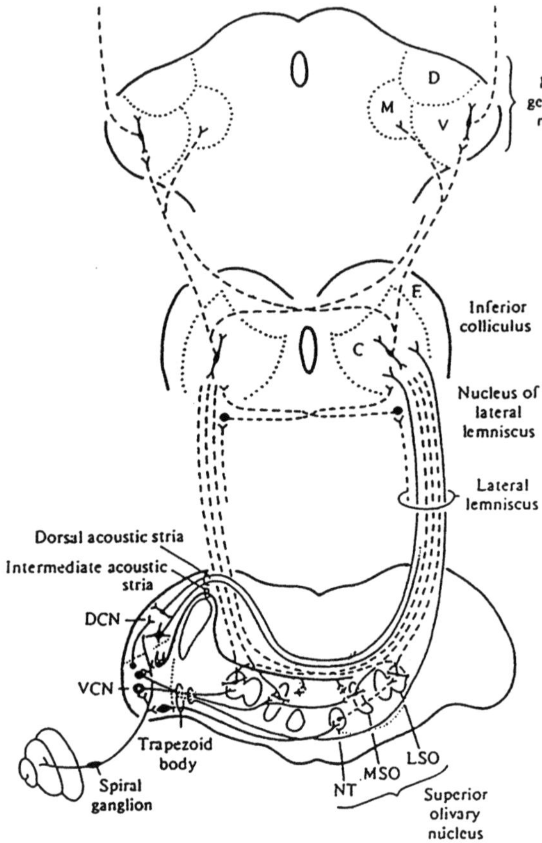

a

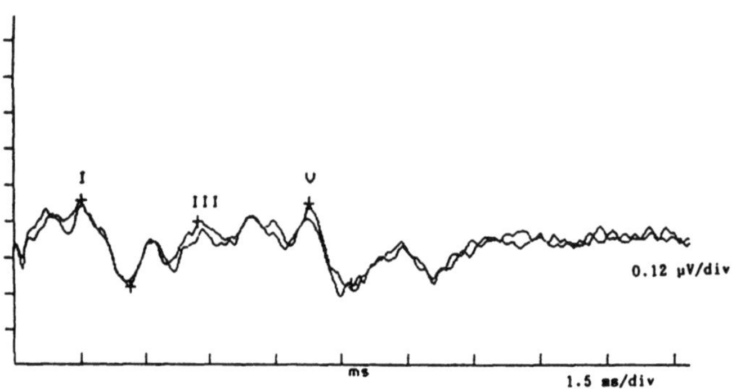

b

Fig. 5. **a** Auditory pathway *VCN* ventral cochlear nucleus; *DCN*, dorsal cochlear nucleus; *NT*, nucleus of trapezoidal body; *MSO*, medial superior olive; *LSO*, lateral superior olive; *C*, central and *E*, external nuclei of inferior colliculus; *M*, medial, *D*, dorsal, and *V*, ventral nuclei of medial geniculate nucleus. **b** BEARs to 90 dB hearing level clicks presented at 21.7/s recorded from a normal neonate showing the main components (waves *I*, *III*, *V*). Positivity is shown as an upward deflection. (Adapted from Regan 1989)

ponents. Wave I latency reaches mature values by 6–8 weeks postterm. In contrast, wave V demonstrates a rapid decrease over the first 12 weeks and then a slower decrease to reach mature values by 3–5 years (Salamy 1984; Eggermont 1989; Stapells and Kurtzberg 1991). These changes with maturation require age-matched-normative data to interpret the response correctly. Cortical auditory-evoked potentials (CAEPs) have also been used and abnormal CAEPs have been associated with abnormal language skill (Stapells and Kurtzberg 1991).

Values of EPs in the Asphyxiated Newborns

Cerebral Function

Like EEG, MEPs can assess the severity of hypoxic encephalopathy particularly in the first few days of life when CT scan and ultrasound often appear normal, and can follow the progression of the encephalopathy and the impact of acquired postnatal insults. CBF has been coupled with electrical activity. Several works about cerebral ischemia have identified a threshold of deficiency in neuroelectric-monitoring capabilities, that is, a gap between the last measurable electrical change and the developement of irreversible tissue damage (Shapiro 1986). In adult humans and in animals, critical levels below which the neuronal transmission failed have been defined to approximately 15–20 ml/100 g min, corresponding to an oxygen delivery of about 2.5 ml/100 g min. (Trojaborg and Boysen 1973; Brantson et al. 1984). In experimental animals, the threshold for neuronal death is approximately 10 ml/100 g min (Astrup 1982). In newborns the cerebral blood flow requirement for brain viability is lower than in adults. There are data indicating that mean cerebral blood flow as low as 5 ml/100 g min in newborn infants can be associated with normal neurological outcome (Altman et al. 1988). Thus, intact VEPs have been observed in ventilated preterm infants with cerebral blood flow below 10 ml/100 g min as estimated by ^{133}Xe clearance after intravenous injection (Greisen and Trojaberg 1987). Using fVEPs in preterm infants the critical levels of PaO_2 at which electrical failure occurred was demonstrated at PaO_2 below 2.5 kPa (Graziani et al. 1972; Pryds and Greisen 1990).

In addition, cerebral function can be evaluated by MEPs in the iatrogenically paralysed infant.

Effects of Sleep

Visual-Evoked Potentials

Sleep state affects the fVEPs. With quiet sleep there is a decrease in amplitude of the N2. The P2 disappears in both quiet and active sleep

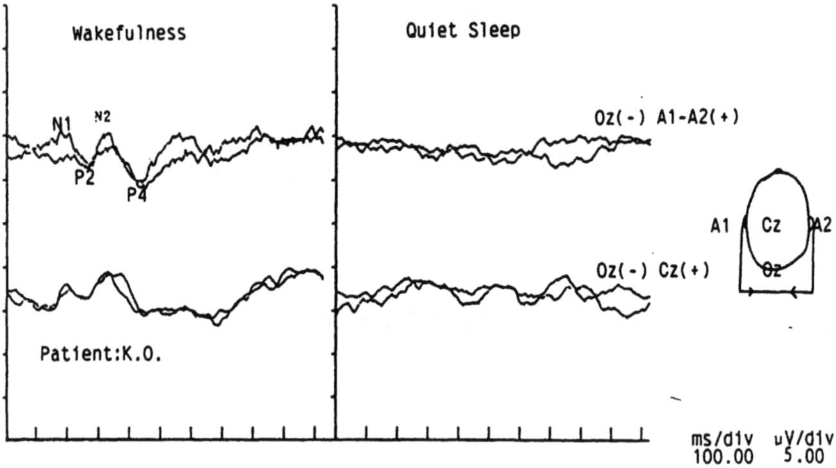

Fig. 6. fVEPs recorded from severely asphyxiated neonates in the waking state and in quiet sleep: missing components P2 (P200) and P4 (P450). Negative-upward convention

(Whyte et al. 1987). We examined the effects of arousal state changes in normal ($n = 4$) compared with mild to moderate ($n = 11$) and severe-asphyxiated newborns ($n = 12$; Scalais et al. 1992a):

- The P3 amplitude was decreased in wakeful severe versus mild- to moderate-asphyxiated newborns ($p < 0.001$, Student's t test)
- Amplitude reduction with quiet sleep was observed for N2 ($p < 0.05$) and for both N2 ($p < 0.001$) and P3 ($p < 0.01$) in mild to moderate asphyxia (Mann-Whitney U test)
- Components P2 and P3 more frequently disappeared with quiet sleep in severe versus mild to moderate asphyxia ($p < 0.00002$, binomial test)

Figures 1 and 6 show the differences observed between VEPs performed in the waking state and the sleep state in normal and in severely asphyxiated neonates, respectively. Owing to the significant variations observed in VEPs during arousal state changes, it is important to compare VEPs results in the same sleep state. Recording VEPs in both arousal states yields more information.

Somatosensory-Evoked Potentials

Differences between waking and slow-wave sleep have been observed in SSEPs performed in term neonates, but little difference has been observed between the waking state and the rapid eye movement sleep (Desmedt and Manil 1970). Figure 7 shows that sleep can affect cortical (N20) and sub-

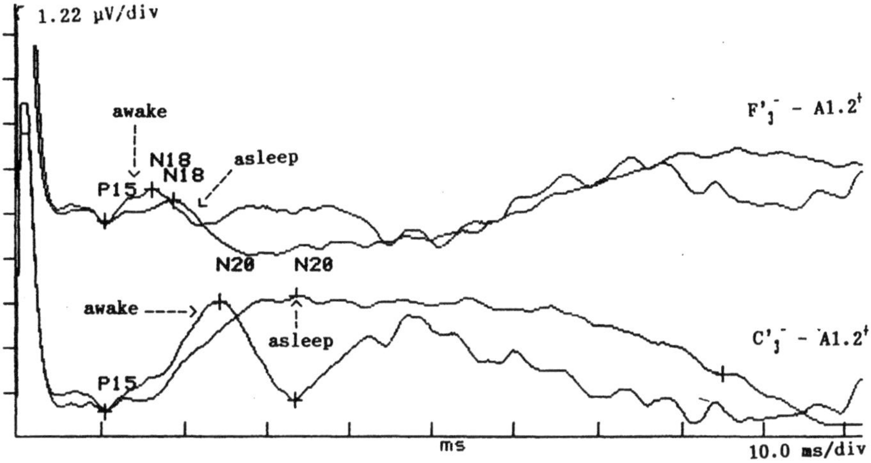

Fig. 7. Subcortical and cortical components of the SSEPs to electrical stimulation of the right median nerve at the wrist in the waking state and in sleep. In sleep there was a decrease in N20 amplitude and an increase in N18 and N20 latency

cortical (N18) components. However, sleep facilates SSEPs recording because of the concomitant reduction in artifacts. Performing SSEPs in sleep state with our technique, cortical SSEP activity was recorded in all normal and term infants. Cortical SSEP activity was also recorded in all normal newborn infants in a recent study (George and Taylor 1991). In contrast, in previous reports cortical activity could not be elicited in all normal neonates (Willis et al. 1984; Laureau et al. 1987, 1990; Taylor 1991).

Brain-Evoked Auditory Responses

Sleep state does not affect BEARs.

Effects of Drugs

Barbiturates and other sedative drugs can produce severe depression of the EEG. In contrast the EPs show a greater degree of resistance to barbiturates. Figure 8 demonstrates that despite marked EEG effects (very depressed) secondary to phenobarbital (Pb), SSEPs were still recorded in a 31-week infant with seizures. The effects of sedation were studied in five preterm infants using fVEPs. Pb (15 mg/Kg IV) associated with valium (0.5–1 mg/kg IV) attenuated transiently the VEP amplitude, whereas the latency was unaffected (Pryds and Greisen 1990).

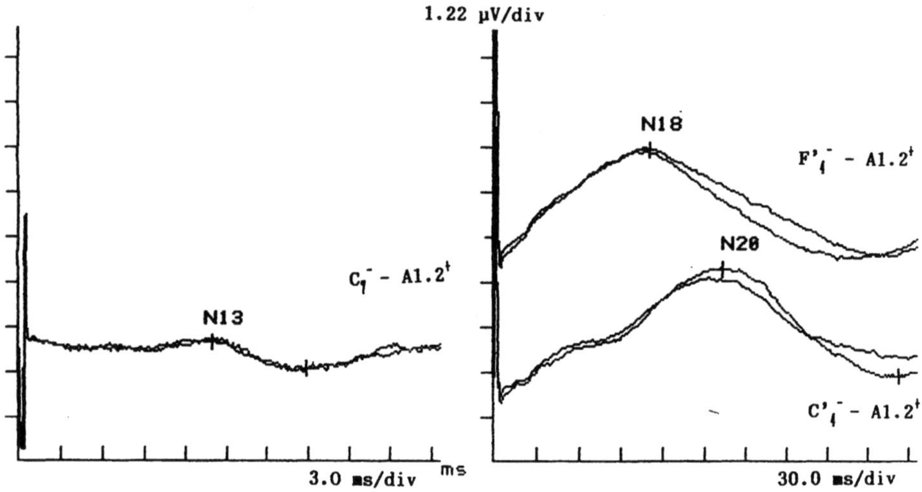

Fig. 8. Early components of the SSEPs to electrical stimulation of the right median nerve at the wrist in a 31-week gestation 3-day-old infant. He received Pb (15 mg/kg IV) for seizures. EEG was very depressed

Classification of the Abnormalities

Several methods of recognizing and grading degrees of hypoxic–ischemic events have been applied to neonates. In term infants abnormal neurological behavior and a number of schemes have been used to document these abnormalities (Sarnat and Sarnat 1976; Fenichel 1983; Levene et al. 1985). In preterm infants the immature brains have shown separate, subtle, or even no abnormal clinical or physical signs (Levene 1991). Quantification of EEG data in terms of the duration of abnormal findings has been used to assess the severity of the hypoxic insult to the brain in term neonates, and to increase the accuracy of early prognosis in asphyxia (Sarnat and Sarnat 1976; Holmes et al. 1982). In preterm infants a scoring system has also been applied to give prognosis (Tharp et al. 1981; Ellison et al. 1982). Similarly, in preterm and term neonates, we have graded abnormalities of EPs in order to establish a correlation between neurophysiological data at birth and the neurodevelopmental status after 18 months' corrected age (Scalais et al. 1992b,c).

VEPs abnormalities were classified as:

- Grade 0: normal
- Grade I: increased latency (P2) or unusual morphology, poorly reproducible
- Grade II: low amplitude ($<5\,\mu V$) and missing components
- Grade III: absence of major components (N2)

SSEPs abnormalities were classified as:

- Grade 0: normal
- Grade I: increased latency of cortical waves (N20) or abnormal morphology
- Grade II: absence of cortical wave on one side
- Grade III: absence bilaterally of cortical wave

BEARS abnormalities were classified as:

- Grade 0: normal
- Grade I: increased interpeak latencies and/or abnormal amplitude ratio V/I
- Grade II: waves I and III identifiable
- Grade III: wave I only

Neurological status at 18 months' corrected age was classified as:

- Normal:
- Mild: abnormal tone (1)
- Moderate: locomotion impaired (2)
- Severe: locomotion impossible (3)

Psychodevelopmental indices (PDI) were classified as:

- Normal: PDI > 85
- Mild: PDI 70–80
- Moderate: PDI 51–69
- Severe: PDI < 50

using the Griffith developmental scale or Brunet-Lezine.

Prediction of Outcome Following Cerebral Hypoxia–Ischemia

First we investigated the prognostic capability of MEPs in a group of 40 infants with documented cerebral hypoxia–ischemia (Scalais et al. 1992b). SSEPs and VEPs demonstrated a good correlation with neurodevelopmental outcome in infants. In our experience SSEPs proved to be superior to VEPs with an accuracy of 93% for SSEPs and 80% for VEPs. The combination of both SSEPs and VEPs gave a higher predictive power: grade 0 VEPs and SSEPs were associated with normal neurological status and PDI in 88% ($p < 0.01$) of the infants. Grade II–III VEPs and SSEPs were associated with abnormal outcome ($p < 0.001$).

BEARs were less sensitive. An abnormal BEAR predicted neurological deficit at 18 months' corrected age, but a normal test did not predict a normal outcome. BEARs were less sensitive because brain-stem structures are more resistant to anoxia (Guerit 1991). However, peripheral auditory abnormalities were frequent in the acute phase of asphyxia. An absence of any response, including wave I, with complete recovery on the following days has been observed. Secondly, we also studied MEPs in 61 preterm infants (mean gestational age: 31.3 ± 2.3 weeks) suffering from hypoxia–

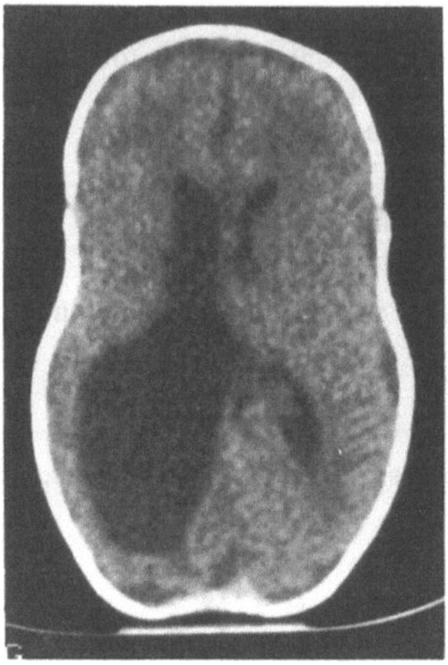

a

Fig. 9. a CT scan for the patient described in (b) showing cortical atrophy with right ventricle enlargement. b Neurophysiologic data of a 31-week gestation and neurological status at 18 months

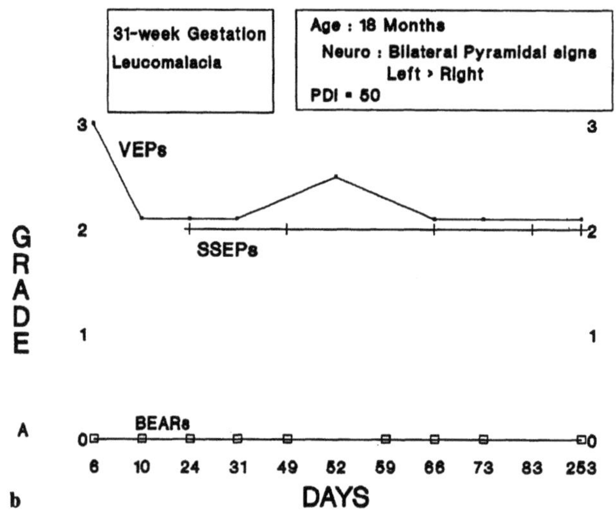

b

ischemia to investigate the correlation between the MEPs in the neonatal period and subsequent neurodevelopmental outcome (Scalais et al. 1992c) VEPs and SSEPs demonstrated good correlation with both neurologic outcome and PDI. BAERs did not predict accurately the neurodevelopmental outcome. The combination of both VEPs and SSEPs also gives a higher predictive power: grade 0 VEPs and SSEPs were associated with normal neurological status in 86% ($p < 0.01$) of infants and normal PDI in 94% ($p < 0.01$) of infants. Grade II-III VEPs and SSEPs were not associated with normal outcome ($p < 0.01$).

Finally, by comparing the predictive power of both ultrasound and EPs in preterm infants we observed that both US and EPs were effective in predicting neurological outcome; however, EPs seem to give earlier prognostic information (François-Adant et al. 1992).

Case Studies

Case 1. Male infant born at 31 weeks' gestation with a history of severe antenatal leukomalacia. CT scan (Fig. 9a) done at 9 weeks of age showed cortical atrophy with progressive right ventricle enlargement. The neurophysiologic evolution is depicted (Fig. 9b) from birth to 41 weeks' corrected age and at 8 months. SSEPs were grade II, VEPs improved from grade III to grade II, and BEARs were grade 0. Neurological status at 18 months was severely abnormal (bilateral pyramidal signs more involved on the left side, cortical blindness) and PDI was less than 50.

Case 2. Female infant born at 40 weeks' gestation by spontaneous vaginal delivery. Several hours after birth she presented right focal seizures and apnea. The EEG showed marked asymmetry with frequent spikes and sharp waves on the left. A CT scan (Fig. 10a) done at 5 days of age showed an

a

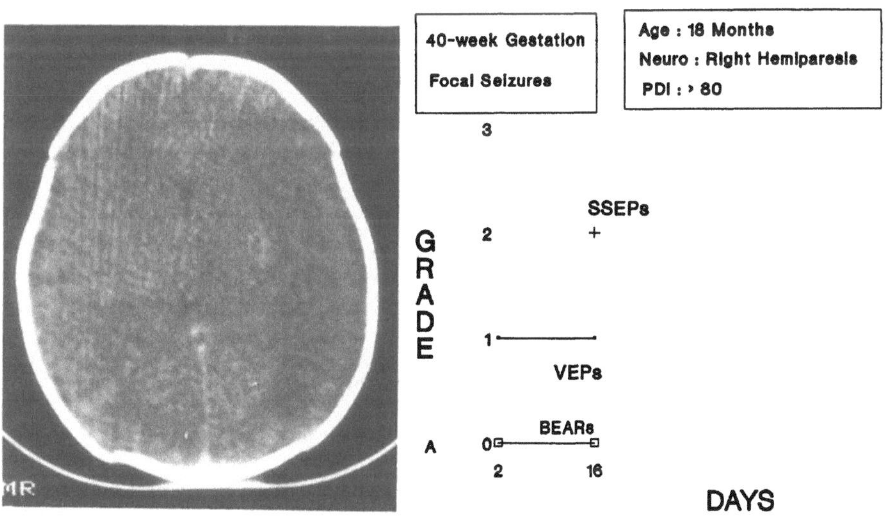

b

Fig. 10. a CT scan for the patient described in (b) showing an infarction in the left hemisphere following the distribution of the middle cerebral artery. b Neurophysiologic data of a 40-week gestation, 2- and 16-day-old infant and neurological status at 18 months

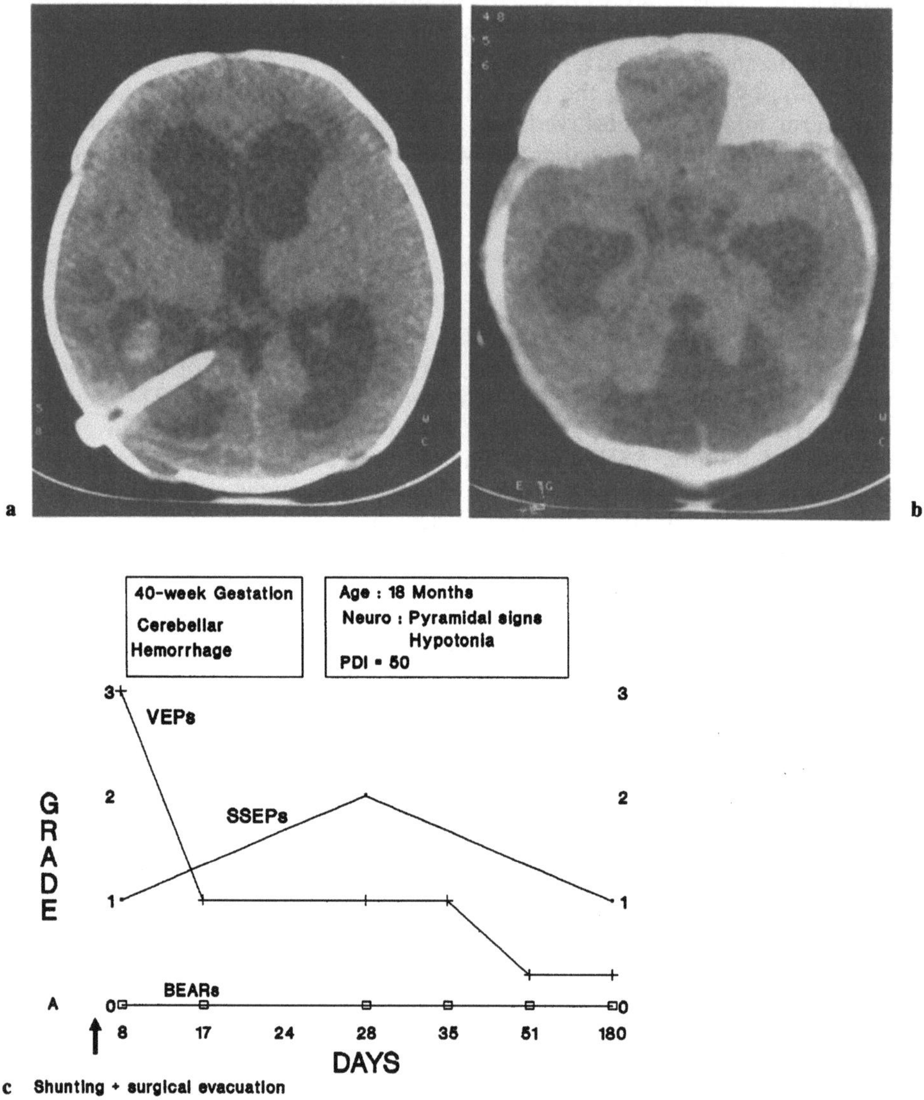

Fig. 11. a,b CT scan for the patient described in **(c)** showing ventricular dilation, right parietal hypodensity, and cerebellar lesions. **c** Neurophysiologic data at 40-week gestation and neurological status at 18 months

area of hypodensity in the left side region. SSEPs were classified grade II (the cortical wave after right median nerve stimulation was not recorded on the left hemispheric side), VEPs grade I, and BEARs grade 0. Neurological status at 18 months was II (impaired locomotion because of right hemiparesis) and PDI was above 70 (Fig. 10b).

Comment: The SSEPs were classified grade II for both cases 1 and 2. Neurological status and PDI were severely abnormal in the first case and moderately abnormal in the second case. The combination of SSEPs with VEPs could differentiate between both cases and contributed more information in addition to that already received by the SSEPs.

Case 3. Male born at 40 weeks' gestation with a history of hypoxic–ischemic events, cerebellar hemorrhage, and secondary hydrocephalus. Shunting and surgical evacuation were done. CT scan (Fig. 11a,b) done at 14 days of age demonstrates mainly the destruction of cerebellar structures. Neurophysiologic evolution are shown (Fig. 11c) from birth to 18 months of age. SSEPs improved from grade III to I and finally to 0. VEPs were classified grade I followed by grade II and finally grade I. Neurological status at 18 months was severely abnormal with PDI less than 50.

Comment: In this case cerebellar structures were damaged and MEPs do not contribute any information to the sensory cerebellar pathway.

Case 4. Male infant born at 40 weeks' gestation with a history of hypoxic events, intraventricular, intraparenchymal hemorrhage, and hypoglycemia. CT scan (Fig. 12a) done at 14 days of age showed intraventricular, left intraparenchymal hemorrhage, and hydrocephalus. Neurophysiologic data are presented in Fig. 12b. SSEPs were classified grade II followed by grade I, VEPs were classified grade I, and BEARs grade 0. Neurological status at 12 months was mildly abnormal (mild hemiparesis) and PDI was 83.

Case 5. Male infant born at 40 weeks' gestation with a history of severe asphyxia. SSEPs were classified grade III on day 1 and day 2, VEPs were classified grade II followed by grade III, and BEARs grade I followed by grade II (Fig. 13). The child died.

Comment: Even in the acute phase the MEPs can assess the severity of the encephalopathy.

Conclusion

1. MEPs are useful for estimating the visual, somatosensory, and auditory pathway in hypoxic-ischemic neonates.
2. fVEPs test hemispheric stuctures, SSEPs test peripheral, spinal cord, central brain stem and, hemispheric structures. BEARs test brain-stem auditory pathway.
3. SSEPs give the best predictive power but the combination of SSEPs and VEPs yields a higher predictive reliability. BEARs were the least sensitive.
4. Topographic and temporal limits of the information received by the EPs are important to consider in order to interpret the responses correctly.

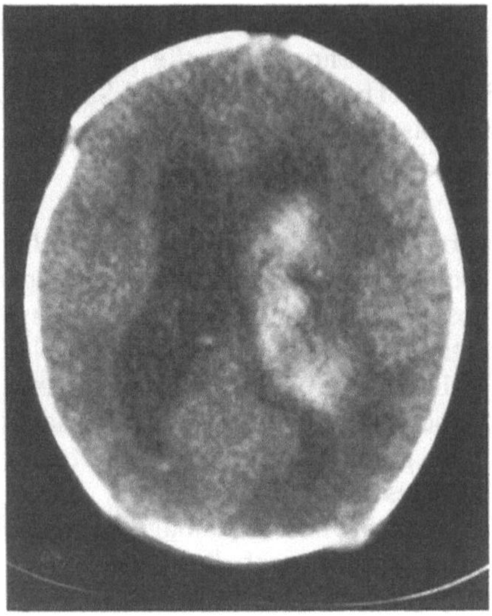

a

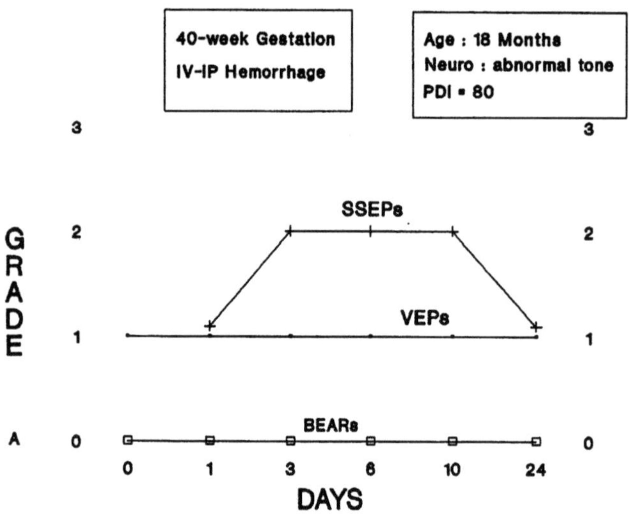

b

Fig. 12. a CT scan for the patient described in (**b**) showing hemorrhagic infarction in the left parietal region with accompanying intraventricular hemorrhage and hydrocephalus. **b** Neurophysiologic data of a 40-week gestation and neurological status at 18 months

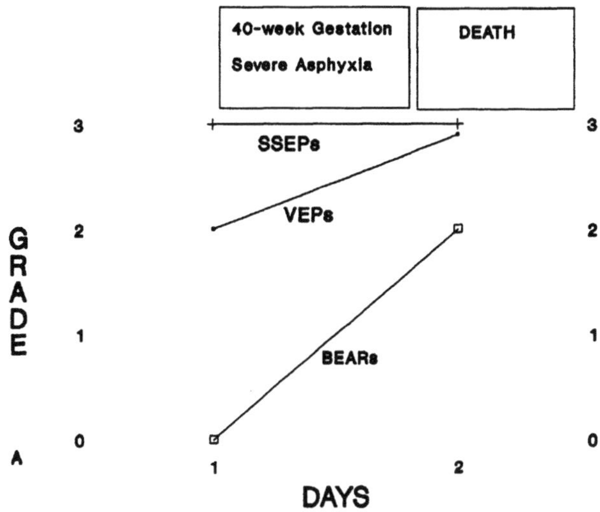

Fig. 13. Neurophysiologic data of a 40-week gestation and neurological outcome

References

Altman DI, Powers WJ, Perlman JM (1988) Cerebral blood flow requirement for brain viability in newborn infants is lower than in adults. Ann Neurol 24:218–226

Astrup J (1982) Energy requiring cell functions in the ischemic brain. J Neurosurg 56:482–497

Brantson NM, Ladds A, Symon L, Wang AD (1984) Comparison of the effects of ischaemia on early components of somatosensory-evoked potentials in brain stem, thalamus, and cerebral cortex. J Cereb Blood Flow Metab 4:468–481

Celesia GG (1985) Visual evoked responses. In: Owen JH, Davis H (eds) Evoked potential, testing, Clinical applications. Grune-Stratton, Orlando

Desmedt JE, Manil J (1970) Somatosensory-evoked potentials of the normal human neonate in REM sleep, in slow-wave sleep and in waking. Electroencephalogr Clin Neurophysiol 29:113–126

Dubowitz LMS, Mushin J, DeVries L, Arden GB (1986) Visual function in the newborn infant: is it cortically mediated? Lancet 1:1139–1141

Duffy FH, Iyer VG, Surwillo WW (1989) Average evoked potentials. In: Clinical electroencephalography and topographic brain mapping. Springer, Berlin Heidelberg, New York pp 203–217

Eggermont JJ (1989) The onset and development of auditory function: contributions of evoked potential studies. J Speech Lang Pathol Audiol 13:5

Ellison P, Franklin S, Jones MG (1982) Serial EEG in the preterm infant as a predictor of neurological sequelae. Ann Neurol 12:213

Fenichel GM (1983) Hypoxic-ischaemic encephalopathy in the newborn. Arch Neurol 40: 261–266

François A, Scalais E, Battisti O, Langhendries JP, Bertrand JM, Withofs L (1992) Predicting neurological outcome in preterms: ultrasound? evoked potentials? Biol Neonate 62:305 (Abstract)

George SR, Taylor MJ (1991) Somatosensory evoked potentials in neonates and infants. Developmental and normative data. Electroencephalogr Clin Neurophysiol 80:94–102

Graziani LJ, Weitzman ED, Pineda G (1972) Visual-evoked potentials during neonatal respiratory disorders in low-birth-weight infants. Pediatr Res 6:203–210

Greisen G, Trojaborg W (1987) Cerebral blood flow, $PaCO_2$ changes, and visual evoked potentials in mechanically ventilated preterm infants. Acta Pediatr Scand 76:394–400
Guerit JM (1991) Les potentiels evoqués. Masson Paris
Holmes G, Rowe J, Hafford J, Schmidt R, Testa M, Zimmerman A (1982) Prognostic value of the electroencephalogram in neonatal asphyxia. Electroencephalogr Clin Neurophysiol 53:60–72
Klimach VJ, Cooke RWI (1988) Short latency somatosensory-evoked responses of preterm infants with ultrasound abnormalities of the brain. Dev Med Child Neurol 30:215–221
Levene MI (1991) Physiological assessment of the asphyxiated newborn infant. In: Minn RH (ed) Problems of intracranial pressure in childhood. Blackwell, Oxford, pp 141–169 (Clinics in developmental medecine, vol 113/114)
Levene MI Kornberg J, Williams THC (1985) The incidence and severity of post-asphyxial encephalopathy in full-term infants. Early Hum Dev 11:21–26
Laureau E, Marlot D (1990) Somatosensory evoked potentials after median and tibial nerve stimulation in healthy newborns. Electroencephalogr Clin Neurophysiol 76:453–458
Laureau E, Majnemer A, Rosenblatt B, Riley P (1987) A longitudinal study of short latency somatosensory evoked responses in healthy newborns and infants. Electroencephalogr Clin Neurophysiol 3:350–355
Majnemer A, Rosenblatt B, Riley P, Laureau P, O'Gorman A (1988) Somatosensory-evoked response abnormalities in high-risk newborns. Electroencephalogr Clin Neurophysiol 71:100–108
Mayer M (1990) Potentiels évoques et électromyographie en pédiatrie. ABC, Masson, Paris
Muttiff SC, Taylor MJ, Kobayashi JS, Whyte HE, MacMillan L, Whyte HE (1991) Serial evoked potentials and outcome in term birth asphyxia. Pediatr Neurol 7:86–90
Pryds O, Greisen G (1990) Preservation of single-flash visual-evoked potentials at very low cerebral oxygen delivery in preterm infants. Pediatr Neurol 6:151–158
Purpura DP (1977) Developmental pathobiology of cortical neurons in immature humane brain. In: Gluck LL (ed) Intrauterine asphyxia and the developing humain brain. Year Book Medical Publishers, Chicago
Regan D (1989) Technical and mathematical considerations. In: Regan D (ed) Human brain electrophysiology. Evoked potentials and evoked magnetic fields in science and medecine. Elsevier, Amskrdam, pp 1–166
Salamy A (1984) Maturation of the auditory brain stem from birth through early childhood. J Clin Neurophysiol 1:293–329
Sarnat HB, Sarnat MS (1976) Neonatal encephalopathy following fetal distress: a clinical and electroencephalographic study. Arch Neurol 33:696–705
Scalais E, Debauche C, Verellen G, Guerit JM (1992) Arousal state changes on visual-evoked potentials (VEPs) in normal or asphyxiated newborns. Electroencephalogr Clin Neurophysiol (in press) (abstract)
Scalais E, François-Adant A, Bachy A, Gerard P, Guerit JM (1992b) Multimodality Evoked Potentials as a prognostic tool in term hypoxic-ischemic newborns. Biol Neonate Abstract 62:300
Scalais E, François-Adant A, Langhendries JP, Battisti O, Bertrand JM, Guerit JM (1992c) Multimodality evoked potentials assessment in hypoxic-ischemic preterm infants. Ann Neurol Abstract 32:480
Shapiro HH (1986) Anesthesia effects upon cerebral blood flow, cerebral metabolism, electroencephalogram, and evoked potentials. In: Miller RD (ed) Anesthesia, 2nd edn. Churchill Livingstone, New York, p 1276
Stapells DR, Kurtzberg D (1991) Evoked potential assessment of auditory system integrity in infants. Clin Perinatol 18:497–518
Taylor MJ (1991) EPs and ERPs in paediatrics. In: Barber C, Taylor MJ (eds) Evoked potentials review. IEPs publication 4:145–150
Taylor MJ, Menzies R, MacMillan LJ, Whyte HE (1987) VEPs in normal full-term and premature neonates: longitudinal versus cross-sectional data. Electroencephalogr Clin Neurophysiol 68:20–27

Taylor MJ, Murphy WJ, Whyte HE (1992) Prognostic reliability of Somatosensory and visual evoked potentials in asphyxiated term infants. Dev Med Child Neurol 34:507–515

Tharp BR, Cukier F, Monod N (1981) The prognostic value of the electroencephalogram in premature infants. Electroencephalogr Clin Neurophysiol 51:219–236

Trojaborg W, Boysen G (1973) Relation between EEG, regional cerebral blood flow, and internal carotid artery pressure during carotid endarterectomy. Electroencephalogr Clin Neurophysiol 34:61–69

Whyte HE, Pearce JM, Taylor MJ (1987) Changes in the VEP in preterm neonates with arousal states, as assessed by EEG monitoring. Electroencephalogr Clin Neurophysiol 68:223–225

Willis J, Duncan C, Bell R (1987) Short latency somatosensory-evoked potentials in perinatal asphyxia. Pediatr Neurol 3:203–207

Willis J, Duncan C, Bell R, Pappas F, Moniz M (1989) Somatosensory-evoked potentials predict neuromotor outcome after periventricular hemorrhage. Dev Med Child Neurol 31:435–439

Magnetic Resonance Spectroscopy

E.B. Cady, C. Boesch, and E. Martin

Introduction

In vivo magnetic resonance spectroscopy (MRS) is noninvasive and can obtain metabolic information concerning developmental changes, regulatory functions, and adaptive mechanisms of the human neonatal central nervous system by discerning relevant metabolites from intact cells. Many compounds exist in sufficiently high concentrations in the cytosol of brain cells to be detectable in vivo with modern instrumentation using ^{31}P-, ^{1}H-, or even ^{13}C MRS. Various MRS methods are available which allow both the monitoring of physiological changes in neonatal brain, either during development (Cady et al. 1983, 1991; Younkin et al. 1984; Hamilton et al. 1986; Martin et al. 1988; Boesch et al. 1988, 1989; Azzopardi et al. 1989a; Van der Knaap et al. 1990) or following perinatal injury, by comparing relative metabolite concentrations (Hope et al. 1984; Hamilton et al. 1986; Azzopardi et al. 1989b; Laptook et al. 1989). Some of these techniques have enabled the study of specific areas of the brain under equilibrium conditions (van der Knaap et al. 1990; Moorcraft et al. 1991a) and more recently the estimation of absolute concentrations (Cady and Azzopardi 1989; Cady 1990, 1991), or measuring enzymatic reaction rates, e.g., of the creatine kinase (CK) reaction (Shoubridge et al. 1982; Rudin and Sauter 1989; Mora et al. 1991).

The biological roles of most phosphorylated compounds, e.g., nucleotide triphosphate (NTP), phosphocreatine (PCr), inorganic orthophosphate (Pi) and sugar phosphates (all involved in energy metabolism), or phosphomonoesters (PME) and phosphodiesters (PDE), containing compounds associated with phospholipid metabolism, are well understood and these substances may serve as biochemical markers. One question to be answered is: Do these markers correlate with the morphological changes observed, e.g., with magnetic resonance imaging (MRI) during early postnatal life or, even more importantly, do they have a predictive value for eventual outcome following perinatal asphyxia? Together with the determination of intracellular pH (pH_i), the relative or absolute concentrations of these compounds should allow both the evaluation of the severity and duration of such cerebral insults, and the monitoring of the effects of therapeutic interventions. First results from recent prognostic investigations indicate that in vivo MRS may give positive answers (Hamilton et al. 1986; Moorcraft et al. 1991b). A fall in pH_i, concomitant with a rise in intracellular lactate (Lac) is one of the major consequences of oxygen deprivation due to, e.g., perinatal asphyxia. The cellular energy potential, indicated by the concentrations of the high-energy phosphates (e.g., NTP and PCr) relative to that of Pi, decreases at a slower rate resulting in cellular-energy failure and breakdown of cellular homeostasis (the inability to maintain osmotic gradients leading to the cessation of biosynthesis). Major transmembranous shifts of Na^+, K^+, and Ca^{2+} causing edematous cellular swelling are then followed by the release of excitatoxic neurotransmitters and a destructive cascade of various metabolic sequences ultimately leading to neuronal death. The combined use of ^{31}P- and ^1H MRS has allowed the observation of the time dependence of hypoxic neuronal metabolism after hypoxia–ischaemia and the estimation of the reparative capacity during recovery in animal models in vivo.

Except for Lac (an end-product of anaerobic metabolism), creatine (Cr) and PCr, and glucose (Gluc) and glycogen (Gly; assignable in ^1H and ^{13}C spectra respectively), the roles and functions of the amino acids *N*-acetyl-aspartate (NAA), glutamine (Gln)/glutamate (Glu), aspartate (Asp), taurine (Tau), and of inositol (Ino) and the choline-containing compounds (Cho) are diverse and often not well understood (Miller 1991; Ross 1991). The three major signals seen in the in vivo ^1H spectrum from neonatal brains are due to (a) NAA – localised largely in neurons together with the excitatory amino acid neurotransmitter precursor NAA-Glu, (b) Cr and PCr – a reservoir for high-energy phosphates, and (c) Cho – a precursor for both the neurotransmitter acetylcholine and the membrane phospholipid phosphorylcholine (PCh). Thus MRS provides a noninvasive clinical method whereby, in addition to the investigation of other disorders, the biochemical responses of brain tissue to hypoxia–ischaemia can be evaluated and sequentially monitored. The ability to perform repeated, noninvasive, in vivo studies on organs (e.g., the brain of the newborn human infant) which are normally inaccessible to other biochemical techniques is an outstanding feature of MRS. The noninvasive and repeatable aspects of MRS are of great importance in paediatric medicine because physiological information can be acquired which would otherwise be unobtainable. Ultrasound, MRI or X-ray-computerised tomography can visualise changes due to haemorrhage. However, the occurrence of biochemical effects due to hypoxia cannot be detected by these modalities and gross structural changes are only observed when loss of brain tissue has taken place.

In "Studies of Normal Human Infants" and "Perinatal Asphyxia," this chapter, some important results of in vivo ^{31}P- and ^1H MRS on both normal and asphyxiated newborn infants, reported by various groups working in this field, are reviewed. The contents of these sections address questions both about the type and extent of metabolic derangement following birth asphyxia which produces changes observable by MRS in vivo and about the predictive value of such information.

Techniques

Nuclear magnetic resonance (NMR – termed MR in medical applications) is a physical effect which provides insight into the living human body, as well as helping us to understand chemical reactions and the conformation of molecules in solution. In vivo, both MRI and MRS are based on this fundamental and versatile phenomenon.

Basic MRS Theory

The MR effect is the ability of certain atomic nuclei in a strong external magnetic field to receive and re-emit electromagnetic fields (radiowaves) of a characteristic resonant frequency (Fig. 1). The most fundamental attribute of the MR effect (formulated in the "Larmor-equation," see Fig. 1) is that the emitted radiofrequency is proportional to the effective magnetic field strength experienced by the resonating nucleus. Therefore, the MR experiment is an extremely subtle measurement of magnetic field strength at the location of the resonating nucleus. In a magnetic field which has been thoroughly homogenised by a process called "shimming," all nuclei should experience the same magnetic field strength and therefore emit radiowaves of the same frequency. The information content of the emitted radiowaves would be poor under these circumstances, but it can be greatly increased by two different types of magnetic field variations, i.e., both those due to the chemical environment within molecules, and instrumentally produced, superimposed magnetic field gradients.

Chemical bonds have slight effects on the local magnetic field within a molecule (Fig. 2). Therefore, nuclei in different chemical environments experience different field strengths and thus emit radiowaves at different

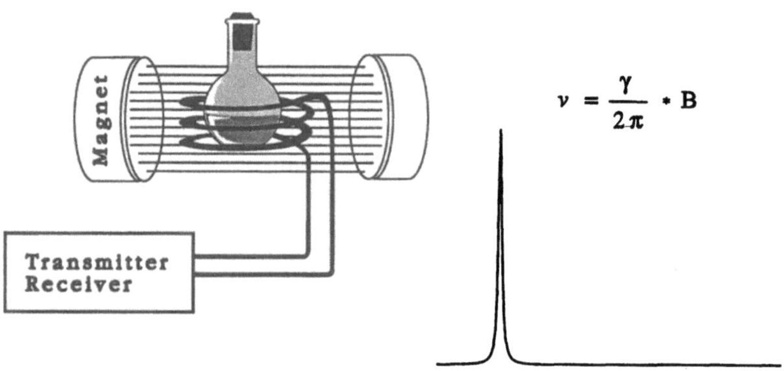

$$\nu = \frac{\gamma}{2\pi} \cdot B$$

Fig. 1. Illustration of the MR phenomenon. A sample, containing only a single type of nucleus, in a homogeneous magnetic field absorbs and emits a single frequency given by the "Larmor-equation", which describes the linear relationship between magnetic field strength and resonance frequency of the radiowaves. ν, radiofrequency; γ, gyromagnetic ratio; B, magnetic field strength

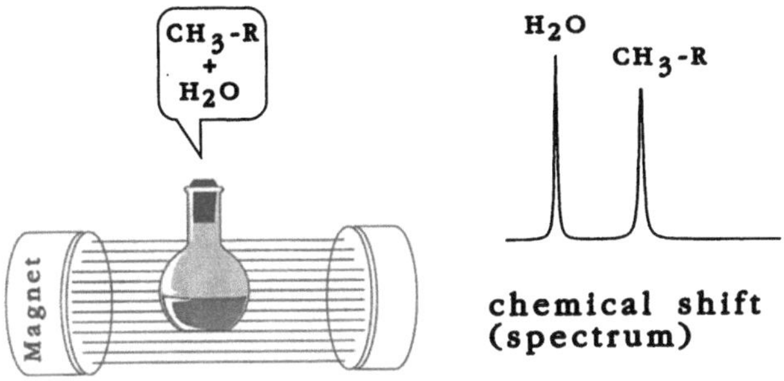

Fig. 2. The molecular–electronic environment produces slightly different magnetic field strengths for nuclei in various chemical bonds (e.g., protons in H_2O and CH_3-R), resulting in a range of emitted frequencies (i.e., a spectrum). *R*, any chemical moiety that might be bound to a methyl group

frequencies. This makes metabolites distinguishable even if this "chemical shift" of frequency is only small (a few parts-per-million, ppm, of the resonance frequency). A plot of the amplitudes of the emitted radiowaves against frequency forms an exceedingly informative "spectrum." Such spectra can be used to characterise the physicochemical environment within unknown molecules (e.g., the three-dimensional structure) or they can be used to measure the amounts of different known chemical components with separated MR signals in a tissue sample (the most common application in vivo). Alterations in chemical structure and therefore of the local magnetic field strength may be due to physiological causes. These may result in chemical shift changes of clinical interest (e.g., the relationship between the Pi chemical shift and pH_i).

The magnetic field can also be altered instrumentally to superimpose "gradients" on the magnetic field in different spatial directions (Fig. 3). Using these time-dependent gradients, it is possible to encode the spatial

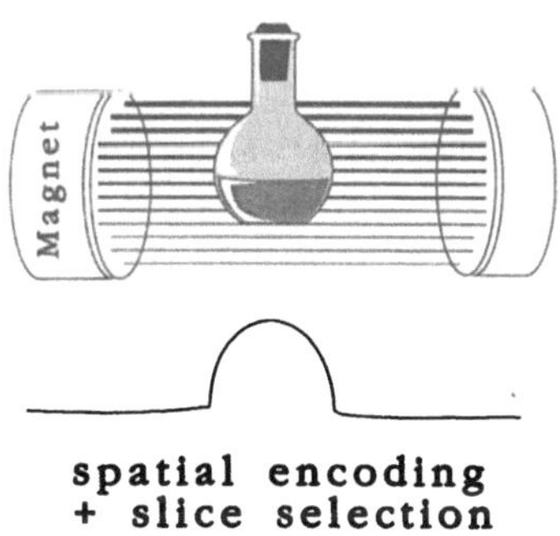

Fig. 3. A spatially linear variation of the magnetic field strength, produced by currents in gradient coils, allows the encoding of the spatial location of nuclei. An acquisition during the application of a gradient gives a projection of the object. The use of radiowaves with a narrow bandwidth (i.e., a selective RF pulse) together with a gradient excites only a slice of the object. These basic principles are used in MRI

distribution of MR signals and thereby produce the well-known MR images (Lauterbur 1973). Usually, these gradients produce larger frequency differences (around 50 kHz) than the chemical shift effect, thus hiding the chemical information in a combined image. However, in volume-selective spectroscopy (Fig. 4) both spatial and chemical information can be obtained simultaneously using special acquisition sequences.

To understand the application of MRS, one has to distinguish three different magnetic or electromagnetic fields which are used in combination and have to be produced or detected instrumentally (Fig. 1):

1. The static external magnetic field is typically generated by a large superconducting magnet with a horizontal aperture for the patient. Additional currents in coils around the aperture homogenise the magnetic field strength in the examination volume. MRI tolerates variations of ~10 ppm within the observed volume, whereas in spectroscopy, inhomogeneities determine the spectral resolution, which should be of the order of 0.1 ppm or better. Spectral resolution increases with magnetic field strength because of the increased frequency separation of the different resonances. Whereas MRI can be performed at 0.02 tesla (but usually between 0.25 and 1.5 tesla), spectroscopy requires magnets of 1.5 tesla or higher strength to resolve resonances sufficiently.
2. Time-dependent magnetic field gradients are spatial, linearly dependent changes of the magnetic field, produced by currents in additional coils driven by large amplifiers. The switching times are typically of the order of milliseconds, resulting in a characteristic acoustic noise due to the forces between the two neighbouring magnetic fields (i.e., the static magnetic field and the gradient field).
3. A set of antennae (coils) emit and detect radiowaves to and from the examined tissue or material. Typically, a MR system is equipped with a whole-body coil, a head coil, extremity coils, and several surface coils. Strong radiowaves of several kW are produced by large amplifiers and emitted by the coils. The signals coming back from the tissue, however,

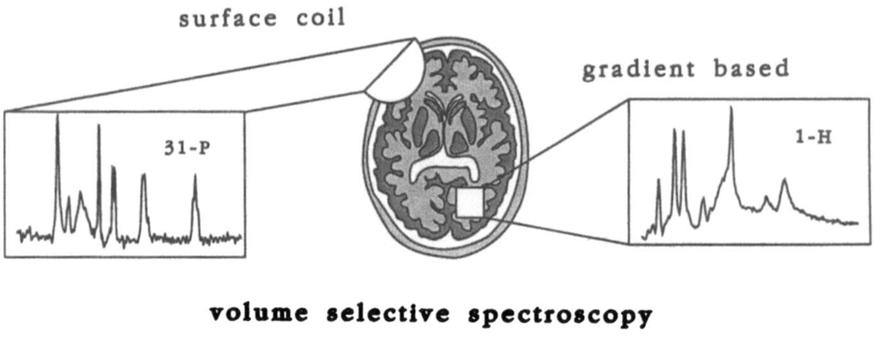

Fig. 4. Volume-selective spectroscopy can be based either on spatial variations of the radiowaves (illustrated by surface coil acquisition of a ^{31}P spectrum), or on spatial encoding by magnetic field gradients, as shown in Fig. 3 (illustrated by the acquisition of a ^{1}H spectrum)

are very small and need amplification before they are detected by a very sensitive radio receiver.

Resonance frequencies of different isotopes are well separated (e.g., at 1.5 tesla; 64 MHz for ^{1}H; 25 MHz for ^{31}P), enabling individual detection of different isotopes, such that signals in a spectrum originate from a specific isotope only, e.g., ^{1}H, ^{13}C, ^{31}P. In living systems, the chemical shift range of ^{1}H is ~10 ppm, ~30 ppm for ^{31}P, and ~200 ppm for ^{13}C. In principle, one could observe MR signals of hundreds of different isotopes, but in living systems only a few can be studied due to the inherently low MR sensitivity of many of these (e.g., ^{14}N, ^{15}N, ^{35}Cl, ^{39}K), the low-natural abundance of the isotope (e.g., ^{13}C, ^{17}O), or the small concentrations of the specific isotope in living tissue (e.g., ^{19}F). ^{1}H in water plays an outstanding role in MRI because of the high overall signal intensity in living tissue (due to high concentration, high-MR sensitivity, and nearly 100% isotopic natural abundance). ^{1}H MRS of metabolites other than water, however, suffers from the lower concentrations of these substances.

Signal/noise ratio (SNR) is a key issue in MRS. Since MRS deals with very low specific energy radiation (radiowaves), its sensitivity is inherently lower than radioactive methods, and the concentrations which are observable have to be much higher than those detectable by such high-energy techniques. This is probably the most serious limitation of MR. Some techniques can be used to increase SNR, but there are serious limitations, e.g., SNR increases with magnetic field strength, however, whole-body magnets above 1.5 tesla are very expensive. Another possibility is to average data together. Since the SNR increases as the square root of the number of averaged acquisitions, an increase of two in SNR requires a fourfold lengthening of experimental time. Patient compliance limits the examination time to roughly an hour, which makes any increase in examination time cumbersome. The minimum detectable concentration depends strongly on the nucleus, the specific MR technique, the apparatus, field strength, homogeneity, the observed volume, and the experimental time. As a rough rule for in vivo studies, minimum detectable concentrations for ^{1}H (metabolites), ^{31}P, and enriched ^{13}C (the natural isotope abundance of which is only 1.1%) have to be in the order of 1 mmol/l. The possible sample volumes for MRS of different nuclei are typically 3–8 cm^3 for ^{1}H and 27–64 cm^3 for ^{31}P and ^{13}C.

There are two relaxation phenomenon which influence the observed signal: spin-lattice relaxation (characterised by a time T_1), and spin-spin relaxation (characterised by a time T_2). T_1 relaxation relates to the amplitude of the signal obtainable following each interrogation. If a subsequent interrogation occurs almost immediately very little signal is obtained. However, if adequate time (~5T_1) is allowed for relaxation, the maximum possible signal is detected. T_2 relaxation concerns the mobility of nuclei. Immobile nuclei have short T_2s and produce broad signals in the spectrum, whereas mobile nuclei (e.g., in solution) have long T_2s and produce narrow resonances. Under certain conditions (full relaxation and detection without delay after excitation), the area under the resonances is proportional to the concentrations of the metabolites – at least of the MR visible portions. Metabolites in solids are relatively immobile and only contribute broad resonances to the MR signal. This is clearly the case for ^{31}P signals from bone and also from certain lipid components of cell membranes.

Data Acquisition Techniques

In order to interpret MR spectra usefully they should originate from a well-defined volume within the body (Fig. 4), e.g., the cerebellum, the wrist flexors, or the liver parenchyma. Various techniques are available to help distinguish signals from different regions, i.e., to perform volume selection. These methods are all based on the inhomogeneity (i.e., the spatial variation) of one of the electromagnetic fields used in MR: either the radiofrequency (RF) field, the magnetic field, or both are varied spatially so that identification of the location of the nuclei is possible. The inhomogeneity of the radiowaves or the magnetic field can be constant or time dependent. Furthermore, it is possible to distinguish between volume selection methods which acquire the data from a single localised volume (voxel), and those which acquire data simultaneously from the entire sample and spectra are localised by post-processing:

1. A surface coil (Ackerman et al. 1980) is formed by one or more electrically conducting loops. Because of both easy implementation and high SNR, this technique was used at the beginning of in vivo spectroscopy and is still used routinely to study easily accessible organs of relatively high homogeneity (e.g., muscles). However, the sensitive volume is very complicated and strongly dependent on experimental parameters (e.g., pulse power, pulse sequence), but can be described roughly as a hemisphere with the same radius as, and concentric with, the coil (Fig. 4).
2. An important source of uncertainty in the signal origin is tissue adjacent to the coil, e.g., cranial muscles during brain spectroscopy. Several schemes have been published to reduce the intensity from tissue near the surface coil e.g., time-dependent gradients (Chen and Ackerman 1990) or plastic sheets dotted with ferromagnetic material (Hennig et al. 1987).
3. Topical NMR (TMR; Gordon et al. 1980) was successfully used during the first years of in vivo spectroscopy. The sensitive volume is fixed relative to the magnet and is determined by shaping the static magnetic field. Only signals originating from the homogeneous region of magnetic field are detected as narrow resonances. Resonances from nuclei outside this region are greatly broadened.
4. Radiowaves of increasing amplitude can be used in connection with surface coils in a technique called "rotating frame spectroscopy" (Hoult 1979), which distinguishes between spectra from different depths perpendicular to the surface coil plane.

5. Analogous to MRI, time-dependent gradients can be used to encode the spatial position of the resonant nuclei. Beginning with volume-selective excitation (VSE; Aue et al. 1984), a large number of different methods have been published. Nowadays, the image-selected in vivo spectroscopy (ISIS; Ordidge et al. 1986), point-resolved spectroscopy (PRESS) (Bottomley 1984), and stimulated-echo acquisition mode (STEAM) (Frahm et al. 1987) techniques are used most frequently. In these methods, "selective RF pulses" excite the sensitive volume. These pulses are radio-waves with a small frequency range which excite only a slice of tissue if they are applied in the presence of a magnetic field gradient. This effect is used in MRI for the selection of a specific image slice and can also be used in volume-selective spectroscopy. Both PRESS and STEAM are "spin-echo" techniques. In these methods, three successive RF pulses are applied at intervals during which T_2 relaxation occurs. The last two RF pulses "refocus" the effects of the T_2 relaxation and generate a spin-echo signal.
6. A combination of surface coil and gradient localisation, called DRESS, gives spectra from single disk-like slices parallel to the plane of the surface coil (Bottomley et al. 1984).
7. Chemical shift imaging acquires data simultaneously from the entire sample. The spatial information is encoded through the signal "phase", such that it can be deconvoluted by mathematical processes (Brown et al. 1982; Maudsley et al. 1983). The spatial resolution of these methods is rather poor (e.g., 16 × 16 voxels) and many technical problems have to be solved (e.g., homogeneity over the whole field-of-view, long acquisition time, motion artifacts, poor voxel definition). However, the simultaneous acquisition of multiple volumes is promising because repeated acquisition and averaging increases the SNR of all the voxels and because voxels can be directly compared.

Safety and Patient Monitoring

The energy of electromagnetic waves is proportional to their frequency. Thus, radiowaves used in NMR have quite low-specific energies, leading to two important consequences: (1) MR is insensitive relative to many other methods and (2) radiowaves used in MR have less energy than that required to break a typical chemical bond and, therefore, are not ionising. This explains the inherent safety of this method.

However, other hazards are very crucial. The strong magnetic field attracts nearby ferromagnetic material and innocent tools can become projectiles. Ferromagnetic or conducting implants can move and could damage surrounding tissue (e.g., aneurysm clips). Devices such as pace-markers or neurostimulators can fail in the magnetic field. These possible hazards have to be considered when working in the fringe field of a magnet since any carelessness can instantly cause injuries. However, in contrast to the inherent ionising effect of X-ray or radioisotope examinations, these hazards can be completely avoided by appropriate management.

Other problems occur during MR examination of critically ill or unstable patients (Roth et al. 1985; McArdle et al. 1986; Nixon et al. 1986; Boesch and Martin 1988; Karlik et al. 1988; Shellock and Slimp 1989). During the relatively long time of an MR examination, one has very limited patient accessibility and visibility. Magnetic materials or monitors must not be positioned near the magnet. RF interference can cause malfunction of physiological monitors. Screens of commonly used monitors can be completely distorted and associated sensors can detect signals induced by pulsed magnetic field gradients or by patient movement in the static magnetic field. In other words, MR examinations take longer and patient monitoring and care is much more difficult than in comparable situations (e.g., ultrasound, X-ray, computerised tomography).

Concepts of patient monitoring vary from MR site to MR site. Generally, one has to decide the location of the person who is responsible for the patient. This decision is substantial, since the room containing the magnet (and the patient) is completely closed during an MR acquisition. A Faraday shield stops the sensitive MR system from picking up environmental RF interference. Opening the door of this shield for a brief examination of the patient can introduce interference and this should be avoided. For critically ill, ventilated, or anaesthetised patients, or very premature infants, it is obvious that the monitoring person has to be located near the patient, i.e., next to the magnet within the Faraday shield. This requires a monitoring system which functions near the magnet. For more stable patients, it has to be decided if the observing person can be placed outside the magnet room with an appropriate number of monitoring facilities.

Experience has shown that more than one monitor should be available for the same physiological parameter to allow double-checking (e.g., pulse-oximetry vs ECG, pulse-oximetry vs expiratory CO_2). Motion of an extremity of the patient in the magnet often causes monitoring artifacts. Over several years, in various generations of equipment (Boesch and Martin 1988; Boesch et al. 1988), ECG, pulse-oximetry, capnography (end-expiratory CO_2), body and incubator temperature, noninvasive blood pressure, and transcutaneous pO_2 and pCO_2 have been used to monitor neonatal and paediatric patients. With these sensors, reliable monitoring of respiration/oxygenation, heart-rate/pulse/blood pressure, and temperature of a patient is possible.

Careful analysis of all possible interactions between MR systems and monitors was necessary prior to the construction of newly designed, MR-compatible monitoring systems (Fig. 5). This has revealed influences in both directions. The MR system produces a static magnetic field, time-dependent gradients, radiowaves, and system control currents. The associated electromagnetic fields can influence the action of sensors and monitors. The monitors on the other hand often emit radiowaves which interfere with the tiny MR signals from the patient, thus producing noise and artifacts in an

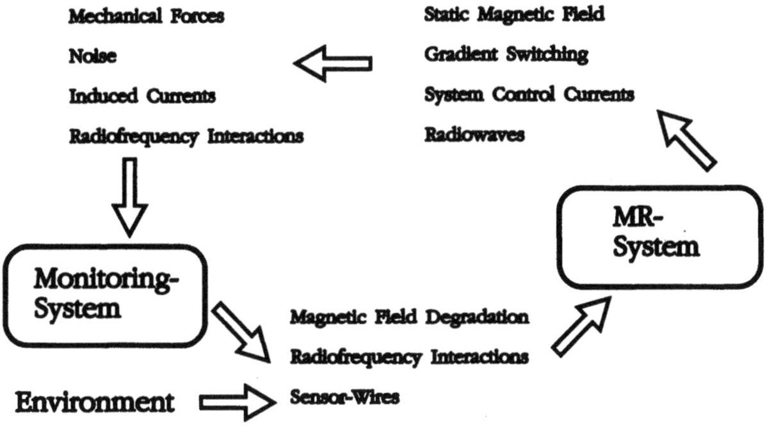

Fig. 5. Schematic representation of possible interactions between MR instrumentation and physiological monitoring systems

image or spectrum (e.g., stripes in MR images). Any ferromagnetic material in a monitor or in a sensor can degrade the magnetic field homogeneity. Sensor cables can allow environmental RF noise to reach the MR receiver via the coil. It has to be considered that an MR system is not only a very hostile environment for technical devices, but it is also an exceptionally sensitive instrument for the detection of radio waves. Making monitors MR compatible requires years of development time and this may, even for big bio-engineering departments in hospitals, be prohibitive. It is not surprising that only a few such monitors are available commercially.

Some fundamental principals underly MR-compatible monitoring. All RF-emitting circuitry must be heavily RF shielded. No magnetic material is usable or the distance from the monitor to the magnet has to be increased (e.g., for transformers). Screens must use liquid crystal or electroluminescent displays. Sensors and cables have to be replaced by optical devices or optical links to avoid induced currents and "hot spots" which can cause burns (Shellock and Slimp 1989). Therefore, many sensing and measurement devices have to be redesigned completely. Some methods are inherently electrical, e.g., ECG and associated electrodes and cabling has to be made of RF-transparent material (e.g., carbon fibre etc.). Otherwise, not only degradations of images and spectra are possible, but also burns.

In the early days of clinical MR, some sites refrained from examining critically ill or unstable patients such as premature babies. Other sites decided to examine babies but performed insufficient patient monitoring. However, technical difficulties produced by an MR system should not lead to careless or inadequate monitoring, or to the rejection of a clinically necessary diagnostic procedure for the evaluation of the newborn.

Dedicated Neonatal Hardware

Generally, MR systems are not planned for use with newborn premature or other small infants. Clinical MR-system design includes attention to medical devices for patient care as well as to other electronic circuits (e.g., RF coils). An overview of physiological monitoring has been given above. In addition to many of the monitoring techniques used for critically ill adult patients, transcutaneous pO_2/pCO_2 is desirable for premature babies. Partially MR-compatible sensors have been developed and successfully used in a small-bore MR system (Boesch et al. 1988). In whole-body magnets, however, the introduction of electrically conductive wires should be avoided completely. This problem has not been solved yet for the pO_2/pCO_2 sensor.

Similar to monitoring devices, ventilation and anaesthesia equipment are not a priori MR compatible. However, the development of compatible devices is less complicated than for physiological monitors and several products are commercially available. Nevertheless, two points have to be considered carefully: (1) anaesthesia equipment should be fully MR compatible. Ventilators, which have to be placed outside the MR suite are not appropriate in general and especially not for the ventilation of premature babies and (2) the ventilators used should allow the application of very small volumes for premature babies.

Standard incubators are generally incompatible with MR, i.e., they do not fit into the bore of either the magnet or the RF coil. There have been several different solutions to this problem: warm blankets (McArdle et al. 1987), an incubator with specially designed coils inside the incubator (Younkin et al. 1984; Thalayasingam et al. 1985; Boesch and Martin 1988), and an incubator with dimensions compatible with a common head coil (Boesch et al. 1988). The temperatures of electrically heated and regulated incubators are easier to control than those of warm blankets. However, the construction of such an MR-compatible incubator is very difficult, since the heater has to be placed far from the centre of the magnet, thereby making the regulation circuit slow in operation. Moreover, to prevent possible arcing, high voltages (commonly used in transmitting volume coils) should be avoided within an incubator when oxygen is applied therapeutically with the possibility of elevated oxygen levels in the incubator.

In commercial systems, mechanical support facilities for patients are not immediately adaptable to neonates and small infants. MR sites that undertake paediatric work find their own solutions. However, as a general rule, fixation of the examined organ is mandatory. Small infants usually tolerate tight wrapping in blankets much better than loose restriction of extremity movement.

In most commercial MR systems, specialised RF coils for premature babies and infants are not available. Without the need of an incubator, examination of small infants in standard extremity coils can give acceptable results. However, either for infants who are too large for this coil or for premature infants inside an incubator, there is generally no commercially available coil and one has to build one's own. In doing this, one has to understand the underlying physics of the problem. With the large head of an adult, the principal source of RF noise is the patient and head coils are optimised for this particular situation. This is not the case for a ~700 g premature baby within an incubator. Here the RF noise contribution of the coil is significant because the patient occupies only a small portion of the volume of the relatively large incubator. This situation requires coils which themselves amplify the RF signal due to their higher quality factor (Q).

The use of small-bore magnets (~30- to 40-cm-bore diameter) mainly for the spectroscopic examination of neonates has been successful in several clinical laboratories (Cady et al. 1983; Younkin et al. 1984; Boesch and Martin 1988). Clear advantages seems to favor this solution: Higher magnetic fields are available, the capital cost is lower, and running expenses are reduced. A smaller fringe field makes installation easier and a shorter magnet allows direct monitoring of the patient. Interactions with other devices are generally less important. However, in the last couple of years, imaging and volume selection using gradients has become more and more common. Nowadays, several serious disadvantages with small-bore magnets have become apparent (Boesch et al. 1991). Limited homogeneous volume and gradient linearity leads to distorted images. Space-consuming RF coils (including an integrated body coil) are not possible. Interactions between the magnetic gradients and the super-conducting magnet are pronounced and this leads to very strong eddy-current effects. Shielded gradients and increasing the distance from the gradient coils to the magnet, which could partially solve the eddy-current problems, further reduce the available bore of the magnet. In summary, small-bore magnets can be used very successfully for spectroscopy of a confined region of a neonate (or of an extremity). MRI, volume-selective MRS using gradients, and follow-up of patients over several years, however, is severely restricted. For these types of experiments, large-bore magnets (>60 cm) are required.

Studies of Normal Human Infants

A substantial amount of information about brain development and metabolism has now been obtained from normal infants using both ^{31}P- and ^{1}H MRS.

^{31}P Studies

^{31}P spectra from human neonatal brain in vivo are presented in Fig. 6. From animal models we have learned that in the neonatal mammal the major contributory chemical species of the brain PME signal (resonating at ~6.7 ppm – the signal from PCr is 0 ppm by convention) is phosphoethanolamine (PEt) – a structural component of the ethanolamine phosphoglycerides. There are also minor signals from PCh and α-glycerophosphate (the common anabolic precursors of the phospholipids) and also from

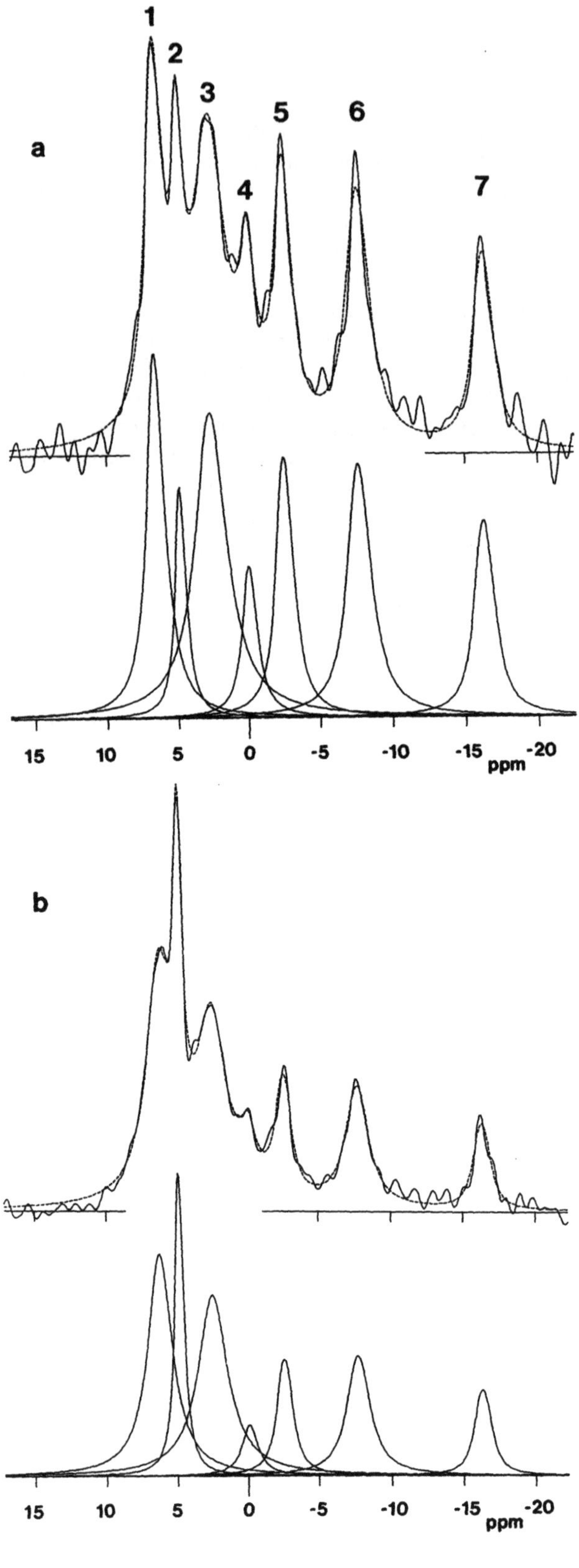

Fig. 6a,b. ^{31}P spectra obtained from the cerebral cortex of a severely birth-asphyxiated infant. Acquisition conditions: 2.4 T Bruker Biospec (^{31}P frequency 40 MHz, 40-cm clear bore); 6-cm surface coil; 256 scans; 90° flip angle at coil centre; pulse repetition time = 2.256 s. Peak assignments: *1*, PME; *2*, Pi; *3*, PDE; *4*, PCr; *5*, *6*, and *7*, γ-, α-, and β-NTP. The *dashed lines* are Lorentzian curves fitted to the spectrum using χ^2 minimisation (Cady 1991). The individual fitted components are shown also. The NTP multiplets were fitted as doublets (γ and α) and a triplet (β) with J-coupling of 16 Hz. **a** On the day of birth the spectrum appeared similar to that obtainable from a GPA-matched normal infant. However, PCr/Pi was reduced (0.37). NTP/P_{tot} was normal (0.095). **b** At age 2 days the severity of the hypoxic–ischaemic injury was indicated both by a further reduction in PCr/Pi (0.18) and by the reduced NTP/P_{tot} (0.039). Note the *ordinate scales* for spectra **a** and **b** are not comparable (Spectra collected in collaboration with S. Roth, R. Aldridge, M. Wylezinska, and E.O.R. Reynolds.)

phosphoserine and phosphoinositol. All these compounds giving rise to the PME signal are predominantly found in the anabolic pathway of membrane phospholipids (e.g., in myelin synthesis and in glial and neuronal proliferations). These substances play important roles in the biosynthesis of nervous tissue. In the newborn, the PME resonance is very strong, indicating high concentrations of these phospholipid precursors. It is during the neonatal period, when myelination and brain growth are rapid, that the PME signal is highest. The PDE resonance at ~2.9 ppm consists of signals from mobile phosphoglycerides, phospholipid bilayers, nonglycerol-containing sphingomyelin in addition to the phospholipid catabolic products glycero-phosphorylcholine (GPC) and glycero-phosphorylethanolamine (GPE; Pettegrew et al. 1987; Cohen 1988).

Although Pi exists in two major forms, i.e., $H_2PO_4^-$ (acid) and HPO_4^- (base), because there is rapid chemical exchange of protons there is only one averaged peak observable in the ^{31}P spectrum (at ~4.9 ppm). PCr is considered to be the storage form of high-energy phosphates in the brain from which adenosine triphosphate (ATP) can be mobilised very rapidly by exchanging its phosphorus atom with the γ-phosphate of ATP in the CK reaction. In addition to very efficient mitochondrial ATP synthesis through oxidative phosphorylation, the CK and adenylate kinase reactions are the major extra-mitochondrial sources for ATP production. The high activity of CK together with the large PCr pool keeps [ATP] virtually constant and adenosine diphosphate [ADP] below the level detectable by MRS. However, [ADP] can be calculated from the equilibrium equation of the CK reaction. A fall in [ATP] indicates a disturbance of the CK equilibrium and a breakdown of energy metabolism. In recent years, many investigators have demonstrated that acute cerebral hypoxia is followed by a decline in [PCr], indicating its utilisation. [Pi] increases simultaneously and the ratio [PCr]/[Pi], proportional to the phosphorylation potential [ATP]/([ADP] × [Pi]) – an index of the energy reserve of the cell – is often used to characterize brain-energy state (Chance et al. 1978). [PCr]/[Pi] is indirectly proportional to [ADP], which in turn is the major trigger for CK activity and also the principal controller of cellular respiration.

Normal postnatal developmental changes in the various phosphorus metabolites observed by in vivo MRS have to be considered if observations of acute and rapid metabolic derangements following hypoxic–ischaemic injury are to be interpreted correctly. If [NTP] is assumed constant during brain development, a relative increase in [PCr] is observed, with a [PCr]/[NTP] ratio of 0.73 ± 0.12 (mean ± SD; n = 10) in the normal neonate and 1.34 ± 0.11 (n = 10) in the normal 1-year-old infant. There is a parallel increase in the ratio [PCr]/[Pi] (the phosphorylation potential) from 0.94 ± 0.18 at birth to 1.28 ± 0.17 at age 1 year. During the same period, a steady decrease in the PME signal (continuing until adolescence) can be observed. PEt diminishes probably as a result of the formation of ethanolamine plasmalogens which are actively synthesised during myelogenesis and these are most abundant in mature myelin. However, no change in the chemical shift of the PME resonance occurs with age, indicating that PEt and PCh are still the major compounds of PME in older children. The PDE peak increases with progressing developmental age, probably as a result of increased amounts of membrane bilayers and phospholipids in myelin (Boesch et al. 1989). However, it has to be remembered both that structurally bound molecules have very short T_2s, and that their signals decay rapidly making them difficult to detect by in vivo MRS. As a result of developmental changes, the ratio of PME/PDE drops from 1.5 in the neonate to about 0.4 in the older child (Cady et al. 1991).

As stated in the previous section, the equilibrium reaction between the two anionic forms of Pi has a pK_a of ~6.8 and hence the chemical shift of this resonance is dependent on intracellular $[H^+]$. From the chemical shift difference (δ) between PCr and Pi in the ^{31}P spectrum, pH_i can be estimated with an accuracy of ~0.1 pH units using the Henderson-Hasselbalch equation (Petroff et al. 1985): $pH_i = 6.75 + \log_{10}[(\delta - 3.72)/(5.69 - \delta)]$. In several separate studies of normal infants, the brain pH_i measured in vivo (~7.1) has shown no changes during postnatal development (Hope et al. 1984; Azzopardi et al. 1989a; Boesch et al. 1989).

Absolute concentrations (i.e., in mmol/l wet of bulk tissue) of phosphorylated metabolites in the brains of newborn human infants have recently been estimated using a technique in which the ^1H signal from tissue water was used as an endogenous, internal concentration reference (Thulborn and Ackerman 1983). The values measured by MRS in normal human infants of a gestational age plus postnatal age (GPA) of 35–37 weeks agreed well with those obtained from neonatal mammalian brain using invasive techniques as shown in Table 1 (Cady and Azzopardi 1989; Cady 1990, 1991). In many instances neonatal metabolite concentrations are somewhat different from those measured in the adult human brain (Cady 1992). These differences probably relate to brain development. The MRS [NTP] agrees closely with high-pressure liquid-gas chromatographic measurements of NTPs (ATP, guanosine-, cytidine-, and uridine triphosphate) in

Table 1. Brain metabolite concentrations in normal human infants and mammals

Metabolite	Human infants[a] (n = 5, mmol/l wet)		Mammal[b] (mmol/kg wet)
	Mean	SD	
PME	9.3	1.5	[PEt] ~8.0
Pi	2.4	0.6	[Pi] ~2.2
PDE	25.4	11.0	[mobile (GPC+GPE)] ~2.2
PCr	2.7	1.1	[PCr] ~2.4–3.0
NTP	3.3	0.6	Σ[NTP] = ~3.7

[a] Human data obtained in collaboration with D. Azzopardi, R. Aldridge, M. Wylezinska, S. Roth, and E.O.R. Reynolds. GPA 35–37 weeks.

[b] Mammalian data – for source references see text.

adult rat brain extracts (3.7 ± 0.1 mmol/kg wet; Chapman et al. 1981). [PCr] agrees well with values of 2.4 to 3.0 mmol/kg wet measured in newborn mammalian brain (Lolley et al. 1961; Miller and Shamban 1977; Vannucci and Duffy 1977). For [Pi], MRS gives a value close to the 2.2 ± 0.2 mmol/kg wet measured in adult rat brain (Veech et al. 1973). In neonatal brain, PME includes a major contribution from PEt (Gyulai et al. 1984; Brenton et al. 1985; Pettegrew et al. 1986; Burri et al. 1988) which has a relatively high concentration (Agrawal and Himwich 1970). In the newborn rabbit [PEt] ~8 mmol/kg wet (Cohen and Lin 1962) and there is evidence of a similar value in the human fetus (Okumura et al. 1960). Several other metabolites may also contribute to the PME (Glonek et al. 1982). The PDE peak is complex and includes resonances from GPC and GPE, in addition to a broader signal from semimobile, cross-linked PDEs in membrane bilayers (Bates et al. 1989; Murphy et al. 1989). In Table 1, [PDE] by MRS ≫ [mobile PDEs] in adult guinea pig brain (Glonek et al. 1982). The excess may be due to the semimobile fraction (Cerdan et al. 1986). There is evidence from studies of premature infants (Cady and Azzopardi 1989; Cady et al. 1990) that brain metabolite concentrations may increase with GPA. Hence, it may be important to consider GPA when interpreting concentrations in sick infants.

^{1}H Studies

Without special editing methods, ^{1}H spectra show resonances mainly due to water and fatty acids, which both have a very high concentration (e.g., ^{1}H in water ~110 M) compared to most metabolites which occur in concentrations of a few mmol/l. To detect the small metabolite signals which resonate close to water and fat, some MR methodological tricks are necessary. (One possible scheme to suppress the strong water signal is a frequency-selective pulse on the water resonance and a subsequent dephasing of the signal by magnetic gradients.) In the human brain, the fatty-acid signals are small and, with good localisation, no suppression within the volume-of-interest (VOI) is necessary. However, bone marrow and subcutaneous tissue show large fat signals. To avoid any contamination from these tissues, nearly perfect pulse shapes are necessary, only exciting the nuclei within the VOI. Additionally, outer-volume presaturation is possible, i.e., saturation of all nuclei outside the VOI. In neonatology, it is crucial to consider the effect of motion on the suppression of these large signals. One single acquisition out of, e.g., 256 (a typical number of averages), during which movement of the head occurs can render the spectrum useless, e.g., additional "resonances" can be produced if patient movement shifts the VOI so as to include bone marrow or subcutaneous fat.

The ^{1}H spectrum of a normal term infant shows several resonances (Fig. 7). Most of these are, even now, not well understood or assigned only tentatively. Moreover, it is important to distinguish between several possible acquisition regimes. In many of the volume-selective schemes, spin-echo signals are acquired after a certain delay (echo time). This can be used to edit (simplify) the spectra. After an echo time of 135 or 270 ms, the large signal from fat has decayed and even the signal from water is much smaller compared to an acquisition immediately after a single RF pulse. In Fig. 7, a spectrum acquired after the very short echo time of 7 ms is shown. Due to this brief echo time, many resonances are visible. Obviously, it is desirable to have the maximum information content in a spectrum and the increased SNR of such short echo times. Nevertheless, the spectrum is very crowded and susceptible to artifacts from large resonances due to water and fat. The biochemical functions of MR-visible metabolites such as Gluc, Lac (a product of anaerobic glycolysis), and PCr and Cr (partners in the high-energy phosphate equilibrium) are relatively clear. However, the roles of NAA, Cho, Tau, and Ino are less obvious and still under investigation. NAA is frequently used as a neuronal marker in the ^{1}H spectrum, however, this is an oversimplification. Even if the neuronal-marker hypothesis holds for most ^{1}H studies (which would have strong implications for spectrum interpretation), there is evidence against this hypothesis from some experiments.

In acquisitions with long echo times, resonances with short T_2s become negligible and a relatively simple spectrum (NAA, PCr and Cr, and Cho) is obtained. The Lac resonance at ~1.33 ppm also decays relatively little during long echo times. Because the Lac CH_3 resonance is a doublet, it is modulated by coupling with the neighbouring CH proton during echo formation. Since this modulation is absent for certain echo times and for certain conditions, usually 135 or 270 ms are chosen as the echo time.

During human-brain development, the relative areas of the resonances change, i.e., the relative concentrations of the observed metabolites alter (Van der Knaap et al. 1990; Hueppi et al. 1991). In Fig. 7, spectra of both a premature baby and an adult are shown. Hueppi et al. used a biochemically determined combined PCr + Cr concentration as an internal standard and observed an increase of the NAA concentration during cerebellar development. Ino and Cho did not change significantly during the age range covered by the study, whereas Tau increased up to term. The chromatographically determined PCr + Cr concentration showed a significant increase during development.

Although ^{1}H MRS studies are in their infancy, this modality holds great promise for increasing the understanding of the sequelae of birth asphyxia, e.g., the possible role of anaerobically generated lactate as a cause of brain injury subsequent to hypoxia–ischaemia.

Perinatal Asphyxia

The results presented here have been obtained using ^{31}P MRS. Although ^{1}H studies of the sequelae of perinatal asphyxia using recently developed localisation methods are proceeding at many centres worldwide, these investigations are at an early stage and, at the time of writing, discussion is

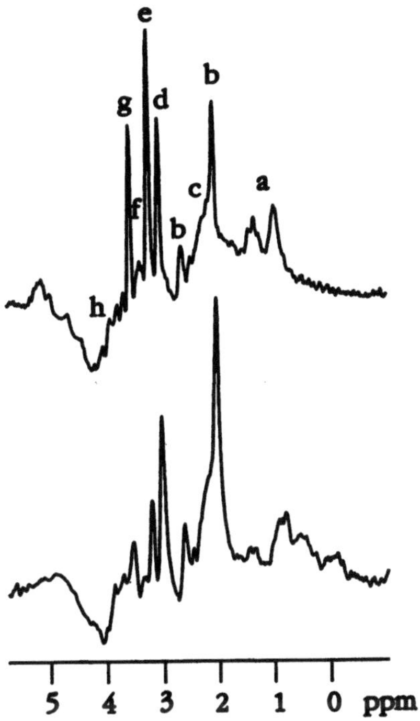

Fig. 7. An ^{1}H spectrum of the human cerebrum at 1.5 tesla (^{1}H frequency = 64 MHz), using STEAM localisation (Frahm et al. 1987) with an echo time of 7 ms. The *upper trace* is a spectrum from a premature baby (34 weeks' GPA), the *lower trace* is a spectrum from an adult (aged 32 years). The *vertical scalings* are arbitrary. The following resonances can be distinguished: *a*, methyl resonances (fatty-acid chains, proteins/peptides), in some cases contamination from fat; *b*, NAA at ~2.0 ppm and ~2.6 ppm; *c*, unresolved resonances from Glu and Gln amongst other compounds; *d*, Cr and PCr at ~3.0 ppm; *e*, Cho at ~3.2 ppm; *g*, Ino at ~3.3 ppm. The intense water signal *h*, at ~4.7 ppm has been reduced both by saturation and by postacquisition suppression (Kreis 1991). Resonances near the water (i.e., above ~3.5 ppm) can be affected by the saturation technique. Between the Ino and Cho resonances, an additional resonance *f*, at ~3.3 ppm (tentatively assigned to Tau) is visible. Since some protons from Gluc resonate at this position also (Gyngell et al. 1991), unique assignment is not possible (at least not under hyperglycaemic conditions). The Lac resonance, if present, appears at 1.33 ppm. (Spectra collected in collaboration with C. Fusch, N. Herschkowitz, P.S. Hueppi, and S. Posse)

not possible. The ^{31}P studies involved simple, surface-coil methods. Therefore, localisation was relatively crude and the sensitive volume approximated a hemisphere of similar radius to the coil. However, on several grounds (i.e., analysis of surface-coil spatial sensitivity) the poorly developed cranial musculature, undetectable PCr and NTP in spectra acquired from an infant with severely deranged metabolism whose skeletal muscle spectra exhibited normal levels, and a ^{31}P MRS [PCr] that compares well with that in neonatal mammalian brain – see "^{31}P Studies" this chapter and Table 1 – contamination of the spectra by signals from superficial tissues was very small (Tofts et al. 1984). A 7-cm-diameter coil was used most often but for smaller infants the diameter was 5 cm. With coils of these dimensions most of the signal originates from a superficial layer of brain ~3-cm thick.

Relative Concentrations

Within 24 h of delivery, ^{31}P spectra from birth-asphyxiated infants are often similar to those obtained from normal infants of a comparable GPA (Fig. 6; Hope et al. 1984, 1986; Azzopardi et al. 1989b). In severe cases, subsequent spectra exhibit abnormalities. In particular the PCr/Pi ratio falls as shown in Fig. 6 (Cady et al. 1983; Hope et al. 1984, 1986; Hamilton et al. 1986; Azzopardi et al. 1989b). In infants who have suffered very severe birth asphyxia, NTP/P_{tot} (P_{tot} is the total mobile phosphate defined as PME + Pi + PDE + PCr + 3 × NTP) may fall also (Fig. 6). Both these ratios (i.e., PCr/Pi and NTP/P_{tot}) have proven to be useful prognostic indicators (see "Prognostic Value of MRS" this chapter). In some severe cases, the Pi peak may eventually dominate the spectrum and, occasionally, both PCr and NTP may become undetectable. In the infants who survive, metabolite ratios often return to normal values after a few weeks, but overall signal amplitude may be reduced (Delivoria-Papadopoulos and Chance 1988; Azzopardi et al. 1989b). The observations of a diminished signal are supported by the results of absolute quantitation studies (Cady

Table 2. Phosphorus metabolite ratios and pH_i in normal infants and infants with a clinical history of perinatal asphyxia[a]

	Normal infants[b] (n = 30)	Birth asphyxia[b] (n = 40)	Postnatal asphyxia[b] (n = 5)
Gestation (weeks)	33 (26–42)	41 (31–42)	33 (31–38)
Age (days)	5 (1–96)	3 (1–10)	5 (4–23)
PCr/Pi	0.99 ± 0.18	0.67 ± 0.35	1.04 ± 0.20
NTP/P_{tot}	0.09 ± 0.01	0.08 ± 0.03	0.10 ± 0.02
PCr/P_{tot}	0.09 ± 0.01	0.08 ± 0.02	0.10 ± 0.01
Pi/P_{tot}	0.10 ± 0.01	0.16 ± 0.10	0.10 ± 0.01
PME/P_{tot}	0.29 ± 0.03	0.29 ± 0.04	0.27 ± 0.02
PDE/P_{tot}	0.20 ± 0.02	0.20 ± 0.03	0.22 ± 0.01
PCr/NTP	1.02 ± 0.20	1.21 ± 0.68	1.06 ± 0.18
Pi/NTP	1.05 ± 0.23	3.28 ± 4.48	1.05 ± 0.29
PME/NTP	3.18 ± 0.61	4.62 ± 2.89	2.97 ± 0.68
PDE/NTP	2.21 ± 0.48	3.26 ± 2.38	2.35 ± 0.47
pH_i[c]	$7.11 \pm 0.13^{n=22}$	$7.16 \pm 0.19^{n=36}$	$7.09 \pm 0.14^{n=5}$

[a] Data at time of lowest observed PCr/Pi.
[b] Medians (ranges) or means ±SD.
[c] pH_i not calculated when Pi unresolved. (From Azzopardi et al. 1989[b]).

and Azzopardi 1989; Cady 1990) which will be described in "Absolute Concentrations" this chapter. A fundamental conclusion from all these studies is that metabolite ratios can be strongly time dependent following birth asphyxia. Studies carried out within 24 h of birth or when aged >1 week may not provide useful assessments of tissue damage. To obtain a realistic evaluation of injury severity, frequent sequential studies are necessary.

The results of studies both on normal infants (Azzopardi et al. 1989a) and on infants suspected of hypoxic-ischaemic brain injury (Azzopardi et al. 1989b) who were investigated because of evidence of birth asphyxia or postnatal asphyxia are presented in Tables 2 and 3. The data shown are those determined when PCr/Pi was at its lowest value. In Table 2, the infants are classified clinically but no attempt has been made to take GPA-related developmental changes (Azzopardi et al. 1989a; Boesch et al. 1989; Van der Knaap et al. 1990) into account. In Table 3, by presenting the results as standard deviation scores (SDS) corrected for GPA, asphyxiated infants are compared with normal infants and developmental changes are accounted for. Table 3 shows that several metabolite ratios (e.g., PCr/Pi, PCr/P_{tot}, and NTP/P_{tot}) were significantly less (Mann-Whitney test) in the asphyxiated infants, and that Pi/P_{tot} and Pi/NTP were significantly higher.

A more recent study on 52 birth-asphyxiated infants has related PCr/Pi, measured within 7 days of birth, to both outcome and head growth (Roth et al. 1992). Table 4 gives the mean occipito-frontal head circumference (OFC)

Table 3. Age-dependent SDS for phosphorus metabolite ratios and pH_i in the brains of infants suspected of birth asphyxia or postnatal asphyxia[a]

Phosphorus metabolite ratios and pH_i	Birth asphyxia ($n = 40$)	Postnatal asphyxia ($n = 5$)
PCr/Pi	−2.14 ± 2.19*	−0.03 ± 0.76
NTP/P_{tot}	−1.17 ± 1.82**	−0.01 ± 1.03
PCr/P_{tot}	−1.79 ± 2.59*	0.33 ± 1.04
Pi/P_{tot}	5.97 ± 8.91*	0.69 ± 0.99
PME/P_{tot}	0.63 ± 1.58	−0.79 ± 1.26
PDE/P_{tot}	−0.60 ± 1.44***	0.39 ± 0.50
PCr/NTP	0.65 ± 3.14	0.15 ± 0.66
Pi/NTP	8.96 ± 17.94*	0.35 ± 1.06
PME/NTP	2.49 ± 4.71**	−0.26 ± 1.11
PDE/NTP	1.43 ± 4.48	0.17 ± 0.77
pH_i[b]	$0.23 \pm 0.97^{n=36}$	$-0.28 \pm 0.22^{n=5}$

*$p < 0.001$; **$p < 0.01$; ***$p < 0.05$.
[a] Values obtained when PCr/Pi was lowest. Mean values for SDS ± SD vs the normal infants.
[b] pH_i not calculated when Pi unresolved. (From Azzopardi et al. 1989b).

Table 4. Neonatal OFC, OFC SDS on discharge or death at a median age of 16 (range 1–94) days and at 12 (range 11–13) months, and OFC growth velocity, according to PCr/Pi SDS

PCr/Pi SDS	Discharge or death[a]			Age 1 year			
	n	OFC (cm) mean (SD)	OFC SDS mean (SD)	*n*[b]	OFC (cm) mean (SD)	OFC SDS mean (SD)	OFC growth velocity ($cm\,yr^{-1}$) mean (SD)
>−1.99	25	34.9 (2.7)	−0.8 (1.5)	21	46.2 (2.0)	−0.9 (2.2)	10.4 (3.1)
−2.00 to −3.99	11	34.9 (2.5)	0.6 (1.5)	7	45.5 (2.0)	−1.5 (1.6)	9.5 (2.7)
<−4.00	16	35.7 (1.7)	0.0 (1.5)	5	41.2 (2.1)	−4.8 (2.0)*	5.6 (2.1)**

*$p < 0.002$; **$p < 0.009$ compared with normal values using one way analysis of variance. (From Roth et al. 1992).
[a] OFC on admission was substituted in 12 infants with missing values.
[b] Fifteen infants died and four values were missing.

growth velocity for three PCr/Pi SDS ranges. For PCr/Pi SDS < −4.0, OFC growth velocity is significantly reduced. Figure 8 shows the relationships between PCr/Pi SDS, neurodevelopmental outcome, and OFC SDS.

Intracellular pH

Many experimental MRS studies of neonatal brain involving hypoxia–ischaemia (e.g., Hope et al. 1988) have demonstrated a marked intracellular acidosis during the insult. In survivors, pH_i often returns to normal following the restoration of normoxic blood flow. However, MRS studies of asphyxiated infants are, of necessity, carried out some time after the period of hypoxia–ischaemia. Therefore, clinical investigations are limited to the investigation of the sequelae of asphyxia. In studies on 40 birth-asphyxiated infants, rather than the observation of a residual acidosis, pH_i was slightly alkalotic (Azzopardi et al. 1989b). Recent studies on neonatal mammalian brain have demonstrated a slight alkalosis in the presence of ~6 m*M* lactate (Nakada et al. 1991), and the possibility of dissociation between pH_i and LAC concentration (Hida et al. 1991).

Absolute Concentrations

It has already been mentioned in "Relative Concentrations", this chapter, that a few weeks after the hypoxic–ischemic episode, many surviving infants exhibit an overall reduction in signal amplitude (Delivoria-Papadopoulos and Chance 1988; Azzopardi et al. 1989b). MRS absolute-quantitation techniques (e.g., Cady 1992) provide both objective measurements of bulk-tissue metabolite concentrations and methods, whereby the observation of a reduced signal can be objectively assessed.

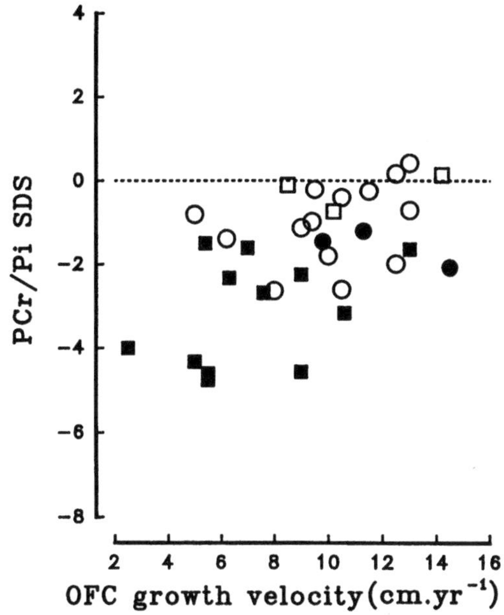

Fig. 8. PCr/Pi SDS related to occipito – frontal head circumference (*OFC*) growth velocity during the first year of life in 32 birth-asphyxiated infants. ○, normal outcome; □, minor impairment; ●, major impairment; ■, multiple impairments. (From Roth et al. 1992)

In addition to measuring absolute-metabolite concentrations in the brains of normal infants (see "^{31}P Studies" this chapter), the same technique has been employed to investigate 29 infants with a clinical history of birth asphyxia. The results have been related to neurodevelopmental outcome assessed at age 1 year when 13 of the infants were progressing normally and 16 infants were either abnormal or had died. Table 5 summarises the results and groups the results according to clinical outcome. [PCr] was significantly reduced in all the groups and the decrease was particularly marked ($p < 0.001$) in the infants who died. Other metabolite concentrations showed no significant relationship with outcome, although [Pi] tended to be higher in the infants who eventually died. The results given in Table 5 indicate that [PCr] may be a somewhat more sensitive prognostic index than the commonly used PCr/Pi ratio. This may be related to the observation that [Pi] is not significantly elevated in severe asphyxia. (This contrasts dramatically with the observation that in relative terms this resonance is greatly increased in many spectra acquired from asphyxiated infants.) Some of the factors which influence [Pi] operate in conflicting directions and this may explain the absence of a significant [Pi] increase. PCr and NTP hydrolysis will increase [Pi], whereas both oedema and membrane damage (leading to Pi washout) will reduce [Pi]. In contrast to the [Pi], many of the influencing factors (i.e., PCr hydrolysis, oedema, PCr washout) operate such as to

Table 5. Brain metabolite concentrations in birth-asphyxiated infants and neurodevelopmental outcome at age 1 year – preliminary results

Neurodevelopmental status	Concentrations (mmol/l)					
	PME	Pi	PDE	PCr	NTP	PCr/Pi
Normal progress ($n = 26$)	7.4 ± 2.2	2.7 ± 3.1	20.1 ± 7.9	1.8 ± 0.8*	2.6 ± 0.6	1.5 ± 1.6
Major/minor impairments ($n = 15$)	9.4 ± 3.1	2.3 ± 1.6[a]	17.0 ± 7.4	1.4 ± 0.9**	2.4 ± 0.7	0.9 ± 0.7[a]
Died ($n = 10$)	7.1 ± 3.2	5.9 ± 6.1	19.7 ± 9.2	0.9 ± 0.4****	2.3 ± 0.8	0.3 ± 0.3***

Values are means ±SD. Spectra acquired at age ≤7 days from infants born at ~ full term. Data collected in collaboration with D. Azzopardi, S. Roth, R. Aldridge, M. Wylezinska, and E.O.R. Reynolds. Neurological assessments by A.L. Stewart and J. Baudin.
*$p < 0.05$; **$p < 0.02$; ***$p < 0.005$; ****$p < 0.001$ when compared with normal controls.
[a] $n = 14$.

reduce [PCr] and this may explain the significant decreases. It appears that after severe asphyxia, [PCr] may be consistently reduced. However, [Pi] may be high, approximately normal or low depending on the time of study. More detailed work is required to elucidate the time dependence of changes in absolute metabolite concentrations following birth asphyxia.

Prognostic Value of MRS

Sufficient results are now available from studies of both relative and absolute concentrations of phosphorylated metabolites, measured within a few days of birth, to assess the prognostic value of MRS. Figure 9 shows the relationships between the minimum-observed values for PCr/Pi, survival, and neurodevelopmental outcome aged 1 year (Azzopardi et al. 1989b). Surviving infants who were progressing normally or with only minor impairments had minimum PCr/Pi values within or close to the 95% confidence limits for normal controls, as shown in Fig. 9a. In those infants with major neuromotor impairments, minimum PCr/Pi values were often low, especially when the impairments were multiple (Fig. 9b). Of the infants who died, a very large proportion had minimum PCr/Pi values below the confidence limit as shown in Fig. 9c. Values for NTP/P_{tot} (determined at the same time as minimum PCr/Pi) below the confidence limit were almost always associated with a fatal outcome. Assessment of the Griffiths general quotient (GQ; Griffiths 1954) for survivors at age 1 year indicated a correlation with the minimum PCr/Pi as shown in Fig. 10. Those infants having low PCr/Pi values within a few days of birth performed badly at assessment. (It should be noted that both Figs. 9 and 10 include some data from infants studied solely because of increased echodensities on ultrasound scans.) No significant relationship between pH_i and outcome has so far been reported. In a more recent study solely of birth-asphyxiated infants (Roth et al. 1992), a PCr/Pi SDS < -2 SD had a sensitivity of 82%, a specificity of 86%, and a positive predictive value of 88% for multiple impairments or death. In the

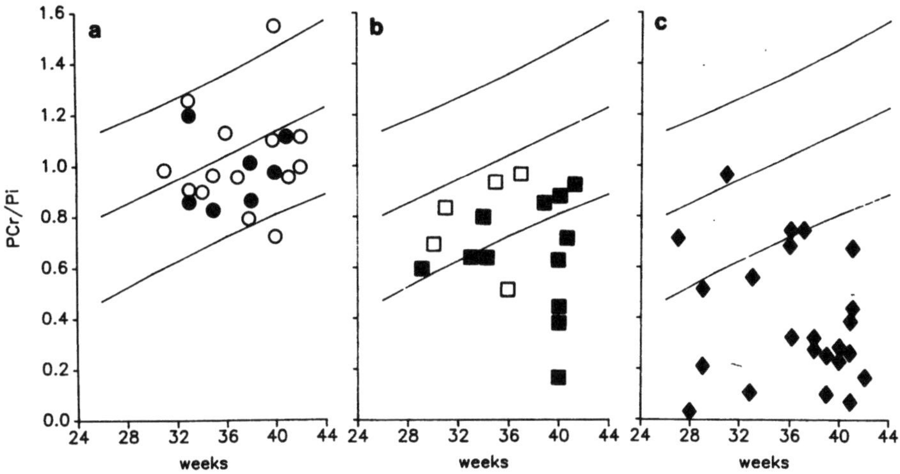

Fig. 9a–c. Relationships between minimum PCR/Pi in the neonatal cerebral cortex, survival, and neurodevelopmental outcome aged 1 year. The *regression line* and 95% confidence limits for normal values vs GPA are given. It should be noted that this figure includes data from 16 infants who were studied solely because of increased echodensities on ultrasound scans attributable in many cases to periventricular leukomalacia or intraparenchymal haemorrhage. **a** ○, normal progress (SDS vs normal infants = −0.29 ± 1.21 − SD − NS; ●, minor impairments (SDS = −0.54 ± 0.52, NS). **b** □, major neuromotor impairment (SDS = −1.06 ± 1.22, $p < 0.05$); ■, multiple major impairments (SDS = −2.60 ± 1.44, $p < 0.001$). **c** ◆, died (SDS = −3.82 ± 1.89, $p < 0.001$). (From Azzopardi et al. 1989b)

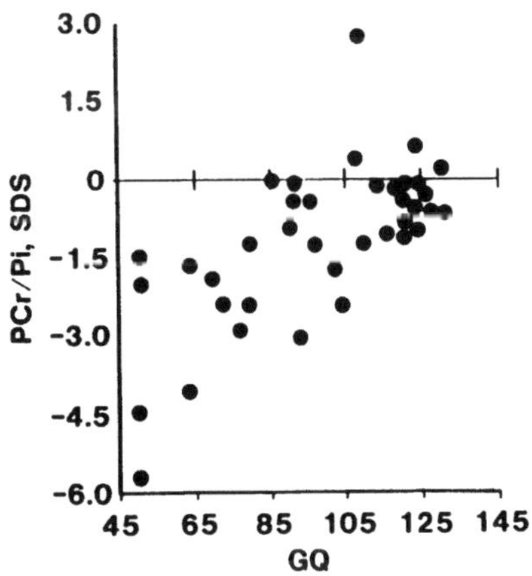

Fig. 10. Relationship between minimum PCr/Pi (expressed as a SDS vs normal infants) and Griffiths general quotient (*GQ*) at age 1 year in 38 survivors. SDS ⩽ 50 recorded as 50 ($r = 0.67$, $p < 0.001$). This figure includes data from 10 infants studied solely because of increased echodensities on ultrasound scans attributable in many cases to periventricular leukomalacia or intraparenchymal hemorrhage. (From Azzopardi et al. 1989b)

same study, values below −2.5 SD gave sensitivity, specificity, and positive predictive values of 72%, 92%, and 91%, respectively.

The data presented in Table 5 indicate that [PCr] may prove a useful prognostic indicator. [PCr] was particularly reduced in the group with fatal outcome, whereas PCr/Pi was not so significantly different from normal. Further investigation of the use of [PCr] as a prognostic index is required. If the errors of measurement for absolute concentrations can be reduced, [PCr] may eventually prove a better prognostic index than Pi/PCr.

Conclusions

^{31}P MRS already has the capability of providing useful prognostic information following birth asphyxia. It has also the potential, as a research tool, for elucidating some of the physiological and metabolic mechanisms responsible for brain-tissue injury due to hypoxic–ischaemic insult. ^{1}H MRS will, no doubt, prove to be an equally useful technique. MRS remains a rapidly developing field of human endeavour. Methods for both data acquisition and data analysis are improving all the time. Once universally accepted and commercially implemented, MRS techniques will be standard in available medical equipment and the contribution to the understanding and clinical evaluation of perinatal asphyxia will be significant.

References

Ackerman JJH, Grove TH, Wong GG, Gadian DG, Radda GK (1980) Mapping of metabolites in whole animals by ^{31}P NMR using surface coils. Nature 283:167–170

Agrawal HC, Himwich WA (1970) Amino acids, proteins and monoamines of developing brain. In: Himwich WA (ed) Developmental neurobiology. Thomas, Springfield, pp 298–299

Aue WP, Mueller S, Cross TA, Seelig J (1984) Volume-selective excitation. A novel approach to topical NMR. J Magn Reson 56:350–354

Azzopardi D, Wyatt JS, Hamilton PA, Cady EB, Delpy DT, Hope PL, Reynolds EOR (1989a) Phosphorus metabolites and intracellular pH in the brains of normal and small for gestational age infants investigated by magnetic resonance spectroscopy. Pediatr Res 25: 440–444

Azzopardi D, Wyatt JS, Cady EB, Delpy DT, Baudin J, Stewart AL, Hope PL, Hamilton PA, Reynolds EOR (1989b) Prognosis of newborn infants with hypoxic–ischemic brain injury assessed by phosphorus magnetic resonance spectroscopy. Pediatr Res 25:445–451

Bates TE, Williams SR, Gadian DG (1989) Phosphodiesters in the liver: the effect of field strength on the ^{31}P signal. Magn Reson Med 12:145–150

Boesch C, Martin E (1988) Combined application of magnetic resonance imaging and spectroscopy in neonates and infants: installation and operation of a 2.35 Tesla system in a clinical setting. Radiology 168:481–488

Boesch C, Luescher K, Brunner P, Kaelin P, Zuerrer M, Dangel P, Martin E (1988) Incubator and monitoring and anesthesia equipment for MR examination of premature and intensive care patients at 2.35 T. Radiology 169P:384

Boesch C, Gruetter R, Martin E, Duc G, Wüthrich K (1989) Variations in the in vivo P-31 MR spectra of the developing human brain during postnatal life. Radiology 172:197–199

Boesch C, Fusch C, Gruetter R, Martin E (1991) Magnetic-resonance examinations of neonates and infants in small-bore systems. In: Lafeber HN (ed) Fetal and neonatal physiological measurements. Excerpta Medica, Amsterdam, pp 25–31

Bottomley PA (1984) Selective-volume method for performing localized NMR spectroscopy. US patent 4 480 228

Bottomley PA, Foster TH, Darrow RD (1984) Depth-resolved surface-coil spectroscopy (DRESS) for in vivo ^{1}H, ^{31}P, and ^{31}C NMR. J Magn Reson 59:338–342

Brenton DP, Garrod PJ, Krywawych S, Reynolds EOR, Bachelard HS, Cox DW, Morris PG (1985) Phosphoethanolamine is major constituent of phosphomonoester peak detected by ^{31}P NMR in newborn brain. Lancet 1:115

Brown TR, Kincaid BM, Ugurbil K (1982) NMR chemical shift imaging in three dimensions. Proc Natl Acad Sci USA 79:3523–3526

Burri R, Lazeyras F, Aue WP, Straehl P, Bigler P, Althaus U, Herschkowitz N (1988) Correlation between ^{31}P NMR phosphomonoester and biochemically determined phosphorylethanolamine and phosphatidylethanolamine during development of the rat brain. Dev Neurosci 10:213–221

Cady EB (1990) Absolute quantitation of phosphorus metabolites in the cerebral cortex of the newborn human infant and in the forearm muscles of young adults using a double-tuned surface coil. J Magn Reson 87:433–446

Cady EB (1991) A reappraisal of the absolute concentrations of phosphorylated metabolites in the human neonatal cerebral cortex obtained by fitting Lorentzian curves to the ^{31}P NMR spectrum. J Magn Reson 91:637–643

Cady EB (1992) Determination of absolute concentrations of metabolites from NMR spectra. In: Rudin M, Seelig J (eds) In vivo magnetic resonance spectroscopy. Springer, Berlin Heidelberg New York, pp 249–281

Cady EB, Azzopardi D (1989) Absolute quantitation of neonatal brain spectra acquired with surface coil localization. NMR Biomed 2:305–311

Cady EB, Costello AM de L, Dawson MJ, Delpy DT, Hope PL, Reynolds EOR, Tofts PS, Wilkie DR (1983) Non-invasive investigation of cerebral metabolism in newborn infants by phosphorus nuclear magnetic resonance spectroscopy. Lancet 1:1059–1062

Cady EB, Roth S, Azzopardi D, Reynolds EOR (1990) Developmental changes in the absolute concentrations of phosphorylated metabolites in the cerebral cortex of the newborn human infant – Results from frequency domain spectrum analysis. Proc Society of Magnetic Resonance in Medicine, 9th Annual Meeting, New York, p 1003

Cady EB, Hennig J, Martin E (1991) Magnetic resonance spectroscopy. In: Haddad J, Christmann D, Messer J (eds) Imaging techniques of the CNS of the neonates. Springer, Berlin Heidelberg New York, pp 117–146

Cerdan S, Subramanian VH, Hilberman M, Cone J, Egan J, Chance B, Williamson JR (1986) ^{31}P NMR detection of mobile dog-brain phospholipids. Magn Reson Med 3:432–439

Chance B, Nakase Y, Bond M, Leigh JS, McDonald G (1978) Detection of ^{31}P nuclear magnetic resonance signals in brain by in vivo and freeze-trapped assays. Proc Natl Acad Sci USA 75:4925–4929

Chapman AG, Westerberg E, Siesjö BK (1981) The metabolism of purine and pyrimidine nucleotides in rat cortex during insulin-induced hypoglycemia and recovery. J Neurochem 36:179–189

Chen W, Ackerman JJH (1990) Spatially localized NMR spectroscopy employing an inhomogeneous surface-spoiling magnetic field gradient. NMR Biomed 3:147–165

Cohen MM, Lin S (1962) Acid-soluble phosphates in the developing rabbit brain. J Neurochem 9:345–352

Cohen JS (1988) Phospholipid and energy metabolism of cancer cells monitored by ^{31}P magnetic resonance spectroscopy: possible clinical significance. Mayo Clin Proc 63: 1199–1207

Delivoria-Papadopolous M, Chance B (1988) ^{31}P NMR spectroscopy in the newborn. In: Guthrie RD (ed) Recent advances in neonatal intensive care. Churchill Livingstone, Edinburgh, pp 153–179

Frahm J, Merboldt KD, Hanicke W (1987) Localized proton spectroscopy using stimulated echoes. J Magn Reson 72:502–508

Glonek T, Kopp SJ, Kot E, Pettegrew JW, Harrison WH, Cohen MM (1982) P-31 nuclear magnetic resonance analysis of brain: the perchloric acid extract spectrum. J Neurochem 39:1210–1219

Gordon RE, Hanley PE, Shaw D, Gadian DG, Radda GK, Styles P, Bore PJ, Chan L (1980) Localisation of metabolites in animals using ^{31}P topical magnetic resonance. Nature 287:736–738

Griffiths R (1954) The abilities of babies. University of London Press, London

Gyngell ML, Michaelis T, Hörstermann D, Bruhn H, Hänicke W, Merboldt KD, Frahm J (1991) Cerebral glucose is detectable by localized proton NMR spectroscopy in normal rat brain in vivo. Magn Reson Med 19:489–495

Gyulai L, Bolinger L, Leigh JS, Barlow C, Chance B (1984) Phosphorylethanolamine – the major constituent of the phosphomonoester peak observed by ^{31}P – NMR on developing dog brain. FEBS Lett 178:137–142

Hamilton PA, Hope PL, Cady EB, Delpy DT, Wyatt JS, Reynolds EOR (1986) Impaired energy metabolism in brains of newborn infants with increased cerebral echodensities. Lancet 1:1242–1246

Hennig J, Boesch C, Gruetter R, Martin E (1987) Homogeneity spoil spectroscopy as a tool of spectrum localization for in vivo spectroscopy. J Magn Reson 75:179–183

Hida K, Suzuki N, Kwee IL, Nakada T (1991) pH – lactate dissociation in neonatal anoxia: Proton and ^{31}P NMR spectroscopic studies in rat pups. Magn Reson Med 22:128–132

Hope PL, Costello AM de L, Cady EB, Delpy DT, Tofts PS, Chu A, Hamilton PA, Reynolds EOR (1984) Cerebral energy metabolism studied with phosphorus NMR spectroscopy in normal and birth-asphyxiated infants. Lancet 2:366–370

Hope PL, Costello AM de L, Cady EB, Delpy DT, Tofts PS, Chu A, Reynolds EOR (1986) Cerebral metabolism in newborn infants studied by phosphorus nuclear magnetic resonance spectroscopy. In: Rolfe P (ed) Neonatal physiological measurements. Butterworth, London, pp 382–389

Hope PL, Cady EB, Delpy DT, Ives NK, Gardiner RM, Reynolds EOR (1988) Brain metabolism and intracellular pH during ischaemia: effects of systemic glucose and bicarbonate administration studied by ^{31}P and ^{1}H nuclear magnetic resonance spectroscopy in vivo in the lamb. J Neurochem 50:1394–1402

Hoult DI (1979) Rotating frame zeugmatography. J Magn Reson 33:183–197

Hueppi PS, Posse S, Lazeyras F, Burri R, Bossi E, Herschkowitz N (1991) Magnetic resonance in preterm and term newborns: ^{1}H-spectroscopy in developing human brain. Pediatr Res 30:574–578

Karlik SJ, Heatherley T, Pavan F, Stein J, Lebron F, Rutt B, Carey L, Wexler R, Gelb A (1988) Patient anesthesia and monitoring at a 1.5-T MRI installation. Magn Reson Med 7:210–221

Kreis R (1991) Post-acquisition water suppression with digital low-frequency filters for ^{1}H MRS in man. Proc Eur Soc Magn Reson Med Biol, 8th Annual Meeting, Zürich, p 205

Laptook AR, Corbett RJ, Uauy R, Mize C, Mendelsohn D, Nunnally RL (1989) Use of ^{31}P magnetic resonance spectroscopy to characterise evolving brain damage after perinatal asphyxia. Neurology 39:709–712

Lauterbur PC (1973) Image formation by induced local interactions: examples employing nuclear magnetic resonance. Nature 242:190–191

Lolley RN, Balfour WM, Samson FE (1961) The high-energy phosphates in developing brain. J Neurochem 7:289–297

Martin E, Boesch C, Grueter R, Kikinis R, Kewitz G, Boltshauser E, Werner B, Eich G (1988) Magnetresonanz in der padiatrischen Forschung und Klinik. II. Untersuchungen zur Entwicklung und Pathologie des Gehirns bei Neugeborenen, Säuglingen und Kleinkindern. Helv Paediatr Acta 43:75–86

Maudsley AA, Hilal SK, Perman WH, Simon HE (1983) Spatially resolved high-resolution spectroscopy by "four-dimensional" NMR. J Magn Reson 51:147–152

McArdle CB, Nicholas DA, Richardson CJ, Amparo EG (1986) Monitoring of the neonate undergoing MR imaging: technical considerations. Radiology 159:223–226

McArdle CB, Richardson CJ, Nicholas DA, Mirfakhraee M, Hayden CK, Amparo EG (1987) Developmental features of the neonatal brain: MR imaging. Radiology 162:223–229

Miller AL, Shamban A (1977) A comparison of methods for stopping intermediary metabolism of developing rat brain. J Neurochem 28:1327–1334
Miller BL (1991) A review of chemical issues in ^{1}H NMR spectroscopy: N-acetyl-L-aspartate, creatine and choline. NMR Biomed 4:47–52
Moorcraft J, Bolas NM, Ives NK, Ouwerkerk R, Smyth J, Rajagopalan B, Hope PL, Radda GK (1991a) Global and depth-resolved phosphorus magnetic resonance spectroscopy to predict outcome after birth asphyxia. Arch Dis Child 66:1119–1123
Moorcraft J, Bolas NM, Ives NK, Sutton P, Blackledge MJ, Rajagopalan B, Hope PL, Radda GK (1991b) Spatially localized magnetic resonance spectroscopy of the brains of normal and asphyxiated newborns. Pediatrics 87:273–282
Mora B, Narasimhan PT, Ross BD, Allman J, Barker PB (1991) P-31 saturation transfer and phosphocreatine imaging in the monkey brain. Proc Natl Acad Sci USA 88:8372–8376
Murphy EJ, Rajagopalan B, Brindle KM, Radda GK (1989) Phospholipid bilayer contribution to ^{31}P NMR spectra in vivo. Magn Reson Med 12:282–289
Nakada T, Houkin K, Hida K, Kwee IL (1991) Rebound alkalosis and persistent lactate: multinuclear (^{1}H, ^{13}C, ^{31}P) NMR spectroscopic studies in rats. Magn Reson Med 18:9–14
Nixon C, Hirsch NP, Ormerod IEC, Johnson G (1986) Nuclear magnetic resonance; its implication for the anesthetist. Anaesthesia 41:131–137
Okumura N, Otsuki S, Kameyama A (1960) Studies on free amino acids in human brain. J Biochem 47:315–320
Ordidge RJ, Connelly A, Lohman JAB (1986) Image-selected in vivo spectroscopy (ISIS): a new technique for spatially selective NMR spectroscopy. J Magn Reson 66:283–294
Petroff OAC, Prichard JW, Behar KL, Alger JR, den Hollander JA, Shulman RG (1985) Cerebral intracellular pH by ^{31}P nuclear magnetic resonance spectroscopy. Neurology (NY) 35:781–788
Pettegrew JW, Kopp SJ, Dadok J, Minshew NJ, Feliksik JM, Glonek T, Cohen MM (1986) Chemical characterization of a prominent phosphomonoester resonance from mammalian brain. ^{31}P and ^{1}H NMR analysis at 4.7 and 14.1 Tesla. J Magn Reson 67:443–450
Pettegrew JW, Kopp SJ, Minshew NJ, Glonek T, Feliksik JM (1987) ^{31}P nuclear magnetic resonance studies of phosphoglyceride metabolism in developing and degenerating brain: Preliminary observations. J Neuropathol Exp Neurol 46:419–430
Ross BD (1991) Biochemical considerations in ^{1}H spectroscopy. Glutamate and glutamine; myo-inositol and related metabolites. NMR Biomed 4:59–63
Roth JL, Nugent M, Gray JE, Julsrud PR, Berquist TH, Sill JC, Kispert DB (1985) Patient monitoring during magnetic resonance imaging. Anesthesiology 62:80–83
Roth SC, Azzopardi D, Edwards AD, Baudin J, Cady EB, Townsend J, Delpy DT, Stewart AL, Wyatt JS, Reynolds EOR (1992) Relation between cerebral oxidative metabolism following birth asphyxia, and neurodevelopmental outcome and brain growth at one year. Dev Med Child Neurol (in press)
Rudin M, Sauter A (1989) Dihydropyridine calcium antagonists reduce the consumption of high-energy phosphates in the rat brain. A study using combined $^{31}P/^{1}H$ magnetic resonance spectroscopy and ^{31}P saturation transfer. J Pharmacol Exp Ther 251:700–706
Shellock FG, Slimp G (1989) Severe burn of the finger caused by using a pulse oximeter during MR imaging. AJR 153:1105
Shoubridge EA, Briggs RW, Radda GK (1982) ^{31}P NMR saturation transfer measurements of the steady state rates of creatine kinase and ATP synthesis in the rat brain. FEBS Lett 140:288–292
Thalayasingam S, Chu A, Delpy DT (1985) An infant incubator for use during NMR spectroscopy and imaging. Proc Soc Magn Reson Med, 4th Annual Meeting, London, pp 927–928
Thulborn KR, Ackerman JJH (1983) Absolute molar concentrations by NMR in inhomogeneous B_1. A scheme for analysis of in vivo metabolites. J Magn Reson 55:357–371
Tofts PS, Cady EB, Delpy DT, Costello AM de L, Hope PL, Reynolds EOR, Wilkie DR, Gould SJ, Edwards D (1984) Surface coil NMR spectroscopy of brain. Lancet 1:459
Van der Knaap MS, van der Grond J, van Rijen PC, Faber JAJ, Valk J, Willemse K (1990) Age-dependent changes in localized proton and phosphorus MR spectroscopy of the brain. Radiology 176:509–515

Vannucci RC, Duffy TE (1977) Cerebral metabolism in newborn dogs during reversible asphyxia. Ann Neurol 1:528–534

Veech RL, Harris RL, Veloso D, Veech EH (1973) Freeze-blowing: a new technique for the study of brain in vivo. J Neurochem 20:183–188

Younkin DP, Delivoria-Papadopoulos M, Leonard JC, Subramanian VH, Eleff S, Leigh JS, Chance B (1984) Unique aspects of human newborn cerebral metabolism evaluated with phosphorus nuclear magnetic resonance spectroscopy. Ann Neurol 16:581–586

^{133}Xe Clearance

G. Greisen

Many attemps to quantify cerebral blood flow (CBF) in newborn infants have been made using different principles and techniques (Greisen 1988), but there is still no satisfactory method. However, the use of a radioactive isotope of the noble gas xenon, ^{133}Xe, has provided some useful information.

Methodology

Measurement of tissue perfusion by ^{133}Xe clearance is based on two main assumptions; (1) xenon is inert: not being metabolized in the tissue, and (2) xenon is freely diffusable: tissue concentrations equilibrate instantaneously with venous blood, in proportion to the blood–brain partition coefficient (λ). The CBF method may hence use a modification of the Fick principle which states that the rate of change of indicator concentration in tissue (dC_t) equals flow (f) multiplied by the arteriovenous concentration difference, derived as the difference between the arterial xenon concentration (C_a) and the tissue concentration multiplied by the partition coefficient (i.e., the venous xenon concentration): $dC_t = f \cdot (C_a - C_t \cdot \lambda)dt$. Tissue concentrations are measured by external detection of the gamma radiation emitted by the ^{133}Xe in the brain.

In the case of bolus injection of xenon in a carotid artery or in the aortic arch, the arterial concentrations (C_a) may be assumed to be zero after the first few seconds. This means that the differential equation solves into a simple negative exponential function $C_t = k \cdot e^{-f/\lambda \cdot t}$ showing that the rate of clearance (f/λ) reflects only the flow rate and the partition coefficient. The proportionality constant (k) depends on the amount of xenon injected, which does not affect the flow estimation. The partition coefficient in white matter in newborn infants is low due to the near absence of myelin (Greisen 1986). The rate of clearance from 15 to 60 s after the injection represents only the most highly perfused regions of the brain.

In the case of intravenous injection of xenon, the initial input is more gradual, since the xenon passes through the lungs. Furthermore, the amount of administered xenon has to be larger which results in significant recirculation. In this case, arterial xenon concentration is assumed to be proportional to the xenon concentrations in the alveolar air, which may be measured by a detector over the chest (Greisen and Pryds 1988). A final modification is due to the fact that the brain is not homogeneously perfused, therefore 15 min of

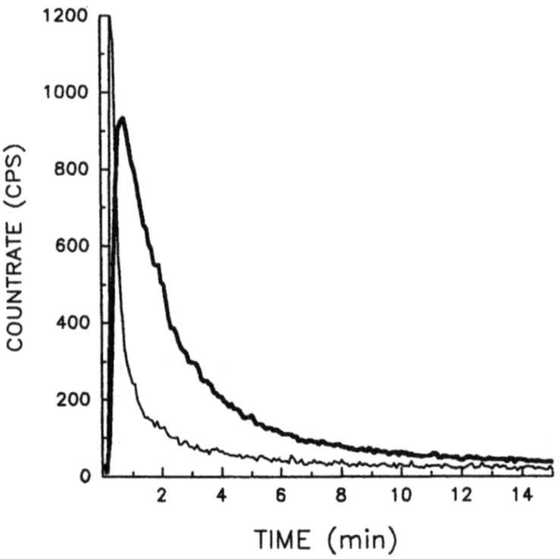

Fig. 1. Count rates over the head (*heavy line*) and chest (*thin line*) for 15 min following intravenous injection of 30 MBq/kg ^{133}Xe in a severely asphyxiated term infant, 23 h after birth. A very rapid clearance, corresponding to a very high global CBF of 60.8 ml/100 g/min is noted

cranial clearance curve (Fig. 1) is fitted by a biexponential function, and the weighted mean of the two exponential clearance rates is taken as a measure of global CBF since the detector samples from a brain volume of about 200 ml.

Review of Clinical Research

In an early paper on CBF in a group of infants with respiratory distress of whom most were moderately asphyxiated at birth, Lou et al. (1977) suggested that perinatal asphyxia may disturb "autoregulation." Hypoxia had previously been shown to abolish the normal relationship between CBF and blood pressure and EEG (Freeman and Ingvar 1968), and among the eight infants studied by the intraarterial ^{133}Xe technique within a few hours after birth, a direct relationship between arterial blood pressure and xenon clearance was seen (Fig. 2). A plateau of constant CBF over the range of normal blood pressure would have been seen if normal cerebrovascular reactivity to variations in blood pressure had prevailed. Furthermore, the low values of CBF were noted (10–50 ml/100 g/min compared to normal "cortical" flow of 70 ml/100 g/min in human adults) and it was speculated that CBF was reduced by preceding hypoxia/ischemia in parallel to the depression of clinical neurological function.

Use of the intravenous ^{133}Xe technique (avoiding the need for arterial catheterization) in larger numbers of newborn infants (Greisen 1986) indicated that CBF is normally much lower in newborn infants compared to

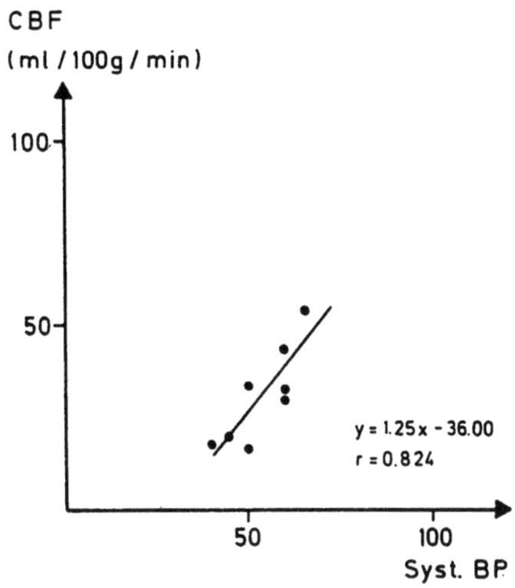

Fig. 2. Simultaneous systolic arterial blood pressure and CBF in eight infants shortly after birth. The CBF values refer to the most highly perfused areas of the brain, "cortical" blood flow. (From Lou et al. 1977)

adults, whereas relatively high CBF was found in a few, severely asphyxiated term infants (Friis-Hansen 1985, Fig. 3). A similar "luxury perfusion" (Lassen 1966) has been described following head trauma, particularly in young individuals (Obrist et al. 1984), presumably as a consequence of the complete loss of cerebrovascular tone.

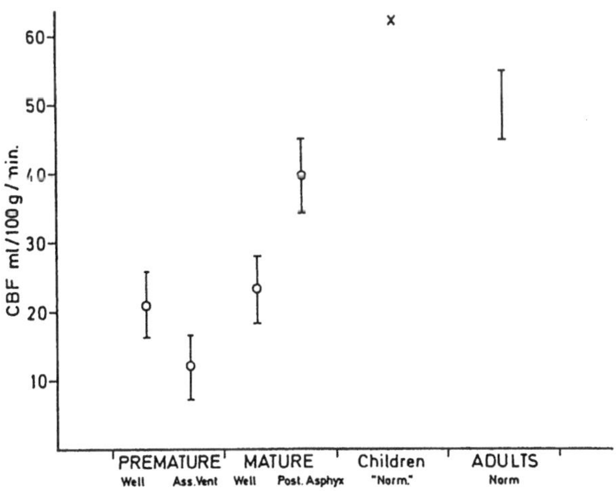

Fig. 3. Human global CBF as a function of age. In preterm infants the values are conspicuously low, rising gradually fivefold to a peak around 2 years of age, well above the value of adults. After severe perinatal asphyxia the CBF may be abnormally high, i.e. luxury perfusion. (From Friis-Hansen 1985)

In a recent larger study of term newborn infants, Pryds et al. (1990) have reconciled these findings. Nineteen infants who were mechanically ventilated and treated with phenobarbitone (15–20 mg/kg) for severe perinatal asphyxia (Apgar score ≤5 at 5 min and pH ≤7.00 in the umbilical artery at birth) were studied 3–22 h after birth with continuous amplitude-integrated EEG monitoring and repeated measurement of mean arterial blood pressure, arterial carbon dioxide tension (P_aCO_2), visual-evoked potentials, and CBF by the intravenous ^{133}Xe technique. As a reference, 12 term infants without asphyxia were studied, who were similarly mechanically ventilated and sedated with phenobarbitone (15–20 mg/kg) for respiratory distress, persistent pulmonary hypertension, or surgical illness.

Five of the asphyxiated infants had no spontaneous electrical activity; electrocerebral silence was confirmed by the absence of visual-evoked responses. Luxury perfusion (a threefold higher CBF compared to the reference group), as well as a direct relationship between CBF and arterial blood pressure and the absence of the normal relationship between CBF and P_aCO_2 were demonstrated in this group (Fig. 4). In summary, a strong indication of a complete loss of cerebrovascular tone and reactivity was found. All five infants died; computed tomography and/or neuropathology demonstrated severe global brain damage.

Of the remaining 14 infants, five developed signs of brain damage (focal periventricular hypodensities, or cerebral infarction on computed tomography, or cerebral palsy were documented at the follow-up). In these five infants, EEG showed burst suppression patterns; in four of the infants this was complicated by epileptiform activity. One of the five infants had abnormally high CBF, whereas the presence of CO_2-reactivity combined with a passive reaction to changes in blood pressure could be demonstrated in this group by statistical analysis. This partial lack of vascular reactivity presumably is a result of a milder degree of injury.

The remaining nine infants (none of whom showed subsequent signs of brain damage) had slightly low levels of CBF compared to the reference group, but had almost normal CBF-reactivity to blood pressure and P_aCO_2.

Normal vascular reactivity was demonstrated in the group of 12 infants without asphyxia.

Clinical Application

The test–retest coefficient of the intravenous ^{133}Xe technique is 10%–15%. This is sufficiently reliable to demonstrate the very high CBF values of luxury perfusion in individual infants for clinical purposes, whereas the estimate of individual CBF reactivity to changes in arterial blood pressure and P_aCO_2 is too imprecise, unless the changes in blood pressure and P_aCO_2 are very large.

The ^{133}Xe technique is technically simple: two single NaI crystal detectors are required, with the appropriate hardware for data aquisition. The software for the analysis of the clearance curves must, however, be adapted

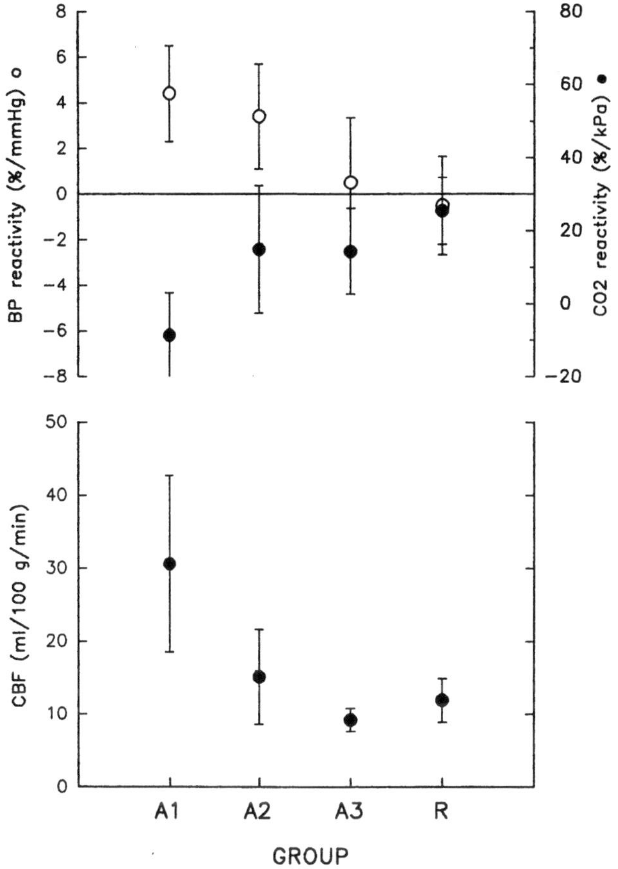

Fig. 4. Global CBF (•, *lower panel*) and its reactivity to changes in P_aCO_2 (•, *upper panel*) and mean arterial blood pressure (*BP*, ○, *upper panel*) in three groups of severely asphyxiated infants (*A1*, *A2*, and *A3*) and a reference group of term infants similarly mechanically ventilated for cardiorespiratory or surgical illness (*R*). The asphyxiated infants were grouped according to the extent of resulting brain damage: severe and global (*A1*), focal (*A2*), and none (*A3*). The *horizontal axis* for the vascular reactivities indicates the normal value of 0%/mmHg for arterial blood pressure and 30%/kPa for P_aCO_2. (Redrawn from Pryds et al. 1990)

to newborns (Greisen and Pryds 1988) and some experience with artifacts is necessary. The patient radiation is very moderate: the whole-body radiation equivalent is 0.2 mSv (20 mrem), corresponding to five chest X-ray examinations with modern techniques (Arrøe 1991).

Future

More precise information on the time course of the loss of cerebrovascular tone and reactivity following asphyxia is needed.

The combination of EEG depression, indicating neuronal hypoxic/ischemic injury and luxury perfusion, indicating a recent injury to the cerebrovascular system as such, defines a condition with a very bad prognosis as regards the brain. It is of particular interest that the combination appears at a very early stage in the clinical course, at a stage where intervention is likely to be relevant. The combination may hence be used as a point of entry into future therapeutic trials, e.g., of excitatory amino-acid receptor-blockers.

Although perinatal asphyxia results in a global insult to the brain, there are cases in which computed tomography and/or EEG demonstrate focal brain damage. ^{133}Xe (or better ^{127}Xe) may be used in combination with a fast-rotating gammacamera to yield absolute values of regional CBF. In particular, it is presently uncertain how much seizures contribute to the ultimate brain damage. By quantification of the regional flow reactivity before, during, and after seizures, it should be possible to investigate the effects of seizures per se.

References

Arrøe M (1991) The risk of X-ray examination of the lungs in neonates. Acta Paediatr Scand 80:489–493

Freeman J, Ingvar DH (1968) Elimination by hypoxia of cerebral blood flow autoregulation and EEG-relationship. Exp Brain Res 5:61–71

Friis-Hansen B (1985) Perinatal brain injury and cerebral blood flow in newborn infants. Acta Paediatr Scand 74:323–331

Greisen G (1986) Cerebral blood flow in preterm infants during the first week of life. Acta Paediatr Scand 75:43–51

Greisen G (1988) Methods for assessing cerebral blood flow. In: Levene MI, Bennett MJ, Punt J (eds) Fetal and neonatal neurology and neurosurgery. Churchill Livingstone, London, pp 151–161

Greisen G, Pryds O (1988) Intravenous ^{133}Xe clearance in preterm neonates with respiratory distress. Internal validation of CBF as a measure of global cerebral blood flow. Scand J Clin Lab Invest 48:673–678

Lassen NA (1966) The "luxury perfusion syndrome" and its possible relation to acute metabolic acidosis localised within the brain. Lancet 2:1113–1115

Lou H, Lassen N, Friis-Hansen B (1977) Low cerebral blood flow in hypotensive perinatal distress. Acta Neurol Scand 56:343–52

Obrist WD, Langfitt TW, Jaggi JL, Cruz J, Genarelli TA (1984) Cerebral blood flow and metabolism in comatose patients with acute head injury. J Neurosurg 61:241–53

Pryds O, Greisen G, Lou H, Friis-Hansen B (1990) Vasoparalysis associated with brain damage in asphyxiated term infants. J Pediatr 117:119–25

Near-Infrared Spectrophotometry

G. Greisen

Light absorption in tissue is very limited in the near-infrared range; thus brain tissue may be transilluminated effectively using standard technology. Measurements of brain oxygenation and circulation can be obtained at the cotside. Although the techniques are still under development and the application in newborn infants is still limited, the use of near-infrared spectrophotometry (NIRS) is likely to expand and hence a brief discussion of the methodology, as well as of the early results is relevant here.

Methodology

Light of three to six wavelengths between 760 and 910 nm produced by laserdiodes is delivered to the infant scalp by optical fibers. Transmitted light is picked up from another area of the scalp, at a 4- to 8-cm distance, by another optical fiber and detected by a photomultiplier tube with a sensitivity of $1-10^8-10^{10}$ emitted photons (8–10 optical densities). This allows 6- to 8-cm brain to be transilluminated. Most of the optical density is due to (multiple) scattering within the tissue, to invariant absorption of photons in the tissue, as well as to imperfect coupling between fibers and scalp.

Monitoring of the Changes. The small variant part of the absorption is assumed to be due to varying tissue concentrations of three absorbers: deoxyhemoglobin (Hb), oxyhemoglobin (HbO_2), their sum being proportional to variations in cerebral blood volume (CBV), and oxygenated cytocrome aa_3 (Cyt), the terminal member of the mitochondrial respiratory chain. Since the changes in optical densities at three to six wavelengths are monitored, it is possible to calculate the corresponding changes in these three absorbers, which may be displayed on-line (Wyatt et al. 1986). In Fig. 1 a gradual drop in arterial blood pressure in a severely asphyxiated infant is associated with a parallel drop in (Hb + HbO_2), whereas Hb remains virtually constant. This directly suggests a loss of normal cerebral vasodilation in the face of a blood-pressure fall.

The optical path length has to be known to calculate concentrations of HbO_2, Hb, and Cyt according to Lambert-Beer's law. Optical pathlength, however, is not simply the distance between the fibers – actually photons are scattered several times per millimeter of tissue. The propagation of light may be likened to a diffusion process; the tissue volume interrogated by

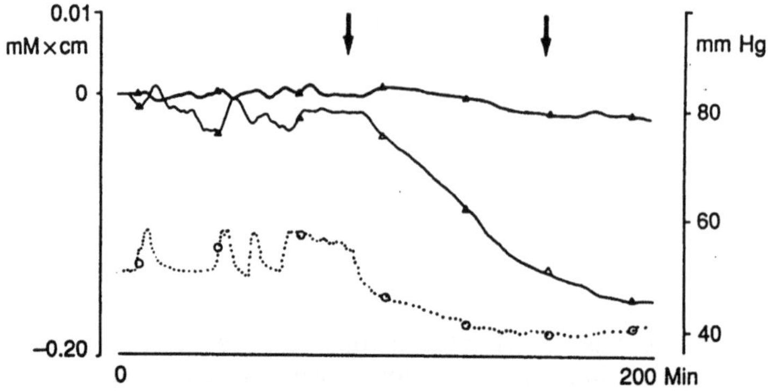

Fig. 1. Monitoring of changes in Hb (▲), Hb + HbO_2 (△), which is proportional to changes in CBV, and mean arterial blood pressure (○) in a severely asphyxiated infant. The parallel drop in blood pressure and CBV suggests a loss of normal cerebral vasoreactivity. P_aCO_2 remained constant throughout. At the two *arrows* CBF was measured by ^{133}Xe clearance obtaining 51 and 30 ml/100 g min, respectively. The NIRS traces are scaled in m*M* × cm since the optical pathlength was not measured (From Skov et al. 1990)

NIRS is roughly the shape of a banana; the effective pathlength in the human neonatal brain has been measured after death as (on average) 4.4 times the distance between the optodes (Wyatt et al. 1990a).

Sources of Errors. Several sources of error may invalidate the results. Firstly, optical density is a function of scatter and absorption as well as of their interaction, hence the relationship between optical density and the concentration of absorbers is not strictly linear. Secondly, the pathlength factor may vary among infants, e.g., as a function of brain maturation. Thirdly, very minor changes in interoptode distance result in major changes in optical density; such may occur as a result of changes in CBV in newborn infants with compliant skulls. Finally, there may possibly be other variant absorbers.

Quantification of Blood Flow, Volume, and Venous Saturation. Small induced changes in arterial oxygen saturation (S_aO_2) may be used to quantify CBV in ml/100 g brain weight (Wyatt et al. 1990b) and cerebral blood flow (CBF) in ml/100 g min (Edwards et al. 1988). The precondition is that the change in S_aO_2 does not influence CBV, CBF, or the cerebral metabolic rate of oxygen in itself. In this case the change in HbO_2 reflects the change in S_aO_2 as a function of CBF and CBV. The values for global CBF obtained in newborn infants compare well to those obtained by the ^{133}Xe clearance (see "^{133}Xe Clearance," this volume; Skov et al. 1991a), although it was not possible to use the NIRS CBF techniques in all the infants tested. The values obtained for CBV have not yet been compared to a reference method.

When infants are tilted head down, a rapid increase in optical density over the brain is seen in most, but not all, infants. If it is assumed that the increased optical density is due exclusively to the pooling of cerebrovenous blood in the brain, then the proportional increase in HbO_2 to (HbO_2 + Hb) estimates the cerebrovenous oxygen saturation, S_vO_2 (Skov et al. 1991b). The S_vO_2 is a very basic physiological variable, being firmly regulated in health and a good indicator of (global) cerebral oxygen sufficiency in disease. At present the only alternative way of measuring S_vO_2 is by direct blood sampling from the sagital sinus or the internal jugular vein.

Cytochrome. The problems of quantification of $cytochrome_{aa3}$ oxygen saturation are an order of magnitude greater than those related to hemoglobin. Firstly, in amplitude, the cytochrome signal is at most a tenth of the hemoglobin signal, this creates a dramatic demand on the precision of the NIRS method. Secondly, the in vivo $cytochrome_{aa3}$ spectrum is determined by animal experimentation with its associated indeterminacies. Thirdly, there is no reference method. Finally, since in physiology and pathophysiology the oxygenation of Hb and Cyt is closely related, it has until now been difficult to prove that changes in the Cyt signal are more than a result of a complex cross-talk between the Hb signal and artifacts due to variations in pathlength.

NIRS in Asphyxiated Infants

Abnormally high CBV in infants after perinatal asphyxia has been demonstrated by NIRS during the first 48 h of life (Wyatt et al. 1990b). This increase was associated with a reduced reactivity in CBV to changes in P_aCO_2 (Wyatt et al. 1989). Within a larger group of asphyxiated infants these abnormalities were most marked in the asphyxiated infants who subsequently died, and tended to be more marked in those who developed major impairment compared to those developing normally (McCormick et al. 1991). These CBV abnormalities very closely parallel the abnormalities of CBF and CBF regulation discussed elsewhere in this book (e.g., see "^{133}Xe Clearance," this volume). Since CBF and CBV do vary in proportion in most situations, though not all, the findings lend mutual support.

Recently, NIRS has provided evidence that the highly increased CBF following the most severe grades of perinatal asphyxia represent blood flow in excess of oxygen needs, a state of "luxury perfusion" (Skov et al. 1993). In a group of mechanically ventilated infants including nine asphyxiated infants the S_vO_2 was closely related to the CBF, i.e., very high CBF following asphyxia was associated with high S_vO_2 indicating a flow in excess of oxygen consumption.

Seizures often complicate the clinical course in infants who have suffered severe asphyxia. If S_aO_2 and arterial blood pressure remain normal, e.g., in paralyzed and mechanically ventilated infants, and if cerebrovascular reactivity is normal, then increased CBF may compensate for the increased metabolic needs of the affected brain areas, at least for some time. NIRS has provided preliminary evidence that this is not always the case (Wyatt et

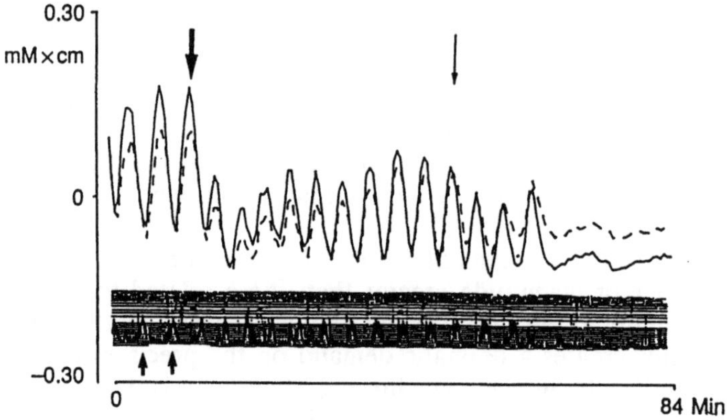

Fig. 2. Monitoring of changes in Hb (---), Hb + HbO_2 (——), which is proportional to changes in CBV, and amplitude integrated EEG, by a cerebral function monitor (Criticon, Tampa, Florida, US) in a severely asphyxiated infant. No clinical seizures were seen. Note that CBV tends to decrease with the EEG paroxysms (the first two of the 15 paroxysms are indicated by *short arrows* below the EEG trace). It should be recognized, however, that the syncrony is not complete. Diazepam (*thick arrow*) and phenytoin (*thin arrow*) were administered. The NIRS traces are scaled in mM × cm since the optical pathlength was not measured (From Skov et al. 1990)

al. 1987; Skov et al. 1990). Figure 2 demonstrates an example of decreasing (Hb + HbO_2), indicating decreasing CBV, with subclinical seizures in a severely asphyxiated infant. Better documentation of timing, of regional EEG activity, as well as of arterial blood pressure and arterial carbon dioxide tension (P_aCO_2) is needed, however, to be able to interpret these findings.

The Future

NIRS is still a research technique. Its application requires considerable insight into its theory, as well as experience of its use in ill newborn infants. It is, however, possible to include in the instrument itself, automatic data quality control and data analysis, as well as instructions for the change in fractional inspiratory oxygen (FiO_2) and P_aCO_2, to provide results of potential clinical relevance. If such instruments prove reliable, larger scale production will reduce costs to a level where they could become available in any neonatal intensive care unit.

Hence, it is reasonable to predict that in vivo NIRS will become an important tool for clinical research, and may even become of documented value for the care of patients within 10 years.

References

Edwards AD, Wyatt JS, Richardson C, Delpy DT, Cope M, Reynolds EOR (1988) Cotside measurements of cerebral blood flow in ill newborn infants by near-infrared spectrocopy. Lancet 2:770–771

McCormick DC, Edwards AD, Roth SC, Wyatt JS, Elwell CE, Cope M, Delpy DT, Reynolds EOR (1991) Relation between cerebral haemodynamics and outcome in birth-asphyxiated newborn infants studied by near-infrared spectroscopy (NIRS). European Society of Pediatric Research, Zurich, p 175 (abstract)

Skov L, Pryds O, Friis-Hansen B (1990) Nær-infrarød spectrofotometri: en ikke-invasiv, kontinuerlig metode til monitorering af den cerebrale tilstand hos nyfødte. Ugeskr Læger 152:1646–1650

Skov L, Pryds O, Greisen G (1991a) Estimating cerebral blood flow in newborn infants: Comparison of near-infrared spectroscopy and ^{133}Xe clearance. Pediatr Res 30:570–573

Skov L, Pryds O, Greisen G, Lou H (1991b) Cerebral mixed-venous saturation and cerebral blood flow. IVth International Conference on Fetal and neonatal physiological measurements. Noordwijkerhout, Erasmus University, Rotterdam, p 304

Skov L, Pryds O, Greisen G, Lou H (1993) Estimation of cerebrovenous saturation in newborn infants by near infrared spectroscopy. Pediatr Res (in press)

Wyatt JS, Delpy DT, Cope M, Wray S, Reynolds EOR (1986) Quantifications of cerebral oxygenation and haemodynamics in sick newborn infants by near-infrared spectrophotometry. Lancet 2:1063–1066

Wyatt JS, Cope M, Delpy DT, Wray S, Reynolds EOR (1987) Cerebral oxygenation and haemodynamics during neonatal seizures assessed by continuous near-infrared spectrophotometry. Early Hum Dev 15:307–307

Wyatt JS, Edwards AD, Azzopardi D, Richardson CE, Cope M, Delpy DT, Cady EB, Reynolds EOR (1989) Cerebral haemodynamics during failure of oxidative phosphorylation following birth asphyxia. Pediatr Res 26:511

Wyatt JS, Cope M, Delpy DT, van der Zee P, Arridge S, Edwards AD, Wray SC, Reynolds EOR (1990a) Measurement of optical pathlength for cerebral near-infrared spectrocopy in newborn infants. Dev Neurosci 12:140–144

Wyatt JS, Cope M, Delpy DT, Richardson CE, Edwards AD, Wray S, Reynolds EOR (1990b) Quantitation of cerebral blood volume in human infants by near-infrared spectroscopy. J Appl Physiol 68:1086–1091

Hexamethypropylene Amine Oxime Single-Photon Emission-Computed Tomography in Perinatal Asphyxia and Ischemic–Hemorrhagic Lesions

J. Haddad, A. Constantinesco, A. Facello, B. Brunot, and J. Messer

Introduction

Perinatal asphyxia is a principal recognized cause of perinatal death and abnormal neurodevelopmental outcome in both mature and preterm neonates. Ischemia and hemorrhage are the main components of central nervous system (CNS) injuries following perinatal asphyxia. The initiating stimulus which leads to lesions is a lack of cerebral blood flow (CBF) autoregulation (Volpe 1987; Lou et al. 1979a). Besides conventional brain imaging, such as ultrasound (US), magnetic resonance imaging (MRI), or computed tomography (CT) scan, the assessment of cerebral hemodynamics seems to be interesting since CBF impairment constitutes the background of ischemia and hemorrhage. In the last few years, increasing interest has been focused on techniques using radionuclides to image and measure CBF. Among them, single photon emission computed tomography (SPECT) is far less expensive and not as complicated as the others (e.g., positron emission tomography-PET scan). Moreover, it has proven to be useful in evaluating cerebral perfusion (CP) or regional cerebral blood flow (rCBF) in adults and children (Hellman et al. 1990; Uvebrant et al. 1991). In neonates, few data are available (Denays et al. 1989; Haddad et al. 1991). In this chapter, we will give an overview of the use of SPECT based on our experience. At the moment we are investigating more than 45 neonates using this technique, mainly for anoxic–ischemic brain injuries.

SPECT Guide to Clinical Brain Imaging

SPECT uses a rotating gamma camera for reconstructing cross-sectional images of radiotracer tissue distribution at equilibrium.

Radiotracer

A number of radiotracers have been developed which act as a chemical microsphere, cross the intact blood–brain barrier, and demonstrate regional blood flow. Superior imaging characteristics and availibility to ^{99m}Tc have led us to choose technetium 99m-labeled agents. ^{99m}Tc-labeled hexamethylpropylene amine oxime (HMPAO) is the radiotracer used in our institution. This lipophilic cerebral radiotracer crosses the intact blood–brain barrier by passive diffusion mechanism and is trapped in the brain, thereby allowing sufficient time for SPECT imaging. Although the trapping mechanism of HMPAO has still not been fully elucidated, current theories explain its retention in the brain as due to a conversion of the primary lipophilic molecule to secondary hydrophilic compounds that cannot diffuse back across the blood–brain barrier into the capillaries (Neirinckx et al. 1988). Some investigators have proposed that HMPAO reacts with glutathione to form a nondiffusible hydrophilic complex.

Approximately 5% of the injected dose of HMPAO localizes in the brain in adults (Neirinckx et al. 1987). Brain uptake in neonates is similar to that reported in adults based on our studies (Table 3). More than 70% of the initial brain uptake remains fixed in the brain up to 24 h postinjection allowing enough time to perform SPECT imaging (Neirinckx et al. 1987). However, the ideal moment to realize SPECT ranges between 15–45 min after the injection, although the washout for HMPAO is less than 1% per hour (Tyrell 1985). The estimated absorption radiation dose has not yet been determined in neonates. However, in adults (Tyrell 1985), dosimetry following intravenous injection of ^{99m}Tc-labeled HMPAO revealed that the organs that receive the highest radiation doses are:

– Lacrymal glands	0.258 rad/mCi
– Gallbladder wall	0.19 rad/mCi
– Kidney	0.13 rad/mCi
– Thyroid	0.10 rad/mCi
– Upper large intestinal wall	0.079 rad/mCi

The brain receives 0.026 rad/mCi and the estimated whole-body dose is 0.013 rad/mCi. In neonates the dose of radiotracer that we have injected is 3 mCi according to Denays et al. (1989). Thus, absorbed radiation dose is limited and less than that due to several X-rays received during the course of a neonate in an intensive care unit.

SPECT Imaging

The intracerebral distribution of the radiotracer is measured with a rotating gamma camera. The camera, equipped with a low-energy high-resolution collimator, rotates step-by-step or continuously through 180° or 360°, and as close as possible to the head of the patient. Some 30–60 projections are currently obtained within 20–30 min in a 64 × 64 matrix. After optimal spatial prefiltering, the images are reconstructed by filtered back-projection into axial (parallel to the orbitomeatal line), coronal, and sagittal slices with a thickness ranging from 5 to 10 mm.

In our institution, for HMPAO SPECT we use a Phillips rectangular gamma camera equipped with a low-energy high-resolution collimator connected to a Medasys A3 processing system. The injected activity is 3 mCi. The start-angle position is lateral (−270°) and the camera rotates step-by-step through 180°; 30 frames of 60″ are obtained and images of ^{99m}Tc-labeled HMPAO distribution in brain are reconstructed into axial, coronal and sagittal slices.

Patient Handling

For SPECT examination, the following conditions are required:

- An intravenous line inserted in an upper or lower limb of the neonate at least 2 h before the examination.
- Since HMPAO SPECT will depict instantaneously the distribution of the radiotracer at the moment of injection, the radiotracer must be injected in a quiet environment to avoid disturbance of CBF. During the injection the following parameters are noted: transcutaneous PCO_2 and PO_2, mean arterial blood pressure (Critikon, Dinamap), blood glucose level, hematocrit level, and pH. These factors are known to act on CBF.
- Absolute immobility of the neonate is required during data acquisition in order to avoid movement artifacts. In prematures the examination often occurs after feeding. In neonates and early infancy, sedation is often used. Chloral hydrate is given via a nasogastric tube (50 mg/kg). However, it is important to stress that sedation must be performed after the injection of the radiotracer in order to avoid distubance of CBF by any drug.
- The infant is brought to the SPECT facilities and placed in a specially designed cot. The infant must be supine, the head in the field of view of the gamma camera, and positioned with the head axis corresponding to the axis of rotation. Besides hearing, touching, and seeing the infants, surveillance is maintained by an ECG and a pulse oxymeter (Critikon).

Interpretation of HMPAO SPECT Scan

In our institution, we have used two methods to interpret HMPAO SPECT scans.

Qualitative Evaluation

It is based on the visual evaluation of CP in each plane of reconstructed images by means of a gray to white scale or a color scale. Regional CP is observed in each frontal, parietal, occipital lobe and in basal ganglia and cerebellum. In these areas, side-to-side differences in brain isotope uptake are determined by visual estimation and must be observed in a minimum of two adjacent planes. To allow a better understanding of the visual evaluation of HMPAO SPECT scan, a comparative morphological brain imaging (US or MRI) is needed. However, to achieve a comparison between CP among different children, a semiquantitative interpretation is required in the absence of quantitative assessment.

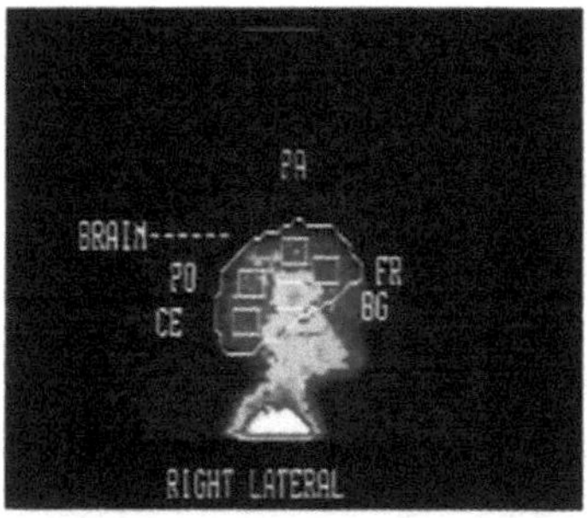

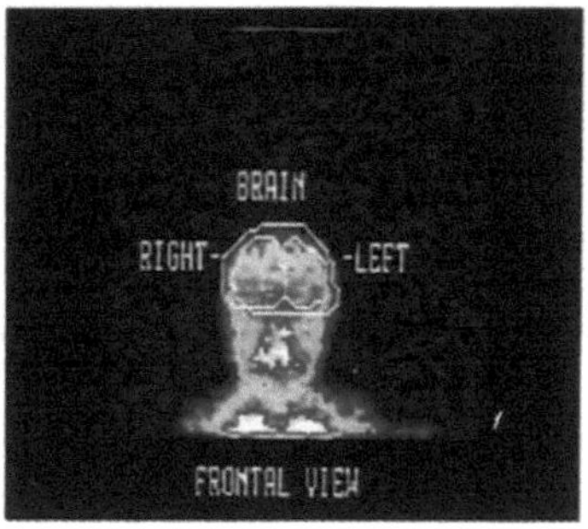

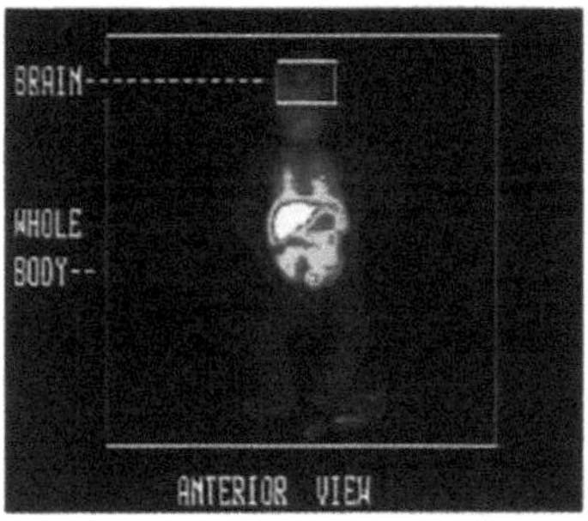

Fig. 1. Assessment of semiquantitative data of HMPAO SPECT

Semiquantitative Interpretation (Fig. 1)

For this purpose, we have choosen the first and the 16th frames from among the 30 frames obtained during SPECT acquisition.

On the first projection (right sagittal view), we have performed the following counts per pixel in selected regions of interest (ROIs): (1) counts of all the cerebrum and (2) counts in rectangular ROIs of 25 pixels arranged in the basal ganglia, cerebellum, parietal, parieto-occipital and frontal areas.

Thus two indexes are drawn: (1) the index of cerebral blood flow (I.CBF) is the ratio of the total counts of the cerebellum and the HMPAO-injected dose (the activity of the syringe is measured before and after the injection) related to body weight:

$$\text{I.CBF} = \frac{\text{total counts of cerebellum}}{\text{HMPAO-injected dose (mCi)/body weight (g)}}$$

I.CBF constitutes a measure of the cerebral perfusion of the whole brain, and (2) the index of regional cerebral blood flow (IrCBF) which is the ratio of counts in each rectangular ROI and counts of all the cerebrum (example):

$$\text{IrCBF of cerebellum} = \frac{\text{counts of selected ROIs of cerebellum}}{\text{total counts of cerebrum}}$$

IrCBF shows the distribution of CBF or CP over the five choosen ROIs and will allow us to distinguish areas of predilection of CBF within the brain.

On the 16th projection corresponding to a frontal view, we have performed the following counts per pixel: (1) counts of all the cerebrum and (2) counts in the right and left hemispheres.

Thus the index of right CP (I.RCP) or left CP (I.LCP) is obtained by dividing counts in each hemisphere by counts of all the cerebrum:

$$\text{I.RCP or I.LCP} = \frac{\text{counts of right or left hemisphere}}{\text{total counts of the cerebrum}}$$

I.RCP or I.LCP will allow as to distinguish (side-to-side) hemispheric differences. In addition to the tomographic-imaging acquisition, we perform a static imaging (coronal view) including the total body. A ROI is placed on the brain. Thus the counts in this latter ROI divided by the total counts of all the body allow us to determine the percentage of HMPAO uptake into the brain. The brain HMPAO uptake is closely correlated to the blood flow allocated to the brain from cardiac output. Thus, the percentage of HMPAO brain-uptake constitutes a marker of CBF.

Validity of HMPAO SPECT Technique

Serial studies have validated the use of HMPAO SPECT to assess CBF. A supplement of the *Journal of Cerebral Blood Flow and Metabolism* in 1988 published the reports of a symposium organized by Niels Lassen in Copenhagen on HMPAO (Lassen and Blasberg 1988). Clinical validation studies compare HMPAO SPECT data with CBF measurements obtained with the same human individuals using ^{133}Xe SPECT or ^{15}O PET techniques. This comparison involved both normal subjects and patients with various neurological illnesses including cerebrovascular disease. By and large, a good correlation was obtained between HMPAO SPECT and rCBF measured by other methods. Although those studies were performed on adults, recent studies have validated this technique in children and neonates (Uvebrant et al. 1991; Denays et al. 1989).

Application to Perinatal Asphyxia

In order to illustrate data of HMPAO SPECT in perinatal asphyxia, we have individualized two separate populations of neonates:

1. Neonates suffering from perinatal asphyxia in whom HMPAO SPECT was performed early in the neonatal period. Perinatal asphyxia criteria were as follows: (a) abnormal cardiotocography, (b) meconium-stained aspiration, (c) neonatal distress Apgar <5 at 5 min, (d) a need for immediate resuscitation and prolonged assisted ventilation, and (e) abnormal neurological signs.
2. Neonates with ischemic-hemorrhagic lesions documented by brain imaging.

Perinatal Asphyxia

Perinatal cerebral asphyxia is initiated by compromised gas exchange which leads to hypoxia with or without hypercapnia (Vanucci 1978). Hypoxia/hypercapnia increases the CBF adequately in order to maintain the brain metabolism stable (Vanucci 1990; Fenichel 1983). The initial effects leading to increased CBF are considered best as compensatory adaptable responses (Richardson 1989). However, with severe perinatal asphyxia, a loss of vascular autoregulation occurs and there is a diminished cardiac output with a subsequently diminished CBF and brain injury (Volpe 1987). In an attempt to detect CBF changes following perinatal asphyxia and to allocate them to a prognosis, we have reviewed HMPAO SPECT data of 14 neonates whose gestational age (GA) ranged from 33 to 40 weeks. They were divided into two groups relating to their neurological signs in the neonatal period and their neurological outcome at 1 year: group 1 (G1, $n = 6$, 33–40 weeks' GA) developed normally and group 2 (G2, $n = 8$, 37–40 weeks' GA), six died in the neonatal period and two are severely retarded.

Neurological Data in the Neonatal Period

In G1, lethargy associated with hypotonia, seizures of a limited duration, and altered primary neonatal reflexes were noted. However, the neurological signs improved in the neonatal period and at discharge neurological examination was normal.

In G2, in addition to the neurological signs described in G1, status epilepticus, coma, hypertonia, apnea, and bradycardia were observed. As mentioned above six neonates died. Aggressive support was withdrawn in view of the poor prognosis drawn from clinical data, EEG, and cerebral blood flow velocities (CBFV) in four of them. In the two remaining, cardiac failure occurred. No neuropathological examinations were performed on the brain. The two surviving children were neurologically abnormal at discharge, as well as at the follow-up.

EEG Data

In G1, the EEG showed moderate abnormalities regarding background activities and in some cases spikes were identified. At the follow-up, the EEG had gained normal activity. In G2, the EEG was severely depressed. The following patterns were observed: electroclinical silence, suppression burst, abnormal background, and spikes. In the two surviving children, abnormal EEG activity was identified until discharge.

Brain Imaging (Ultrasound and MRI) and Doppler

Using brain imaging none of the neonates in G1 showed ischemic or hemorrhagic lesions. In G2, cerebral parasagittal lesions and basal ganglia injuries were identified in the two surviving children at 16 and 12 days old, respectively. CBFV were assessed by means of a transcranial Doppler (2 MHz) in the middle cerebral arteries (Messer 1991). Mean, diastolic, and systolic velocities (cm/s) were noted.

HMPAO SPECT Data

Timing of Examinations

HMPAO SPECT was realized either in the acute (A, 0–12 h), the subacute (S, 24–96 h), or the chronic (C, >96 h) stages following perinatal asphyxia. Twenty-one SPECT examinations were thus performed (Table 1): In G1, four neonates were assessed in S and C and only two in S, and in G2, one neonate was assessed in A and S and died; two neonates in S and C and the remainder were assessed either in A (n = 3) or in S (n = 2) and died.

Visual Interpretation

Qualitative evaluation of HMPAO SPECT allowed us to distinguish several patterns of CBF images as noted in Table 2. In G1, three neonates showed abnormal CBF images, whereas in G2 the latter were abnormal in six neonates. The temporal progression of CBF images after brain asphyxia as assessed by visual evaluation is as follows: there is an increase in CP followed by a return to normal CP. However, in two cases from G2 (cases seven and nine) a decrease in CBF was noted. A hierarchy in rCBF images was detected in cases where hyperemia (an increase in CBF) was evidenced (cases 5, 11, 12, and 14). Indeed, excessive CP reached the parietal and frontal areas (Fig. 2). Furthermore, in case 11 a later MRI showed, areas of high CP as an ischemic pattern (Fig. 3).

Unlike US and MRI, CBF images drawn by HMPAO SPECT allow us to distinguish individual differences in neonates who suffered perinatal asphyxia. Nevertheless, quantitative or semiquantitative data are necessary to achieve a better comparison within the groups.

Table 1. HMPAO SPECT and perinatal asphyxia: distribution of HMPAO SPECT scans between acute, subacute, and chronic stages

Group	n = 21		
	Acute (0–12 h)	Subacute (12–96 h)	Chronic (>96 h)
G1	0	4	→4
		2	
G2	1	→1	→2
		2	
	3	2	

Table 2. Individual qualitative CBF images and semiquantitative I.CBF of G1 and G2

Group	Case *n*	GA	Dates of SPECT	Outcome	Qualitative CBF images	Semi-quantitative I.CBF
G1	1	40	H24 D20	N	↗ →	63–46
	2	36	H72 D28	N	→ →	22–41
	3	36	H72	N	→	40
	4	33	H96	N	→	17
	5	35	H24 D14	N	↗ [a] →	49–29
	6	40	H26 D6	N	↗ →	56–52
G2	7	40	H10	Died	↘	22
	8	38	H9	Died	↗	66
	9	37	H2 H28	Died	↘ ↗	34–37
	10	40	H6	Died	↗	46
	11	40	H72 D16	Retarded	→ [a] →	34–34
	12	40	H41	Died	↗ [a]	82
	13	40	H72 D12	Retarded	↗ →	37–51
	14	40	H48	Died	↗ [a]	83

N, normal; ↗, increased; →, normal; ↘, decreased; H, hours after birth at the moment of HMPAO SPECT scan; D, day after birth at the moment of HMPAO SPECT scan; GA, gestational age.
[a] Hierarchy of regional cerebral perfusion was observed.

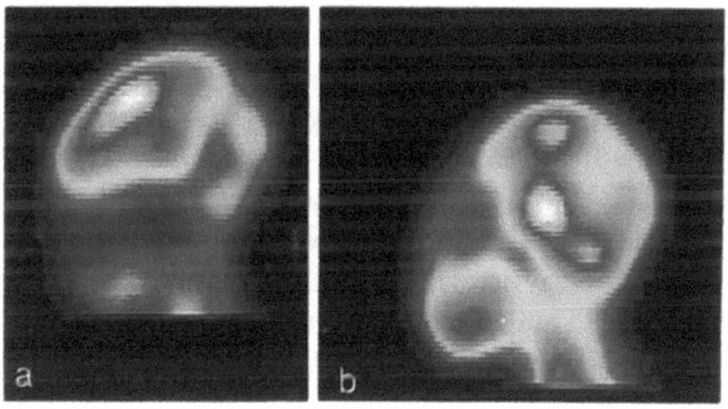

Fig. 2a,b. Case 5. 8-mm slice thickness, sagittal cross-section HMPAO SPECT at **a** 24 h after birth (note cerebral hyperperfusion in frontal and parietal areas – excessive brightness), and **b** 14 days after birth. (Note the return to a normal cerebral perfusion)

Semiquantitative Interpretation

Percentage of HMPAO Brain Uptake. Five neonates, two from G1 and three from G2 were compared to three other normal neonates (control group) in order to assess the percentage of HMPAO brain uptake. As seen in Table 3, there is an important increase in HMPAO brain uptake in S in both G1 and G2 and a return to normal values at a later stage (C). Although the percentage of HMPAO brain uptake was not performed in all cases, the increase in CBF following asphyxia (as revealed by visual evaluation) seems to be confirmed.

Table 3. Semiquantitative data: percentage of HMPAO brain uptake in acute, subacute and chronic phases as a comparison between groups 1, 2, and 3

Group[a]	Acute		Subacute		Chronic	
	%	*n*	%	*n*	%	*n*
G1	0	0	12	2	6	2
G2	11	1	11	2	5	2
G3			5–6	3		

[a] G1 and G2 infants have suffered perinatal asphyxia. G3 is a control group of normal neonates.

Index of Cerebral, Blood Flow; Index of Regional Cerebral Blood Flow; and Indexes of Right or Left Hemispheres. At the different stages A, S, and C, I.CBF, IrCBF, and I of the right or left hemisphere were statistically analyzed in G1 and G2. The methods of statistical analysis used are the Mann-Whitney U test and the Kruskall-Wallis H test. The analysis of variance (ANOVA) test, with a covariant (postconceptional age, PCA) was used after testing the normality of the distribution of parameters using a Kolmogorov-Smirnov test. PCA, is known to influence I.CBF. Indeed, in our population we have measured I.CBF in 12 neonates whose PCA at the moment of the SPECT examination ranged from 33 to 47 weeks. We have found a significant difference ($p < 0.05$, Kruskall-Wallis H test) between I.CBF of neonates whose PCA were less than 37 weeks ($n = 3$), ranged from 37 to 42 weeks ($n = 6$), and were greater than 42 weeks ($n = 3$).

Thus, at the moment of SPECT examination, PCA Tc PCO_2 (mm Hg), diastolic (DV) and mean velocities (MV; cm/sec), pH, hematocrit (%), blood glucose level (mg/l), and mean arterial blood pressure (MABP, mm Hg) were statistically analyzed. The four latter parameters were not subject to any significant differences. They are not mentioned in Tables 4, 5, 6, and 7.

In G1 (Table 4), even though PCA were different, I.CBF (adjusted medium) was higher in S than in C. It seems then that in G1 an increase in CBF occurred following asphyxia, but was not detected after 26 h.

In G2 (Table 5), a comparison was made between A and S in order to follow the progression of CBF. I.CBF was slightly more increased in S than in A and IrCBF of the cerebellum decreased significantly in S. The increase in CBF in G2 was observed 72 h after birth asphyxia and was associated with a lack of CBF/PCO_2 reactivity. Indeed, excessive CP occurred despite low PCO_2. Discrepancies between I.CBF, DV, and MV were noted in this group.

In S (Table 6), we have compared data from G1 and G2. Significant differences regarding Tc PCO_2 were noted although I.CBF (adjusted medium) were similar.

In C (Table 7), there were no statistical differences regarding I.CBF between G1, G2, and G3 (G3 being a control group of three normal neonates). Thus, I.CBF in both G1 and G2 returns to normal values.

Table 4. Semiquantitative data of G1 showing a comparison between subacute and chronic phases

	Subacute $n = 6$ mean SD	Chronic $n = 4$ mean SD	p
PCA (weeks)	36.7 ± 2.8	40.5 ± 2.1	0.033*
$TcPCO_2$ (mm Hg)	39.5 ± 2.6	40.3 ± 2.5	NS
I.CBF	41.1 ± 18.6 (48.7)	41.8 ± 9.6 (30.4)	NS
IrCBF			
BG	11.6 ± 1.5 (11.0)	10.0 ± 0.7 (10.9)	NS
CE	10.1 ± 1.8 (9.1)	8.6 ± 1.6 (10.1)	NS
PO	8.1 ± 1.0 (7.6)	6.6 ± 0.7 (7.3)	NS
PA	9.2 ± 1.3 (9.1)	7.9 ± 0.7 (8.2)	NS
FR	7.9 ± 1.2 (7.8)	6.5 ± 0.4 (6.7)	NS

p, ANOVA test; *, Mann-Whitney U Test; (), adjusted medium of ANOVA test; PCA, postconceptional age; $TcPCO_2$, transcutaneous PCO_2; BG, basal ganglia; CE, cerebellum; PO, parieto-occipital; PA, parietal; FR, frontal

Table 5. Semiquantitative data of G2: acute vs subacute stage

	Acute $n = 4$ mean SD	Subacute $n = 5$ mean SD	p
PCA (weeks)	38.04 ± 1.4	39.4 ± 1.3	NS
$TcPCO_2$ (mm Hg)	32.3 ± 7.6	18.6 ± 3.4	0.018*
DV (cm/s)	11.8 ± 9.3	28.8 ± 12.5	0.032
MV (cm/s)	19.0 ± 7.0	29.0 ± 13.0	0.03
I.CBF	42.1 ± 18.8	54.6 ± 25.7	NS
IrCBF			
BG	13.2 ± 2.2	11.3 ± 1.0	0.095
CE	9.7 ± 0.9	8.9 ± 0.9	0.05
PO	6.7 ± 2.2	7.7 ± 0.9	NS
PA	8.8 ± 2.2	9.5 ± 1.0	NS
FR	7.6 ± 1.7	7.2 ± 1.0	NS

p, Mann-Whitney U Test; *, ; PCA, postconceptional age; $TcPCO_2$, transcutaneous PCO_2; DV, diastolic CBF velocities; MV, mean CBF velocities; BG, basal ganglia; CE, cerebellum; PO, parieto-occipital; PA, parietal; FR, frontal

There were no differences in I of the right and left hemispheres in either groups or stages.

Conclusion

There was an overall agreement between the findings of qualitative and semiqualitative data of HMPAO SPECT (Table 2). It seems that an increase in CBF (hyperemia) occurs following perinatal asphyxia. However, its

Table 6. Semiquantitative data; G1 vs G2 in subacute phase

	G1 $n = 6$ mean SD	G2 $n = 5$ mean SD	p
PCA (weeks)	36.7 ± 2.8	39.4 ± 1.3	0.063
$TcPCO_2$ (mm Hg)	39.5 ± 2.6	18.6 ± 3.7	0.002*
DV (cm/s)	12.7 ± 12.5	28.8 ± 12.5	NS
MV (cm/s)	24.0 ± 13.0	38.0 ± 12.0	NS
I.CBF	41.1 ± 18.6 (48.1)	54.6 ± 25.7 (46.1)	NS
IrCBF			
BG	11.6 ± 1.5 (4.1)	11.3 ± 1.0 (12)	NS[a]
CE	10.1 ± 1.8 (9.42)	8.9 ± 0.9 (9.66)	NS[a]
PO	8.1 ± 1.0 (7.76)	7.7 ± 0.9 (0.08)	NS[a]
PA	9.2 ± 1.3 (9.13)	9.5 ± 1.0 (9.63)	NS[a]
FR	7.9 ± 1.2 (7.76)	7.2 ± 1.0 (7.43)	NS[a]

*, Mann-Whitney U Test; PCA, postconceptional age; $TcPCO_2$, transcutaneous PCO_2; DV, diastolic CBF velocities; MV, mean CBF velocities; (), adjusted medium; BG, basal ganglia; CE, cerebellum; PO, parieto-occipital, PA, parietal; FR, frontal
[a] ANOVA test.

Table 7. Semiquantitative data: G1 vs G2 vs G3 (normal neonates) in chronic phase

	G1 $n = 4$ mean SD	G2 $n = 2$ mean SD	G3 $n = 3$ mean SD	p
PCA (weeks)	40.5 ± 2.1	42.0 ± 0.0	39.3 ± 1.5	NS
$TcPCO_2$ (mm Hg)	40.3 ± 2.5	41.00 ± 0.0	43.0 ± 0.0	NS
I.CBF	41.8 ± 9.6	42.6 ± 12.3	31.6 ± 3.5	NS
IrCBF				
BG	10.0 ± 0.7	10.6 ± 0.1	9.1 ± 0.9	NS
CE	8.6 ± 1.6	9.4 ± 1.0	7.6 ± 0.5	NS
PO	6.6 ± 0.7	8.5 ± 0.1	6.0 ± 0.9	NS
PA	7.9 ± 0.7	8.3 ± 0.1	7.5 ± 1.7	NS
FR	6.5 ± 0.4	7.0 ± 0.5	6.1 ± 0.5	NS

p, Kruskall Wallis H test; PCA, postconceptional age; $TcPCO_2$, transcutaneous PCO_2; BG, basal ganglia; CE, cerebellum; PO, parieto-occipital; PA, parietal; FR, frontal

detection and duration are two main characteristics which might be taken into consideration in differentiating moderate from severe asphyxia. In moderate asphyxia, hyperemia is a transitory phenomenon, whereas in severely asphyxiated neonates, it is detected earlier, lasts longer, and is associated with a lack of CBF/PCO_2 reactivity (Table 8). In addition to hyperemia, a pattern of low CBF was found in two neonates from G2 in A,

Table 8. Characteristics of hyperemia in groups 1 and 2

	G1	G2
Detection (after birth)	24 h	3 h
Duration (after birth)	<36 h	76 h
Vasoparalysis	–	⊕
Outcome	Normal	Died (n = 6) Retarded (n = 2)

although Tc PCO_2, MABP, pH, and hematocrit were within the normal ranges.

The hierarchy of rCBF observed in the qualitative evaluation of hyperemia was not statistically confirmed by semiquantitative data. Nevertheless, trends towards a decrease in rCBF in the cerebellum and basal ganglia and an increase in CBF in the parietal and frontal areas were observed in severely asphyxiated neonates in S.

Discussion

The accuracy of HMPAO SPECT as a measurement of CBF following perinatal asphyxia in our study needs careful consideration. The small number of infants involved limits the results observed. Another limitation arises from the number of SPECT scans realized in each patient. Indeed, to assess the temporal progression of CBF following asphyxia, repeated SPECT scans are needed. However, it seems to be difficult to apply such a policy mainly due to the dose of radiotracer given, the removal of neonates, and in some instances, the severity of the asphyxial event. The selection of patients into G1 or G2 is also subject to discussion. Withdrawal of aggressive support in four cases in G2 could be advanced as a cause of improper comparison between G1 and G2. However, the aim of our study is to define CBF characteristic of a bad outcome by HMPAO SPECT. Since our study and selection were based first on neurological symptoms, a reliable predictor of outcome (Volpe 1987), the influence of withdrawal on selection criteria is null and void. Furthermore, all the neonates were managed uniformly regarding fluid restriction, use of drugs (phenobarbital and hydantoine), and ventilator support. In our population there was an overall agreement between neurodevelopmental outcome and neonatal neurological symptoms.

We have found two patterns of CBF abnormality in a group of full-term infants who had suffered perinatal asphyxia. The most common abnormality reported here is an increase in CBF. Indeed, in our study we believe that excessive HMPAO brain uptake (12%), as well as high indexes of CBF reported in A and S relate to high CBF. A major issue to resolve was to ensure whether high CBF could not account for other factors responsible for cerebral hyperperfusion. pH, blood glucose levels, MABP, and hematocrit did not change in our groups. We were also aware of the role of changes in

CO_2 tension. We have not noted any increase in Tc PCO_2 which could account for an increase in CBF. On the contrary, we have observed a lack of PCO_2/CBF vasoreactivity or vasoparalysis in G2. The final problem to clarify was to ensure that high CBF was not related to seizure activity. Even though EEG was not available during the SPECT procedure, no abnormal movements or clinical seizure activity were seen.

The abnormal high CBF occurred relatively early in G2 (3 h) and was detected as late as 72 h after the asphyxial event. In G1, although none of the neonates had been assessed early, hyperemia was detected between 24 and 26 h after asphyxia in three neonates. Thus our first conclusion relates to the detection and duration of hyperemia. It seems that in severely asphyxiated neonates, the onset of excessive CBF is as early as 3 h and could last as long as 76 h. In moderate asphyxia, unlike other authors (Pryds et al. 1990; Frewen et al. 1991), we have documented a hyperemic stage, not uniformly observed but with a limited duration.

Two possibilities should be considered for the explanation of the increase in CBF following asphyxia. First, it is well known that during asphyxia there is an increase in CBF mainly related to hypoxia and hypercapnia (Fenichel 1983; Richardson 1989). This adaptive response is considered to be a physiological event. Indeed there is an attempt to maintain adequate tissue oxygenation and preventing tissue ischemia by increasing the percentage of cardiac output to the brain (Brann 1986). We were tempted to link this phenomenom to what was observed in G1, since there was a lack of abnormal neurodevelopmental outcome. However, no evidence of such a response was documented either in experimental animal models or in neonates after the asphyxial event.

The second explanation for the increase in CBF is drawn from what we have observed in G2. The combination of hyperperfusion and vasoparalysis is thought to indicate severe cerebral damage (Pryds et al. 1990). The pathophysiology of this hyperemia remains difficult to explain. However, a stage of cerebral hyperemia was initially described by Lassen as a "luxury perfusion" (Lassen 1966). It refers to a dissociation between cerebral metabolic requirements and CBF. Recent results from Frewen et al. (1991) in asphyxiated neonates support the important role of luxury perfusion. Indeed, there was a loss of relationship between CBF and cross-brain oxygen extraction (CBO_2E) in severely asphyxiated neonates. It is thought to be related to the accumulation of lactate and CO_2 in the brain, leading to a decrease in extracerebral fluid pH and subsequent arteriolar vasodilation. Further studies are warranted to clarify this major issue.

From our results, we could stipulate that any asphyxial event in the perinatal period could be responsible for an increase in CBF. The latter phenomenon is thought to indicate severe cerebral damage when it lasts for several days and is associated with a lack of CBF/PCO_2 reactivity. The role of luxury perfusion seems to be a major feature in the pathophysiology of hyperemia. On the contrary, even though the asphyxial event is moderate an increase in CBF could occur. Its significance awaits clarification: long-adaptive response to hypoxia/hypercapnia or limited luxury perfusion.

Two infants from G2 had abnormally low CBF 2 and 10 h after the asphyxial event. A proposed cause for this low CBF is arterial vasospasm (Levene et al. 1989). Although this state has been reported in neonates by Levene et al. (1989) (low CBF velocity) no documented data are yet available. Lou et al. (1979b) have reported critical values of CBF (<20 mg/100 g min) to be associated with neurological handicaps. Recent studies by Altman et al. (1988) refute those findings since premature babies developed normally despite low CBF (5 ml/100 mg min). Despite these discrepancies, it seems to be interesting to argue for vasospasm with regard to the findings of abnormally low CBF after perinatal asphyxia since it is a treatable condition.

Our study cannot support or refute the findings of Lou et al. (1979a) that CBF is pressure passive since MABP were within the normal ranges. Furthermore, vasoactive drugs were administered early after birth and no repeated HMPAO SPECT was available with a variable MABP.

Regarding CBFV, in G2 there was a lack of agreement between DV, mean CBFV, and I.CBF in A. We do think that the small number of patients screened could be an explanation. However, one must not neglect the fact that CBFV are mainly dependent on the diameter of the insonated vessel (Raju 1991). Thus under several circumstances, CBFV could vary and could account for the discrepancies emphasized above. Even though there may be a good relationship between Doppler and HMPAO SPECT data, the latter offers the capacity to evaluate regional CBF. In our study, we were able to identify a pattern of high CP on parietal and frontal areas in three cases. In one case the hyperperfusion noted was the site of cerebral parasagittal lesions on MRI (Fig. 3). With semiquantitative data, we have observed in severely asphyxiated neonates a reduced CBF in cerebellum and to a lesser degree in basal ganglia with a trend towards an increase in CBF in frontal and parietal areas. However, one must be cautious in that those are preliminary findings and need further investigation.

Ischemic and Hemorrhagic Lesions

Ischemic and hemorrhagic lesions are often interrelated in neonates. We will review in this section the interest raised by HMPAO SPECT in assessing such lesions. It is mainly the qualitative evaluation which seems to be the most interesting in such circumstances.

In prematures, periventricular–intraventricular hemorrhages (PV-IVH) are most often encountered. They are classified into grades 1, 2, or 3 and are often complicated by hydrocephalus and parenchymal involvement (Volpe 1989). Although HMPAO SPECT may not have any further benefit, compared to US, in the recognition of PV-IVH, it has been useful in the evaluation of complications. The effect of hydrocephalus on CP is well displayed by HMPAO SPECT images (Fig. 4). Excessive ventricular dilatation leads to a disturbance of marked rCBF and hypoperfusion. HMPAO SPECT data, acquired after ventriculo-peritoneal shunt, show a reappearence of rCBF (Fig. 4). We can state that HMPAO SPECT is useful for monitor-

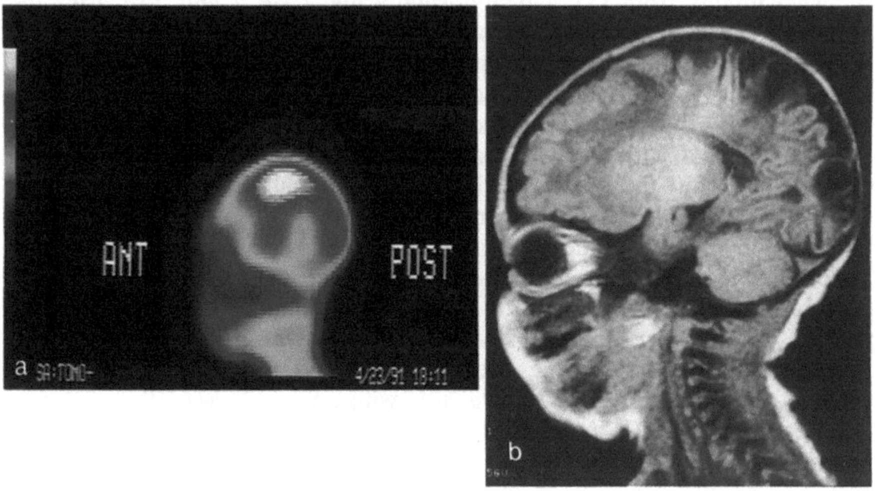

Fig. 3a,b. Case 11. **a** 8-mm slice thickness, left sagittal cross-section HMPAO SPECT. Note the high cerebral perfusion in parietal areas (excessive brightness). **b** MRI at day 16 after birth, left sagittal view, gradient echo (GE) 450/12 ms 90°. Note the ischemic cortical subcortical lesions in the parietal areas corresponding to the high cerebral perfusion noted on HMPAO SPECT (see **a**)

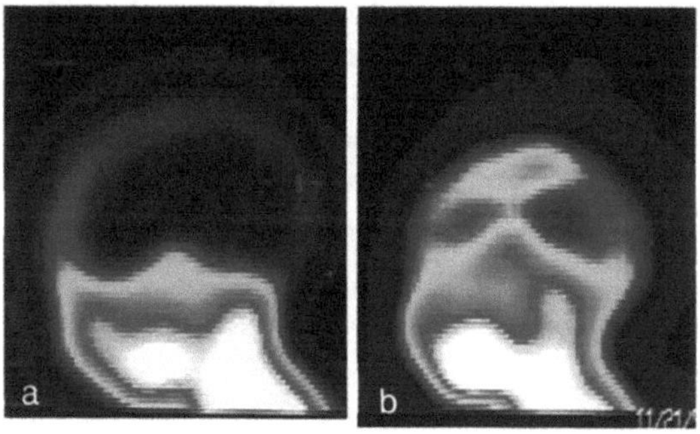

Fig. 4a,b. HMPAO SPECT 8-mm slice thickness, sagittal cross section. **a** Hydrocephalus before ventriculoperitoneal shunt. Note the cerebral hypoperfusion (dark blue). **b** Hydrocephalus after ventriculoperitoneal shunt. Note the return to a normal cerebral perfusion

ing complications brought on by hydrocephalus and this leads to rational patient management.

Parenchymal involvement associated with PV-IVH refers to hemorrhagic infarction (Haddad et al. 1992) or hemorrhagic leukomalacia. In the pathogenesis of both, ischemia is the most important event leading to an impairment of behavioral outcome (Haddad et al. 1991; Volpe 1989). Parenchymal involvement is readily identified by US and located in the periventricular areas. However, the importance of the ischemic damage seems to be under-

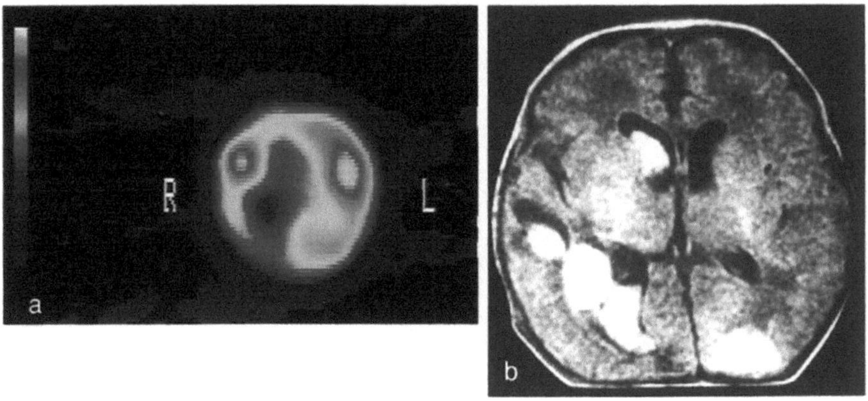

Fig. 5. **a** HMPAO SPECT scan 8-mm slice thickness, axial cross section. Note the important right parietal hypoperfusion extending to the corresponding cortex (dark blue). **b** MRI of the same infant, axial view at the same level as (**a**), GE 450/12 ms 90°. Right intraventricular hemorrhage with bilateral parenchymal participation. Note that the extent of the parenchymal participation on SPECT scan is greater than on MRI

estimated by US, CT, or MRI. HMPAO SPECT shows (Fig. 5) a marked hypoperfusion. Thus the extent of the defect is often greater on HMPAO SPECT than on other brain-imaging methods. It has also been emphasized in adults by Raynaud et al. (1989). This could suggest that there is a hypofunctional area surrounding ischemic lesions which could account for the severe neurodevelopmental outcome.

Furthermore, we have compared HMPAO SPECT data from two neonates suffering from cerebral leukomalacia. In the less severe case, there was a moderate effect on CP (Fig. 6), whereas in the other case an important reduction in CP was noted (Fig. 7). Again the importance of the ischemic damage seems to be well displayed on the HMPAO SPECT scan. Thus HMPAO SPECT constitutes an additional source of information in ischemic-hemorrhagic brain lesions by assessing the effect on CP.

Limits of the Method

Although there are some benefits in assessing brain lesions by HMPAO SPECT, one must not neglect several important limiting factors, including those due to the manner in which the SPECT scan is performed, the technique employed and the interpretation of the information.

SPECT Scan. SPECT examination, as with other imaging techniques (CT scan and MRI), is carried out outside the neonatal unit. Therefore, removal of critically ill neonates remains questionable, unless information expected from SPECT is of major interest with respect to the withdrawal of the intensive care or birth asphyxia, for example.

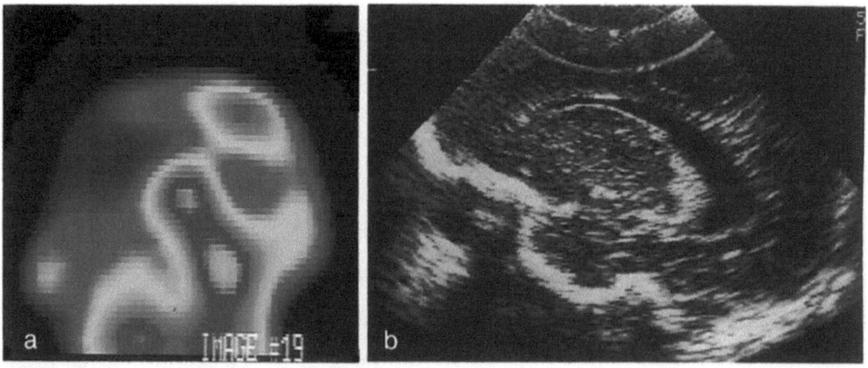

Fig. 6. **a** HMPAO SPECT scan 8-mm slice thickness, sagittal cross section. Note the moderate hypoperfusion on parietal area (dark blue). **b** Ultrasound of the same infant. Note mild periventricular leukomalacia

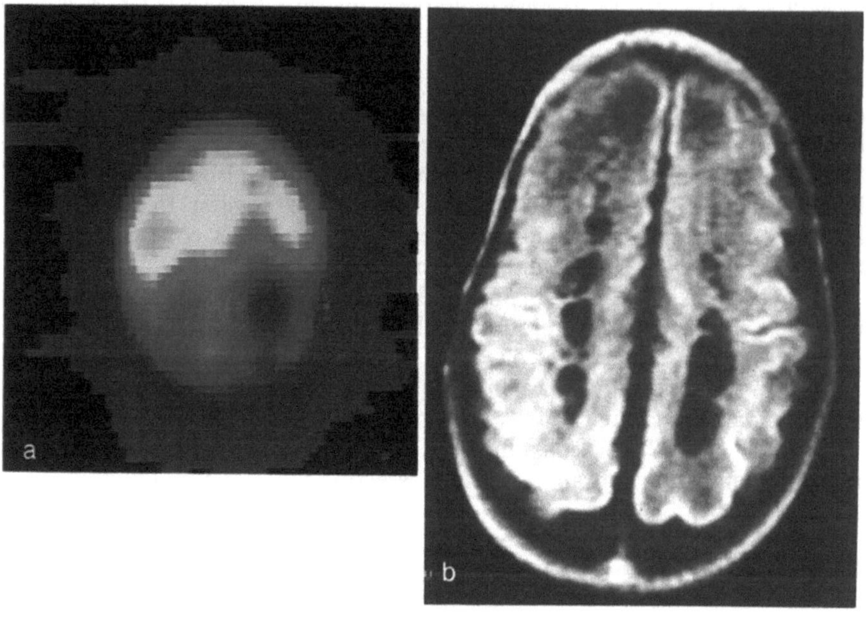

Fig. 7. **a** HMPAO SPECT scan 8-mm slice thickness, axial cross section. Note the important cerebral hypoperfusion of bilateral parietal areas (dark blue). **b** MRI of the same infant, axial view at the same level as (**a**), GE 450/12 ms 90°. Bilateral periventricular leukomalacia. Note that the extent of the defect is greater on HMPAO SPECT than on MRI

Absolute immobility is required to perform SPECT. Chloral hydrate is often used to sedate the infants. It is widely used in MRI examinations and no side effects have, as yet, been noted when the sedative is correctly used. However, it needs careful attention and surveillance of vital parameters.

To perform SPECT examination, care must be taken to ensure that the injection of the radiotracer is done in the absence of any stimuli or drug effects. It is important to obtain the parents' consent since a radionuclide tracer is used.

Technique. For image quality reasons, the administered activity is approximately three times greater, normalized to a weight other than usually given to an adult patient. However, even though the absorbed dose is higher than in an adult, the estimated absorption dose is relatively lower than CT scan or several chest X-rays (see "Radiotracer", this chapter) that a newborn could have in a neonatal intensive care unit. HMPAO SPECT scan performed with conventional (rectangular) gamma camera shows a poor spatial resolution which interferes with interpretation. However, the limitations due to the dose of radionuclide tracer given, as well as to poor spatial resolution seems to be resolved with the development of new SPECT devices with several head detectors.

HMPAO SPECT data will show, instantaneously, the distribution of the radiotracer in the brain. Therefore, it will only report about the functional activity at the moment of injection. It does not allow one to study the evaluation of rCBF over the period following tracer administration.

Interpretation. Qualitative evaluation of rCBF is not sufficient to allow an assessment of CBF. It can detect defects in CP. However, the assessment of small ischemic lesions and poorly perfused areas (e.g., frontal area) is difficult. With advanced technology (multiple-head detectors) qualitative evaluation will gain more interest. An alternative to this limitation is the semiquantitative data. Although it approaches CBF, it cannot yet quantify CBF which constitutes an important criterion in defining perinatal asphyxia. Nevertheless, there was overall agreement between qualitative and semiquantitative data in our study. The absence of healthy newborn controls, for ethical reasons, also constitutes a limitation. Normal findings could be drawn from infants who are neurologically normal at the follow-up.

Conclusions

CP imaging by HMPAO SPECT appears to be interesting in evaluating perinatal asphyxia and ischemic-hemorrhagic lesions. Two main areas of important information were drawn from our study. Regarding perinatal asphyxia, it seems that hyperemia occurs after cerebral asphyxia. However its duration is limited in moderate asphyxia and accounts for an adaptive response or limited luxury perfusion, whereas in severely asphyxiated neonates, it is associated with vasoparalysis if it lasts longer than 3 days and indicates severe brain damage.

Regarding ischemic and hemorrhagic lesions, HMPAO SPECT is an additional source of information which constitutes a functional approach to these lesions. Although the limitations of SPECT examination are numerous, the future is promising with the development of new technologies, such as tridimensional images and multiple-head detectors.

References

Altman D, Powers W, Perelman J, Herscovitch P, Volpe S, Volpe JJ (1988) Cerebral blood flow requirement for brain viability in newborn infants is lower than adults. Ann Neurol 24:218–226

Brann A (1986) Hypoxic-ischemic encephalopathy. Pediatr Clin North Am 33:451–463

Denays R, van Pach Terbke T, Tondeur M, Spehl M, Toppet V, Ham H, Piepz A, Rubinstein M, Noel P, Haumon D (1989) Brain single-photon emission-computed tomography in neonates. J Nucl Med 30:1338–1341

Fenichel G (1983) Hypoxic-ischemic encephalopathy in the newborn. Arch Neurol 40:261–266

Frewen T, Kissoon N, Kronick J, Fox M, Lee R, Bradwin N, Chance G (1991) Cerebral blood flow, cross-brain oxygen extraction and fontanelle pressure after hypoxic-ischemic injury in the newborn. J Pediatr 118:265–271

Haddad J, Constantinesco A, Brunot B (1991) SPECT of the brain perfusion in neonates. In: Haddad J, Christmann D, Messer J (eds) Imaging techniques of the neonate. Springer, Berlin Heidelberg New York, pp 171–182

Haddad J, Messer J, Aranda J (1992) Periventricular hemorrhagic infarction associated with subependymal germinal matrix hemorrhage. Eur J Pediatr 151:63–65

Hellman RS, Tikofsky RS (1990) An overview of the contribution of regional cerebral blood flow studies in cerebrovascular disease: is there a role for SPECT? Semin Nucl Med 20:303–324

Lassen N (1966) The luxury perfusion syndrome and its possible relation to acute metabolic acidosis localized within the brain. Lancet 2:1113–1115

Lassen N, Blasberg R (1988) Hexamethyl-prolylene amine oxime. J Cereb Blood Flow Metab 8:S1–S123

Levene MI, Fenton AL, Evans DH, Archer LNJ, Shortland DB, Gibson NA (1989) Severe birth asphyxia and abnormal cerebral blood flow velocities. Dev Med Child Neurol 31:427–434

Lou HC, Lassen NA, Fries Hansen B (1979a) Impaired autoregulation of cerebral blood flow in the distressed newborn. J Pediatr 94:118–121

Lou HC, Skov H, Pedersen H (1979b) Low cerebral blood flow: a risk factor in the neonate. J Pediatr 95:606–609

Messer J (1991) Cerebral Doppler in neonates. In: Haddad J, Christmann D, Messer J (eds) Imaging techniques of the neonates. Springer, Berlin Heidelberg New York, pp 107–116

Neirinckx RD, Canning LR, Piper IM (1987) Tc99m HMPAO: a new radiopharmaceutical for SPECT imaging of regional cerebral blood perfusion. J Nucl Med 28:191–202

Neirinckx RD, Burke JF, Harrison RC, Forster A, Andersen A, Lassen N (1988) The retention mechanism of 99mTc HMPAO: intracellular reaction with glutathione. J Cereb Blood Flow Metab 8:S4–S8

Pryds O, Greisen G, Lou H, Friis Hansen B (1990) Vasoparalysis associated with brain damage in asphyxiated term infants. J Pediatr 117:119–125

Raju TNK (1991) Cerebral Doppler studies in the fetus and newborn infant. J Pediatr 119: 165–174

Raynaud C, Rancurel G, Tzourio N, Souly JP, Baron JC (1989) SPECT analysis of recent cerebral infarction. Stroke 20:192–204

Richardson B (1989) Fetal-adaptive responses to asphyxia. Clin Perinatol 16:595–611

Tyrell DA (1985) Amersham report N109/28: European phase I and II. Clinical trials

Uvebrant P, Bjure J, Ekholm S (1991) Brain single-photon emission-computed tomography in neuropediatrics. Neuropediatrics 22:3–9

Vanucci RC (1978) Neurologic aspects of perinatal asphyxia. Pediatr Ann 7:15–30

Vanucci RC (1990) Experimental biology of cerebral hypoxia-ischemia. Relation to perinatal brain damage. Pediatr Res 27:317–326
Volpe JJ (1987) Neurology of the newborn. Saunders, Philadelphia, pp 209–233
Volpe JJ (1989) Intraventricular hemorrhage and brain injury in the premature infant: diagnosis, prognosis and prevention. Clin Perinatol 16:387–412

IV. Management

Perinatal Management

F. Pierre

In the 1970s it was a widespread opinion that permanent brain damage resulting from asphyxia followed adverse intrapartum events. As one of the major aims of obstetrical and neonatal care is to reduce the number of children affected by this serious pathology, cardiotography (CTG) and scalp blood sampling were widely introduced in the labor room to identify infants at high risk of suffering asphyxia, in hopes that obstetrical intervention would prevent it.

Ten years later, in the 1980s, every obstetrician knew that CTG in labor was a poor predictor of asphyxia at birth (50% false-positive results) in case of severely abnormal monitoring (Hull and Dodd 1991).

The Dublin study (Mc Donald et al. 1985) confirmed that continuous intrapartum electronic fetal heart rate (FHR) monitoring reduced by 55% the number of infants who suffered seizures or persistent abnormal neurological signs and increased the number of obstetrical interventions, but did not make any difference in rates of cerebral palsy after a 4-year follow-up (Grant et al. 1989).

In a collaborative study involving 64000 births between 1959 and 1966, Nelson and Ellenberg (1986) found 21% of children who developed cerebral palsy with intrapartum asphyxia, and 9% with evident malformations or obstetrical pathologies that could explain the issue. Even so, because intrapartum asphyxia is probably less frequent now than in the 1960s (Nelson and Ellenberg 1986), with improvement of neonatal care in labor rooms and less frequent post-term problems with the introduction of echographic data recording, it probably accounts for a smaller proportion of cases of cerebral palsy. For example, Blair and Stanley (1990) found that intrapartum hypoxia may have played a causal role in less than 2% of 171 cases of cerebral palsy.

What about the majority of unexplained cases left?

In the late 1980s, it seemed clear that most cerebral palsies result from asphyxia before birth. It could be temporary and occur a few weeks before birth, with no detection by CTG and even a good response at birth.

In small-for-gestational-age infants or known maternal pathologies with placental insufficiency, it is logical to try to evaluate the degree of hypoxia before delivery. If this "prenatal asphyxia" related to a degree of placental insufficiency is likely in small-for-gestational-age fetuses or certain maternal pathologies, it is not the same in eutrophic fetuses during the third trimester of a normal pregnancy. Nevertheless, asphyxia may accidentally happen (fetomaternal transfusion, maternal traumatism and collapsus, multiple pregancies in some cases).

In this book, the different methods of fetal and neonatal assessment were discussed. Apart from the clinical aspects, nonstress test (NST; the oldest used in clinical medicine), analysis of Doppler flow velocity waveform (the latest), and biophysical profile (including NST and ultrasound evaluation of fetal reactivity and tone) are the main means of assessment available in clinical practice.

The Role of the Various Means of Fetal Assessment for Perinatal Management

Clinical Fetal Assessment

There are only very limited means of clinical fetal assessment.

Fundal Height

Measurement of fundal height is the easiest way of estimating fetal growth. Several studies had variable conclusions on its efficacy, mainly depending on the fact that patients were seen always by the same observer or by different persons, in successive antenatal clinics. Some studies (Quaranta et al. 1981; Pearce and Campbell 1987) have shown quite good sensitivity in predicting low birth weight for gestation. Others (Persson et al. 1986) find it to be a poor predictor but with a very good negative predictive value (92%).

Pearce and Campbell (1987) concluded that low birth weight diagnosis was as good by fundal height as by ultrasound measurement of the abdominal circumference. In conclusion, fundal height measurement is not very good to evaluate birth weight, but if done by the same observer, it could help to detect growth retardation and could, then, be a good screening test before other investigations.

Intrauterine Growth

The use in clinical practice of a formal risk scoring for intrauterine growth is impossible as it has not yet been defined as it has for prematurity. Nevertheless, if one wants to use, for example, a history of low birth weight infant as a risk factor, its significance will have to be tested in the concerned population.

During pregnancy, in pathologies such as hypertensive disorders of pregnancy a very careful clinical and ultrasonographic evaluation of fetal growth is needed. This does not mean that one cannot encounter small-for-date infants in a no risk population.

Fetal Movement Antenatal Counting

Very few authors have studied the matter of clinical counting of fetal movements by women. These studies, including an interesting one from Neldam, have been well analyzed by Grant and Ellbourne (1991).

We have already described (see chapter on clinical and biophysical profile, this volume) the difficulties of these methods, which can be summarized by: much subjectivity; only the presence of a vigorous fetus is reassuring; and the predictive value, as a test, is considered as poor for a positive test (more than 50% of false diagnosis). As a decrease in fetal movements needs to be reassessed by NST, biophysical profile, or other means, this poor predictive value means that it is not as inexpensive as thought, because of the complementary assessment for 2%–4% of the counting mothers who are referred to the hospital for insufficient movements. Let us also remember that it is very time consuming for the women, that is for specific pathologies (hypertension, fetal growth retardation, post-date pregnancies), or for every pregnant woman (if one wants to detect fetal asphyxia within a random population).

The last point is that recording fetal activity is an indirect way of evaluating central nervous system integrity, and this does not mean that prediction of adverse outcome is done at a time when it could avert or moderate fetal incidence. It would be important to evaluate if it makes a difference in long-term results.

Amnioscopy

Amnioscopy is performed during labor in order to detect meconium-stained amniotic fluid. Encountered in post-date babies in particular, it is also seen in small-for-gestational age infants or in any with placental insufficiency. It seems to be correlated to an increase in perinatal morbidity and mortality, especially in the labor room, with a risk of meconium aspiration.

There is no significant association between acidosis and meconium-stained amniotic fluid (Dooley et al. 1985; Mitchell et al. 1985), and according to the most recent studies (Kariniemi and Harrela 1990), meconium in amniotic fluid seems to be associated with a placental insufficiency but not specifically to fetal gasometry.

Nonstress Test

Nonstress test (NST) is the only mean of fetal assessment that can be done in any obstetrical unit as this monitoring equipment is now available in every labor room, but that does not mean that it has been evaluated (especially for antepartum fetal heart rate (FHR) monitoring).

Schulman's critical analysis in this book shows that the NST was good for identifying dying fetuses but not in saving lives. An explanation for this observation is that the cause of FHR anomalies is more complex than a simple matter of change in acid–base status and that the no-return point, in terms of brain damage, is not clear. Is FHR monitoring based on nervous regulation the right way to detect brain damage?

Some problems are still unsolved.

Study Results

The results of various studies have shown the importance of subjectivity of interpretation (intra and interobserver reliability), especially before 30 weeks of gestation (Borgatta et al. 1988; Dawes et al. 1990). Errors arise in 20% of cases of deceleration at this gestational age and are not always easy to recognize by visual analysis.

For this reason, the development of computer analysis tries to make interpretation more objective (Dawes et al. 1990; Schulman et al., this volume).

Important Intra- and Interfetal Differences in FHR Pattern

For interfetal differences, each fetus is its own best control, which means that you need reference FHR with the same fetus. An early suspicion of the pathology will allow such monitoring before installation of hypoxemia or asphyxia. For example, small-for-date fetuses, have changes in heart rate that are rather late signs of impairment, and so NST is not a good way to detect small-for-date fetuses, but could correctly detect beginning hypoxemia if insufficient growth is known.

Intrafetal differences are seen in the FHR pattern, depending on gestational age. Sadovsky and Nicolaides (1989) showed by correlating FHR patterns with fetal blood gas by cordocentesis in a study on 119 fetuses from 20 to 39 weeks that the FHR baseline and the incidence and amplitude of decelerations decrease with gestation, while baseline variability and incidence and amplitude of accelerations increase. These data suggest that in the interpretation of FHR patterns, adjustments according to the gestational age of this fetus should be made.

Who Should be Monitored and When?

Even if Schifrin (Schifrin et al. 1979) showed on 4000 general obstetric patients that routine NST could be a good screening test to detect high risk patients, this needs more evaluation and, anyway, seems impossible to apply as a routine test, repeated in the third trimester. In addition, one would have to evaluate the number of false positives and their influence on pregnancy outcome.

Devoe (1990) has analyzed in a review of selected clinical reports that NST in high-risk pregnancy is not a good means of fetal assessment, except when there is intrauterine growth retardation, fetal hypoxia, or placental insufficiency. Even then there is a 30%–40% rate of false positives.

In the high-risk selected population in which NST could be effective, test repetition frequency remained controversial. We think it depends very much on the results of other tests; for example, we test more frequently (two to three times a week) when the results of Doppler flow testing are abnormal for the umbilical artery. Of course, such an attitude should be evaluated. It is certain, though, that repeated testing is of no value when the clinician is not prepared to act because of young fetal gestational age.

Contractions Stress Test

The basis of the contractions' stress test (CST) approach is that uterine contractions reduce uterine blood flow with, as a consequence, no compensatory response from compromised fetuses. As pointed out by Schulman (this volume), most of the studies which found this approach to be useful were not controlled, randomized, or prospective. In an obstetrical unit, CST is very difficult because it requires a bed, intravenous perfusion, and a monitor for a few hours. In addition, there are many false positives with a not negligible effect on pregancy outcome and caesarean section incidence.

NST seems to be the correct way of evaluating fetal response to contractions as in the third trimester there are frequently spontaneous contractions that seem to be sufficient to give information on modifications of the NST in real CTG and a detailed interpretation mode.

FHR Patterns and Blood Gas Values

Many authors have recently tried to correlate antepartum FHR patterns with blood gas values obtained by cordocentesis. The attempt has been made to correlate antepartum FHR patterns with fetal hypoxemia. Smith et al. (1988) tried the first during elective caesarean section and demonstrated that decelerative patterns corresponded with fetal hypoxia, but it was impossible to be sure that anesthesia or surgery was not partly responsible for such results.

Visser (Visser and Huisjes 1977; Visser et al. 1980) demonstrated that terminal FHR patterns with absence of heart rate variability and the presence of decelerations after spontaneous contractions are associated with fetal acidemia, but it was not proved. Visser (1990a) later demonstrated correlations between antepartum FHR patterns and blood gas values obtained by cordocentesis. Of interest is that his population was made up of small-for-gestational age, third-trimester fetuses very much concerned by possible obstetrical interventions due to asphyxia: 67 patients between 27 and 38 weeks with cord blood samples with analysis of pH and PO_2. They were all small for gestational age (>2 SD below the normal mean for gestational age) with reduced amniotic fluid volume in 47 cases and oligohydramnios in 20 cases. FHR patterns were analyzed by an observer without knowledge of the case (indication and fetal blood sampling results). It was interesting because the results were the same in terms of predictivity whatever the gestational age was.

One can conclude that changes in FHR patterns are correlated with *hypoxemia* or *acidemia*. When a reactive pattern was present, there was no acidemia and only a few cases of hypoxemia but the samples were mainly from arterial blood, which is less reliable.

Although the implication of a nonreactive FHR pattern is uncertain after short recording periods, it is also indicative of fetal brain damages or asphyxia (Visser 1988; Lawson et al. 1984; Visser 1990a).

No conclusion could actually be made about nonreactive FHR or isolated decelerations; also, some fetuses had abnormal blood gas results. On the other hand, with repetitive decelerations, one could predict an acidemic and hypoxemic fetus. Fetuses should probably be delivered before this point of no return in terms of neurological handicap. But the advantages of earlier delivery (with more prematurity-related incidence) versus the consequences of acidemia later on should be evaluated prospectively.

Conclusion

NST is surely the main mean for fetal assessment, as it is a simple non-invasive method with which every obstetrical unit is well equipped. For evaluation, it could rapidly lead to clinical everyday applications in fetal asphyxia assessment.

The paths of future research are surely:

- Amelioration of FHR interpretation, possibly by computed assistance of the analysis (Maedda 1990; Schulman, this volume). In absence of this help, in clinical practice, one must try to obtain an interpretation by an observer who is unaware of the case and has FHR records classified very strictly like Visser for his small-for-gestational-age, third-trimester fetuses (Visser 1990b).
- A better analysis of each detail in FHR correlated to fetal gasometry to try to detect hypoxemia.

Doppler Assessment

Doppler assessment seems currently to be more a research project than an established means of fetal assessment routinely used in clinical medicine. Many authors have thought that analysis of umbilical Doppler flow velocity waveform could help to detect fetal hypoxia, but as pointed out by Arbeille (this volume), it only detects hemodynamic abnormalities, which does not always mean fetal hypoxia.

The degree of hypoxia does not seem to correlate with the reduction of placental perfusion (Brar et al. 1988; Divon et al. 1989; Nicolaides et al. 1986), but partly depends on deterioration of placental function. Confirming this idea, Degani et al. (1991) demonstrated that placental structure changes responsible for the echogenicity of maturing placenta are not directly related to vascular impedance of the placental vascular bed, but probably reflect the placental function and, thus, the possibility of hypoxia.

In accordance with Arbeille's demonstration, cerebral Doppler index and renal flow should not presently be used to quantify and follow up the evolution of fetal hypoxia.

Screening Tests

As opposed to earlier study conclusions, Newnham et al. (1991) in a recent randomized controlled trial on 505 women with high-risk pregnancies demonstrated that introduction of Doppler waveform study did not result in reduced neonatal morbidity. It even had a small adverse effect on obstetrical management: pregnancies in which Doppler studies have been performed were less likely to have a prelabor diagnosis of fetal distress done.

The availibity of Doppler seemed to alter the use of other means of fetal assessment. In the same study, it was clear that initial use of ultrasound biometry, in addition to clinical evaluation, helped to diagnose small-for-date fetuses and thus diminished the power of Doppler waveform alone to predict this fetal pathology.

The same results were found by Anyaegbunam et al. (1990), who proposed systematic umbilical Doppler evaluation in third trimester of pregnancy to identify small-for-date fetuses. With known hypertension (high-risk population), overall 45% of small-for-gestational age births were not identified, which could be about the same as in simple clinical orientated evaluation (Persson et al. 1986) and simple ultrasound biometry.

Results are even more disappointing for umbilical Doppler as a routine screening test for unselected or medium-risk pregnancies. Newnham et al. (1990) showed in a prospective double blind study that umbilical arterial Doppler was a weak primary screening test.

Routine Clinical Application of Doppler

It has been demonstrated by several authors that the abnormalities in fetal blood flow velocity waveforms precede the appearance of pathological FHR (Arduini et al. 1989). But there is no evidence on the time interval elapsing between Doppler modifications and FHR anomalies. And as seen earlier, it is still unclear which FHR anomalies predict perinatal asphyxia.

Another idea was that oxygen could be a way to distinguish hypoxic small-for-date fetuses and its administration to be potentially beneficial with significant changes induced in fetal hemodynamics (increased cerebral vascular resistance) (Arduini et al. 1989), but Ribbert et al. (1991) demonstrated that it was not proved by controlled trials and it even could unduly prolong small-for-date pregnancies with fetuses apparently in better condition (FHR and Doppler amelioration) with no demonstrated benefit for fetuses with persistent in vitro malnutrition (Ribbert et al. 1991).

Authors like Gudmundsson and Marsal (1991) have been trying for a few years to evaluate blood velocity in umbilical artery by Doppler to detect intrauterine growth retardation. They always compare umbilical blood flow to fetal outcome considering Apgar score and umbilical artery pH after mainly operative delivery. It was concluded that abnormal results were significantly associated with the occurrence of fetal distress and thus could be a reliable guidance in monitoring small-for-date pregnancies, but there is no proof because diagnosis of fetal distress is uncertain (evaluated by CTG with great variations in analysis, depending on obstetrician), and Apgar score after operative delivery depends very much on anesthesia and operative difficulties.

Conclusion

The use of Doppler is very limited and, as seen later, has to be correlated to other tests results, when used in everyday practice. Prospective studies with very strict methodology evaluating not only Apgar score, neonatal pH, but also long-term infant morbidity are needed to know when to use Doppler and how to interpret it.

Biophysical Profile Evaluation (Jackson et al. 1991; Manning et al. 1980, 1990; Manning 1990; Mohide and Keirse 1991; Moore and Piacquadio 1989)

Biophysical profile evaluation has been described in this book and although it seems to be of interest in the future, it needs some modifications (Manning 1990) and its evaluation in different pathologies.

As thought by Manning et al. (1980), it has a better specificity and predictive positive value than other tests, but as shown by us in this book, its reproducibility could be discussed. It uses five interdependant variables; movements can be influenced by many factors in relation with different pathologies in pregnancy (maternal blood glucose level, maternal medications), ultrasonographic evaluation of fetal movements is subjective; and when pathologic, it is time consuming for trained operators.

A response to one of these remarks has been given by increased experience with the use of this test in clinical management. In Manning's experience (1990) NST is not performed anymore if the four dynamic ultrasound monitoring variables are normal (which means a new score of 8 on 8) (see Table 1 in chapter "Clinical and Biophysical Profile"; Pierre, this volume). NST is reserved only to fetuses in whom any other dynamic ultrasound monitoring variables are abnormal. With this modification, biophysical profile score only evaluates fetal movements. Another advantage is that the technique is then 30 min shorter.

With clinical experience, Manning (1990) also modified the definition of oligohydramnios, no longer using the original 1-cm fluid pocket as reference, but 2 cm, based on results of their prospective clinical studies.

Whatever the modifications are, the biophysical profile should not be considered as one means of fetal assessment in itself. On the contrary, it seems to be a cumulative status of different elements which may not have the same significance: NST, fluid volume, fetal tone, gross body movements, fetal breathing movements. Separate study of each element of the biophysical profile score is still needed considering disease-specific testing. For example, it seems that modification of one of the biophysical variables could have a different significance when considered during different pathologies (with hyperglycemia in diabetes mellitus, for example).

If biophysical profile analysis has indeed greatly increased our understanding of fetal behavior, its value in everyday clinical fetal assessment has not yet been evaluated. The summary of a recent publication of Manning (1990) is a good illustration:

> *... Moreover, the method appears to offer the advantage of grading various degrees of fetal compromise. The additional information gained by real-time ultrasound scanning (gestational age determination, fetal morphometrics, fetal anomalies screening) while not an integral part of fetal biophysical profile scoring, nevertheless remains a critical aspect of antepartum fetal assessment... ? These data are collected simultaneously with the fetal biophysical profile score. It is impossible to separate clearly the advantages of fetal biophysical profile scoring in isolation of this additional information collected. It would seem that such an attempt of separation is artificial since this data, in combination provides more key information to the physician to guide fetal management.*

Does the biophysical profile need more variables? Isn't morphological screening a separate job done itself at the beginning of second trimester with its own imperfections? (Constantine and McCormack 1991; Levi et al. 1991). More than a multivariable gross screening test, costly in time and money and impossible to evaluate in itself, wouldn't it seem more logical to find in each of the elements a specific marker of fetal hypoxemia or asphyxia?

Biochemical Assessment of Fetal Well-being

Biochemical assessment will not be discussed as it has a poor predictive power, no demonstrated efficiency, and is expensive. Research has, however, been a great help in the current understanding of the endocrinology and physiology of pregnancy (Alexander et al. 1991).

Association or Comparison of the Various Means of Fetal Assessment?

As seen previously, defenders of each method have tried to prove that one method alone was the best. Later, authors compared one method to another to find which was the best in terms of prenatal results. But which results – pH or Apgar score after labor is completed?

Test Efficiency

Test efficiency has been compared by many authors. Manning et al. (1984) compared the biophysical profile score to NST alone. Sarno et al. (1989) demonstrated that prior to labor, umbilical velocimetry had a very poor sensitivity and predictive positive value as predictor of subsequent fetal distress (meconium staining, caesarean delivery required, Apgar score <7 at

1 min, and mortality, with no reference to long-term morbidity). He concluded that the classical means of FHR tracing or the assessment of amniotic fluid volume in the early intrapartum period were better predictors of fetal ulterior condition.

Making this comparison between different tests in everyday practice, it is tempting to associate different tests in fetal assessment to minimize false-positive results and low positive predictive value of each test taken alone.

Test Association

Different tests have been associated to obtain a better predictive value. For example, Brar et al. (1989) suggested that to obtain a less significant number of false-positive results and low predictive positive value, one has to associate the use of umbilical velocimetry to complement traditional antepartum fetal surveillance schemes. Devoe (1990) examined 1000 consecutive pathologic pregnancies, performed NST, amniotic fluid pocket measurement, and umbilical artery Doppler velocimetry and concluded that the combination of these screening tools performed at the same session may improve the accuracy of prediction of fetal compromise. The association of means seems to be more or less the idea of biophysical score including different criterias: ultrasound evaluation of fetal reactivity, tone, and NST (Manning et al. 1980).

Test Sensitivity

With the same idea, some authors (Gagnon et al. 1988; Ingemarsson et al. 1988; Ohel et al. 1986) tried to make the classical means of assessment such as NST or biophysical profile score more sensitive by a vibroacoustic stimulation of the human fetus. Visser et al. (1989) demonstrated that this type of stimulation by electronic artificial larynx induced "excessive fetal movements, a prolonged tachycardia, non physiological state changes and a disorganisation and change in the distribution of fetal behavioural states and needed more evaluation on its safety and real interest before used in a routine clinical schemes."

Association of Means

As association of means is time consuming, and it is even more difficult to correctly evaluate in terms of prediction of hypoxemia with a growing number of intricated variables; here, gasometry seemed to be a good idea. It is thought to allow progress in the interpretation of easily available test, like NST.

Fetal Gasometry

Gasometry has been used for a long time at birth, but it was impossible to be sure what effect the second stage of labor had on fetal gasometry. This cord blood sample at birth was the consequence of addition of antenatal state and reaction to labor and delivery. With cordocentesis (with a low risk of modifying fetal gasometry if the sampling technique is fast), one can avoid this disadvantage and have true fetal blood gas results (Nicolaides et al. 1986; Sothill et al. 1986).

As seen previously in this text, some authors have tried to correlate blood gas values by cordocentesis to the results of different means of antenatal assessment. The reevaluation of antepartum heart rate patterns have been made by different authors: for small-for-gestational-age, third-trimester fetuses (Visser et al. 1990a) and for many fetal pathologies in the third trimester (nonimmune hydrops fetalis, ultrasound detected malformations, intrauterine growth retardation, Rh incompatibility) (Okamura et al. 1989). All the authors agree with the fact that repetitive late decelerations could predict an acidemic and hypoxemic fetus. But great uncertainty is associated with the meaning of nonreactive NST or isolated decelerations. For example, Okamura has a false-positive rate of hypoxemia prediction of 44% with nonreactive NST.

Other authors have tried to correlate fetal blood gas with blood flow velocity in different fetal arteries (umbilical, cerebral and renal) (Bonnin et al. 1992; Tyrrell et al. 1989; Wladimiroff et al. 1988). A strong relationship between the placental resistance index and the umbilical venous gasometry has been shown by some authors (Bonnin et al. 1992; Vyas et al. 1990), but Vintzileos et al. (1991) suggested that umbilical Doppler does not add to NST analysis in terms of fetal acidosis detection. Research is currently underway by different study groups on the significance of vascular distribution in terms of fetal hypoxia, hypercapnia, and acidosis. As described by Arbeille (this volume), vascular redistribution can be mainly analyzed by measuring umbilical, cerebral, and renal blood flow velocity. This point is still in the experimental stage and needs more studies for correct evaluation.

The relationship between fetal biophysical profile and fetal blood gas values has not been analyzed much and only in very short series (Ribbert et al. 1990). It is uncertain whether it could give more than the NST. Ribbert et al. (1990) and Vintzileos et al. (1987) showed that FHR reactivity and fetal breathing movements are the first, whereas fetal gross body movements and tone are the last biophysical activities to be compromised during acidemia. It means that to diagnose fetal acidemia, it is more useful to obtain information from the sequential examination of each individual item of the biophysical profile score than the total score.

The remaining problems to be solved are:

- The need for exact evaluation of venous fetal gasometry values. As discussed by Thorp (1990), different authors have to be clear on the definition of gasometry and to specify whether it is venous or arterial blood. This problem was discussed earlier in Visser's work (1990a).
- Cordocentesis and its incidence have to be evaluated in different fetal pathologies. It is probably the cause of more lethal complications in intrauterine growth retardation (unpublished series), with emphasis on

decelerations after the procedure. There is hardly any way of knowing if it is an early consequence of the funicular punction or the natural evolution of this fragile fetus.
- Up to now, only very small series of small-for-date fetuses have been available, making its significance uncertain (Ribbert et al. 1990; Bonnin et al. 1992). Larger series have many different fetal pathologies included (Okamura et al. 1989) and they do not give practical information.

In conclusion, fetal gasometry by cordocentesis is an experimental means of fetal assessment which should help to better evaluate the results of more current and practical means, such as NST, or to interpret Doppler results.

Routine Perinatal Management

Antepartum Fetal Assessment in Clinical Practice?

Screening for antepartum fetal asphyxia should be considered at different levels:

- Whether a selection of high-risk women in the first antenatal clinics is efficient is unclear.
- As seen earlier, primary screening by systematic use of tests such as NST or Doppler has no proven value in no-risk pregnancies. Clinical fundal height measurement, controlled if necessary by ultrasonic biometry, is the more efficient screening test to find intrauterine growth retardation (most frequent cause of fetal asphyxia).
- When a pathology with high risk of fetal asphyxia is discovered during pregnancy, some means of fetal assessment may be useful but how often they should be repeated is unclear.

For example, in intrauterine growth retarded fetuses, abnormal velocity waveform patterns of the umbilical artery, precedes the occurrence of antepartum decelerations by several weeks. The duration of this interval is surely different depending on fetuses and the pathology concerned. Even sometimes decelerations occur when Doppler recordings are still normal. So, normal umbilical Doppler is not absolute proof of fetal well-being but in most cases, abnormal Doppler recordings are a strict indication for intensive fetal monitoring. This attitude has still a very limited value in deciding when to deliver the baby (Visser et al. 1990b; Schulman, this volume). Anyway, in intrauterine growth retardation, NST is useful in the third trimester, but the frequency of monitoring will be based on weekly umbilical Doppler control. In the second trimester, NST is not useful, because its interpretation has not been sufficiently studied yet, and, anyway, no intervention could be planned to save the baby. Doppler seems to be the only useful assessment mean in second trimester.

We will not discuss each obstetrical pathology separately but some of the tests seem to be useful in certain conditions: the results of Pearce and McParland (1991) strongly suggest that a combination of umbilical artery

Doppler waveform (absent end diastolic frequencies in umbilical artery) with a 91% sensitivity and a 100% specificity and amniotic fluid determination (reduced amniotic column <3 cm) is a adequate method of monitoring post-date pregnancies, especially in the first stage of labor. In the same study, NST alone was not often performed, with false positives leading to unnecessary intervention. In pregnancy hypertension, blood velocity in uterine arteries by Doppler is currently being explored with great hope that it will permit discovery of those patients who will have significant placental insufficiency, i.e., which patient has high fetal complication risk. In red-cell isoimmunized patients, it is more FHR patterns that have been analyzed and will help to precisely decide when fetal blood sampling and transfusion or intervention is best performed (Sadovsky et al. 1988; Visser 1982). We will not discuss twin pregnancies as there are uncertainties on the significance of Doppler abnormalities.

In conclusion, each specific pathology may require a particular kind of fetal assessment but for most of them, NST remains the major means of assessing hypoxemia if its results are correlated to others such as Doppler, amniotic fluid volume, and fetal activity.

Important Additional Points in Fetal Assessment

Evaluation of Precise Gestational Age

Evaluation of a precise gestational age with a correct datation in the first trimester is very important by clinical means alone or ultrasonographic biometry, if unknown. This point will always be important in the face of fetal asphyxia, as it will avoid making errors in gestational age and it will help in the choice of date for early interventional deliveries.

Ultrasonographic Detection of Congenital Anomalies

Ultrasonographic detection of congenital anomalies is important. As opposed to Manning (1990), we do not think one should not rely on the biophysical profile score to find these anomalies, but on a prenatal diagnosis by mid-trimester screening ultrasound scans (Constantine and McCormack 1992; Levy et al. 1992). This systematic screening should be reinforced by a fetal morphological scan when an obstetrical pathology is discovered. Such a measure will prevent excessive, premature caesarean sections and deliveries of abnormal babies when asphyxia is suspected. Fetal chromosomal analysis by cordocentesis or other means could then be discussed, if one always remember that it does not exclude fine chromosomal structure anomalies or genopathy.

Uncertainties Concerning the Mode and Time of Delivery

Whatever means of assessment is chosen, great uncertainties are still present concerning long-term results. Whether it is clear that small-for-date fetuses in vertex presentation with no hypoxia could deliver vaginally at term; the best mode of delivery for breech "tiny babies" is still unclear. If there is no reason to anticipate delivery time in absence of hypoxia, it is clear that: it is not useful to wait for delivery after 30 weeks of gestation when there is a suspicion of fetal hypoxemia; before 30 weeks of gestation, deciding between delivery before hypoxemia at an earlier age and risk of in utero hypoxemia is not easy and poor outcome might be increased by other neonatal factors. As Visser (1990a) pointed out, this needs further prospective evaluation on long-term morbidity. Whatever the pregnancy age is, caesarean delivery is then the best, as fetuses will not tolerate blood flow modifications during the successive uterine contractions of labor.

As the best long-term neurological results for fetuses seem to be when there was no fetal hypoxemia, the difficulty is to detect when one has to deliver the baby. Doppler analysis could detect flow anomalies that precede hypoxemia diagnosis by NST patterns, but there are too many false positives. The difficulty will be to deliver babies before this point of no return. Anyway, whatever the choice, it should be decided by obstetricians and neonatologists together.

It is also important to remember that whatever the gestational age is, the obstetrician must remember to obtain the best possible fetal respiratory maturation. It means that the mother should be given corticosteroids before 30 weeks of gestation or ambroxol later on.

If premature delivery is the option, informed pediatricians should be present at birth to organize infant care in due time. For example, immediate clearance of the respiratory passage prior to the first fetal breath has been shown to reduce the incidence of meconium aspiration.

Routine Fetal Assessment in the Labor Room

Like Sarno et al. (1989), we think that FHR tracing associated with the assessment of amniotic fluid volume in the early intrapartum period are currently the reasonable predictors of subsequent fetal conditions. It does not mean that Doppler investigations are not interesting for research and will surely be an adjunctive test for the future.

It is clear that there is a relation between abnormal antepartum FHR tracings and subsequent development of intrapartum severe fetal heart rate anomalies and the increased fetal morbidity (Sarno et al. 1990).

Even if some (Hull and Dodd 1991) have recently pointed out the fact that with the development of routine intrapartum monitoring the frequency of perinatal asphyxia or cerebral palsy has not changed, these studies have always considered the infants without separating them into subgroups such as small-for-gestational-age infant, for example. Even if monitoring abnormal patterns does not always mean asphyxia, it is important to monitor these small-for-date fetuses throughout labor.

Some authors (Johnson et al. 1991) are working on fetal monitoring with pulse oximetry, but there are still some unresolved problems, such as the incidence of meconium on the results.

The detection of meconium during labor should not lead to intervention but to strict monitoring (and fetal gasometry) in case of late decelerations.

Anyway, as pointed out by Steer et al. (1989), there is no correlation between intrapartum FHR abnormalities, meconium staining of the amniotic fluid, acidosis, and low Apgar score, and this means that small-for-date fetuses or other babies issued from obstetrical pathologies should be well assessed before vaginal delivery is authorized.

Conclusion: Future Directions of Research

With the recent "discovery" that intrapartum asphyxia is an uncommon cause of cerebral palsy, it seems more probable that most cases of cerebral palsy are incurred before birth without detection. Instead of a very costly diffusion of home-monitoring without any evaluation, some research seems to be reasonable:

- Better analysis of FHR monitoring, which suffers from insufficient use of its real potential (simple mean present in each unit). A better analysis goes through prospective studies on discrete abnormalities and their classifications in terms of significance for hypoxemia and asphyxia. It could need computers as Schulman has pointed out in this book.
- Better interpretation of epidemiologic indices of fetal growth retardation with the idea that all means of current research (aortic, cerebral or renal Doppler assessment, fetal movements) will help to find a neurodevelopmental explanation or consequences for growth retardation (Berg 1989). Contrary to most of the current studies, it is not certain that placental insufficiency problems should be solved by fetal gasometry analysis.
- Prospective evaluation is needed to learn whether if in the long term it is better to deliver very premature babies, with far too many false-positive indices for this decision, or to have the oldest babies, with maybe a greater number of hypoxemic infants among them. But who is sure that the result would not have been the same for these babies if delivered earlier? This point is important to discuss when everyone is competing for the "tiniest baby."

Even if we have lots of "flashing machines", the blind tunnel is still long.

References

Alexander S, Stanwell-Smith R, Buekens P, Keirse MJN (1991) Biochemical assessment of fetal well-being. In: Chalmers I, Enkin M, Keirse M (eds) Effective care in pregnancy and childbirth, vol 1. Oxford University Press, Oxford pp 453–476

Anyaegbunam A, Brustman L, Langer O (1990) A longitudinal evaluation of the efficacy of umbilical Doppler velocimetry in the diagnosis of intrauterine growth retardation. Int J Gynecol Obstet 34:121–125

Arduini D, Rizzo G, Boccolini MR, Caforio L, Massacesi M, Mancuso S (1989) Intrauterine growth retardation and Doppler ultrasound. Fetal Ther 4 [Suppl 1]:63–67
Battaglia C, Larocca E, Lanzani A, Coukos G, Genazzani AR (1991) Doppler velocimetry in prolonged pregnancy. Obstet Gynecol 77:213–216
Berg AT (1989) Indices of fetal growth retardation, perinatal hypoxia-related factors and childhood neurological morbidity. Early Hum Dev 19:271–283
Blair E, Stanley F (1990) Intra-uterine growth and spastic cerebral palsy: association with birth weight for gestational age. Am J Obstet Gynecol 162:229–237
Bonnin P, Guyot B, Bailliart O, Benard C, Blot P, Martineaud JP (1992) Relationship between umbilical and fetal cerebral blood flow velocity waveforms and umbilical venous blood gases. Ultrasound Obstet Gynecol 2:18–22
Borgatta L, Shrout PE, Divon MY (1988) Reliability and reproducibility of non stress test readings. Am J Obstet Gynecol 159:554–558
Brar HS, Platt LD, Paul RH (1989) Fetal umbilical blood flow velocity waveforms using Doppler ultrasonography in patients with late decelerations. Obstet Gynecol 73:363–366
Constantine G, McCormack J (1991) Comparative audit of booking and mid-trimester ultrasound scans in the prenatal diagnosis of congenital anomalies. Prenat Diagn 11:905–914
Dawes GS, Moulden M, Redman CWG (1990) Limitations of antenatal fetal heart rate monitors. Am J Obstet Gynecol 162:170–173
Degani S, Paltieli Y, Lewinsky R, Abramovici D, Sharf M (1991) Fetal and uteroplacental flow velocity waveforms and placental grading. Am J Perinatol 8:47–49
Devoe LD (1990) The nonstress test. Obstet Gynecol Clin N Am 17:111–128
Divon MY, Girz BA, Lieblich R, Langer O (1989) Clinical management of the fetus with markably diminished umbilical artery and diastolic flow. Am J Obstet Gynecol 161: 1523–1527
Dooley SL, Pesavento DJ, Depp R, Socol ML, Tamura RK, Wiringa KS (1985) Meconium below the vocal cords of delivery: correlation with intrapartum events. Am J Obstet Gynecol 153:767–771
Gagnon R, Hunse C, Carmichael L, Fellow SF, Patrick J (1988) Fetal heart rate and fetal activity patterns in growth-retarded fetuses: changes after vibratory acoustic stimulation. Am J Obstet Gynecol 158:265–271
Grant A, Elbourne D (1991) Fetal movement counting to assess fetal well-being. In: Chalmers I, Enkin M, Keirse M (eds) Effective care in pregnancy and childbirth, vol 1. Oxford University Press, Oxford, pp 440–454
Grant A, O'Brian N, Joy MT, Hennessy E, McDonald D (1989) Cerebral palsy among children born during the Dublin randomized trial of intra-partum monitoring. Lancet, ii:1233–1235
Gudmundsson S, Marsal K (1991) Blood velocity waveforms in the fetal aorta and umbilical artery as predictors of fetal outcome: a comparison. Am J Perinatol 8:1–6
Hastie SJ, Brown MF, Whittle MJ (1990) Predictive values of umbilical artery waveforms and repeat cardiotocography in pregnancies complicated by nonreactive cardiotocography. Eur J Obstet Gynecol Reprod Biol 34:67–72
Hull J, Dodd K (1991) What is birth asphyxia? Br J Obstet Gynecol 98:953–955
Ingermasson I, Arulkumaran S, Paul RH, Ingermasson E, Tambyraja RL, Ratnam SS (1988) Fetal acoustic stimulation in early labor in patients screened with the admission test. Am J Obstet Gynecol 158:70–74
Jackson GM, Forouzan I, Cohen AW (1991) Fetal well-being: nonimaging assessment and the biophysical profile. Semin Roentgenol 26:21–31
Johnson N, Johnson VA, Fisher J, Jobbings B, Bannister J, Licford RJ (1991) Fetal monitoring with pulse oxymetry. Br Obstet Gynecol 98:36–41
Kariniemi V, Harrela M (1990) Significance of meconium staining of the amniotic fluid. J Perinat Med 18:345–349
Lawson GW, Dawes GS, Redman CWG (1984) Analysis of fetal heart rate on line at 32 weeks gestation. Br J Obstet Gynecol 91:542–550
Levy S, Hyjazi Y, Schaaps JP, Defoort P, Coulon R, Buekens P (1991) Sensitivity and specificity of routine antenatal anomalies by ultrasound: the Belgian Multicentric Study. Ultrasound Obstet Gynecol 1:102–110

Lowery CL, Henson BV, Wan J, Brumfield CG (1990) A comparison between umbilical artery velocimetry and standard antepartum surveillance in hospitalized high-risk patients. Am J Obstet Gynecol 162:710–714
Maeda K (1990) Computerized analysis of cardiotocograms and fetal movements. Bailliere Tindall Obstet. Gynecol 4:797–813
Manning FA, Platt LD, Sipos L (1980) Antepartum fetal evaluation: development of a fetal biophysical profile score. Am J Obstet Gynecol 136:787–795
Manning FA, Lange IR, Morrison I, Harman CR (1984) Fetal biophysical profile score and the nonstress test: a comparative trial. Obstet Gynecol 64:326–331
Manning FA, Harman CR, Morrison I, Menticoglou SM, Lange IR, Johnson JM (1990) Fetal assessment based on fetal biophysical profile scoring. IV. An analysis of perinatal morbidity and mortality. Am J Obstet Gynecol 162:703–709
Manning FA (1990) The fetal biophysical profile score: current status. Obst Gynecol Clin N Am 17:147–161
McDonald D, Grant A, Sheridan-Pereira M, Boylan P, Chalmers I (1985) The Dublin randomized controlled trial of intra-partum fetal heart rate monitoring. Am J Obstet Gynecol 152:524–539
Mills MS, James DK, Slade S (1990) Two-pier approach to biophysical assessment of the fetus. Am J Obstet Gynecol 163:12–17
Mitchell J Schulman H, Fleischer A (1985) Meconium aspiration and fetal acidosis. Obstet Gynecol 65:352–358
Mohide P, Keirse MJ (1991) Biophysical assessment of fetal well-being. In: Chalmers I, Enkin M, Keirse M (eds) Effective care in pregnancy and childbirth, vol 1. Oxford University Press, Oxford, pp 477–492
Moore TR, Piacquadio K (1989) A prospective evaluation of fetal movement screening to reduce the incidence of antepartum fetal death. Am J Obstet Gynecol 160:1075–1080
Nelson KB, Ellenberg JH (1986) Antecedents of cerebral palsy. N Engl J Med 315:81–86
Newnham JP, Patterson LL, James IR, Diepeveen DA, Reid SE (1990) An evaluation of the efficacy of Doppler flow velocity waveform analysis as a screening test in pregnancy. Am J Obstet Gynecol 162:403–410
Newnham JP, O'Dea MRA, Reid KP, Diepeveen DA (1991) Doppler flow velocity waveform analysis in high risk pregnancies: a randomized controlled trial. Br J Obstet Gynecol 98:956–963
Nicolaides KH, Soothill PW, Rodeck CH, Campbell S (1986) Ultrasound guided umbilical chord and placental blood sampling to assess fetal well-being. Lancet i:1065–1067
Ohel G, Birkenfeld A, Rabinowitz R, Sadovsky E (1986) Fetal response to vibratory acoustic stimulation in periods of low heart rate reactivity and low activity. Am J Obstet Gynecol 154:619–621
Okamura K, Endoh H, Watanbae T, Tanigawara S, Iwamoto M, Murotsuki J, Yajima A (1989) Reevaluation of nonstress test by umbilical venous blood profile using cordocentesis. Fetal Ther 4:146–151
Pearce JM, Campbell S (1987) A comparison of symphysis-fundal height and ultrasound as screening tests for light-for-gestational age infants. Br J Obstet Gynecol 94:100–104
Pearce JM, McParland PJ (1991) A comparison of Doppler flow velocity waveforms, amniotic fluid columns and the nonstress test as a means of monitoring post-dates pregnancies. Obstet Gynecol 77:204–208
Persson B, Stangenberg M, Lunell NO, Brodin U, Holmberg NG, Vaclavinkova V (1986) Prediction of size of infants at birth by measurement of symphysis fundus height. Br J Obstet Gynecol 93:206–211
Quaranta P, Currell R, Redman CWG, Robinson JS (1981) Prediction of small-for-dates infants by measurement of symphysial-fundal-height. Br J Obstet Gynecol 88:115–119
Ribbert LSM, Snijders RJM, Nicolaides KH, Visser GHA (1990) Relationship of fetal biophysical profile and blood gas values at cordocentesis in severely growth-retarded fetuses. Am J Obstet Gynecol 163:569–571
Ribbert LSM, Van Ligen RA, Visser GHA (1991) Continuous maternal hyperoxygenation in the treatment of early fetal growth-retardation. Ultrasound Obstet Gynecol 1:331–335

Sadovsky G, Nicolaides KH (1989) Reference ranges for fetal heart rate patterns in normoxaemic noanaemic fetuses. Fetal Ther 4:61–68

Sadovsky G, Visser GHA, Nicolaides KH (1988) Heart rate patterns in fetal anemia. Fetal Ther 3:216–223

Sarno AP, Ahn MO, Brar HS, Phelan JP, Platt LD (1989) Intrapartum Doppler velocimetry, amniotic fluid volume and fetal heart rate as predictors of subsequent fetal distress. I. An initial report. Am J Obstet Gynecol 161:1508–1514

Sarno AP, Phelan JP, Ahn MO (1990) Relationship of early intrapartum fetal heart rate patterns to subsequent patterns and fetal outcome. J Reprod Med 35:239–242

Schifrin BJ, Foye G, Amato J (1979) Routine fetal heart rate monitoring in the antepartum period. Obstet Gynecol 54:21–30

Smith JH, Anand KJS, Cotes PM, Dawes GS, Harkness RA, Howlett TA, Rees LH, Redman CWG (1988) Antenatal fetal heart rate variations in relation to the respiratory and metabolic status of the compromised human fetus. Br J Obstet Gynecol 95:980–989

Sothill PW, Nikolaides KH, Rodeck CH, Campbell S (1986) Effect on gestational age of fetal and intervellous blood gas and acid base value in human pregnancy. Fetal Ther 1:168–175

Steer PJ, Eigbe F, Lissauer TJ, Beard RW (1989) Interrelationship among abnormal cardiotocograms in labor, meconium staining of the amniotic fluid, arterial cord blood pH and Apgar scores. Obstet Gynecol 74:715–721

Thorp JA (1990) Umbilical artery Doppler velocimetry as a predictor of fetal hypoxia and acidosis at birth. Obstet Gynecol 75:309–310

Tyrrell S, Obaid AH, Lilford RJ (1989) Umbilical artery Doppler velocimetry as a predictor of fetal hypoxia and acidosis at birth. Obstet Gynecol 74:332–337

Vintzileos AM, Gaffney SE, Salinger LM, Campbell WA, Nochimson DJ (1987) The relationship among fetal biophysical profile, umbilical cord pH and Apgar scores. Am J Obstet Gynecol 157:627–631

Vintzileos AM, Campbell WA, Rodis JF, McLean DA, Fleming AD, Scorza WE (1991) The relationship between fetal biophysical assessment, umbilical artery velocimetry and fetal acidosis. Obstet Gynecol 77:622–626

Visser GHA (1982) Antepartum sinusoidal and decelerative heart rate patterns in Rh disease. Am J Obstet Gynecol 143:538–544

Visser GHA (1988) Antepartum fetal heart rate patterns and subsequent handicap. In: Pattel N (ed) Antenatal and perinatal causes of handicap. Baillieres Clin Obstet Gynecol 2: 117–135

Visser GHA, Huisjes HJ (1977) Diagnostic value of the unstressed antepartum cardiotocogram. Br J Obstet Gynecol 84:321–326

Visser GHA, Redman CWG, Huisjes HJ, Turnbull AC (1980) Non-stress antepartum heart rate monitoring: implications of decelerations after spontaneous contractions. Am J Obstet Gynecol 138:429–433

Visser GHA, Mulder HH, Wit HP, Mulder EJH, Prechtl HFR (1989) Vibro-acoustic stimulation of the human fetus: effect on behavioural states organization. Early Hum Dev 19: 258–296

Visser GHA, Sadovsky G, Nicolaides KH (1990a) Antepartum heart rate patterns in small-for-gestational-age third-trimester fetuses: correlations with blood gas values obtained at cordocentesis. Am J Obstet Gynecol 162:698–703

Visser GHA, Bekedam DJ, Ribbert LSM (1990b) Changes in antepartum heart rate patterns with progressive deterioration of the fetal condition. Int J Biomed Comput 25:239–246

Vyas S, Nikolaides KH, Bower S, Campbell S (1990) Middle cerebral artery flow velocity waveforms in fetal hypoxemia. Br J Obstet Gynecol 97:797–803

Wladimiroff JW, Noordam MJ, Van Den Wijngaard JAGW, Hop WCJ (1988) Fetal internal and umbilical artery blood flow velocity waveforms as a measure of fetal well-being in intrauterine growth retardation. Pediatr Res 24:609–612

Treatment of Anoxic-Ischemic Encephalopathy

J. Laugier, E. Saliba, and F. Gold

Anoxic-ischemic encephalopathy (AIE) of the neonate is still frequent and it is the second greatest cause of hospitalization in neonatal intensive care units.

Approximately three to four live-born full-term infants per thousand develop AIE. A quarter of these infants die or have sequelae such as paralysis, epilepsy, and/or mental retardation. This pathology is and has been the subject of many therapeutic trials. Based on the physiopathology of cerebral anoxia-ischemia, two types of treatment have been proposed; firstly to decrease the secondary effects of the cerebral edema and of the resulting intracranial hypertension (ICHT) and, secondly and more recently, to avoid the metabolic effect of hypoxia and its resulting cellular energy deficit. The latter is probably more promising [23].

Treatment of Edema and Intracranial Hypertension

In the 1970s the concept of neuroresuscitation was developed. Its principle is based on the fact that anoxia results in an increase in the quantity of fluid in the cerebral tissue and, therefore, an increase in cerebral volume. This increase, once it has exceeded the capacities of the compensatory phenomena, leads to ICHT which decreases the cerebral blood flow and thus causes secondary ischemic injury.

Anoxia-ischemia also results in the loss of the autoregulation of cerebral perfusion which protects the brain from sudden changes in arterial pressure. Thus, there is no compensatory phenomenon in the cerebral blood flow, explaining why anoxia-ischemia is so dangerous.

Neuroresuscitation techniques seek to decrease the cerebral volume and to reduce intracranial pressure (ICP). Keeping the head straight and slightly raised at 30°, without compression of the jugular vein, facilitates natural venous drainage and decreases ICHT. However, a recent trial in adults showed that this effect is transitory and can be dangerous in cases of hypovolemia and/or hypotension [21].

Hyperventilation is known to decrease ICP very rapidly. Its effect is transitory (a few hours), with a rebound effect necessitating progressive weaning. The aim of hyperventilation should be to maintain the $PaCO_2$ level between 30 and 35 mm Hg (with PaO_2 at 100 mm Hg). The mechanism of hypocapnia is vasoconstriction of the arterioles which lowers the cerebral blood volume and also results in ICP. However, vasoconstriction decreases the cerebral blood flow and oxygen intake, explaining why it is dangerous to obtain a lower level of $PaCO_2$ which may favor ischemia.

Sedation (association of benzodiazepines and morphine derivatives) is necessary to adapt the neonate to the ventilator, avoiding an increase in intrathoracic pressure, which impedes natural venous drainage, and facilitating hyperventilation. This sedation also has two other effects: keeping the neonate still during treatment and examination, and avoiding increased waves of ICP.

In addition to anticonvulsant properties, the following benefits of barbiturates have been described: a decrease in cerebral oxygen consumption, a decrease in cerebral metabolism, and a decrease in ICP [10, 11].

The favorable effect of barbiturates given before anoxia–ischemia has been demonstrated in adult and newborn animals [6]. On the other hand, the results of one controlled study [11] with full-term infants, all of them having suffered anoxia–ischemia, showed that there was no difference between the group subsequently treated with barbiturates and the untreated control group in terms of mortality, frequency of seizures, and final outcome. These results were the same for adults [4]. In another controlled study with very low birth-weight infants [22], the same results were observed in treated and untreated groups; the incidence of intraventricular hemorrhage was not different. In addition, hypotension was more frequent in the treated groups. We will see later the necessity of maintaining cerebral perfusion pressure (CPP) above 30 mm Hg.

Corticoids act against cerebral tumor edema, however, they are not effective for asphyxia edema in the newborn rat or human. They are not necessary for the treatment of anoxia–ischemia and, moreover, they are responsible for an increase in hyperglycemia, hemorrhage, and mortality [1, 7].

Hyperosmolar agents (mannitol 0.25/kg for 20 min) lower ICP by decreasing cerebral cell water content for 3–4 h. This treatment can be repeated every 4–6 h up to a maximum of 1 g/kg/day. Plasmatic osmolality must be checked regularly to maintain 320 mosmol/kg. In asphyxiated rats hyperosmolar agents prevent the increase in cerebral edema, but do not improve the neurologic prognosis [18]. Levene [16] showed in seven human neonates that mannitol decreases ICP and improves CPP. However, without a control group it is not possible to know the effect of this treatment on outcome. Hyperosmolar agents avoid the herniation of temporal lobes, thus lowering mortality but not reducing the cerebral injury itself.

The reduction in water intake (two thirds of requirements) seems to be ineffective. On the other hand, every sizeable rapid change in blood osmolality is very dangerous, requiring frequent surveillance of blood and urine ions [3].

Early seizures (during the first 24 h after hypoxia) are correlated with a greater incidence of neurologic sequelae, but the prognosis is linked more to the cerebral injury itself than to the consequences of the seizures (except status epilepticus). No data show that short seizures lead to injury, unless they are associated with hemodynamic disturbance. Nevertheless, it is necessary to treat hypoglycemia and hypocalcemia, and to prescribe anticonvulsants (diazepam, dilantin, and phenobarbital).

At this point we must answer the three following questions. What is the frequency of ICHT with anoxic–ischemic encephalopathy of the neonate, is it necessary to measure ICP continuously, and what neuroresuscitation treatment should we prescribe?

It is not possible to answer the first question precisely because the few studies with measurements of ICP [14] have concerned only severely asphyxiated neonates, often with status epilepticus. In all these cases, ICHT was above 10 mm Hg. When the autopsy was performed, it showed that considerable cerebral edema was responsible for ICHT.

On the other hand, with less severe asphyxia and open sutures of the skull, compensation for a certain increase in edema is possible without ICHT. The constant measurement of ICP is probably necessary; unfortunately, the most accurate devices (intraventricular or subarachnoid catheters) are invasive, leading to the risk of infection and subdural effusion. To date noninvasive techniques are not sufficiently accurate [14].

A recent survey of American neonatal intensive care units [8] showed a diversity of treatment without demonstrated results. For severe cases of asphyxia, hyperventilation and hyperosmolar agents may be prescribed, while recognizing their limitations. It is also necessary to treat seizures and to prevent the infant becoming agitated and moving. Perhaps the most important point is to maintain an adequate CPP (mean arterial pressure, MAP − ICP) when ICP increases. It is also very important to measure MAP continuously and to maintain it at above 40 mm Hg. It is necessary to avoid hypotension and hypovolemia and to diagnose and treat the often associated myocardial ischemia (continuous infusion of dopamine 10–15 μg/kg/min). The effectiveness of this treatment is not proven, but it has been demonstrated that when CPP is below 20 mm Hg all survivors have sequelae.

In neonates, contrary to the effect in adults, glucose infusion increases cellular resistance to hypoxia and decreases cerebral injury. The lower uptake and metabolism of glucose by cerebral cells in neonates lead to a lower increase in lactates and intracellular $H+$ ions, and thus to less extensive injuries [5].

Neuroprotector Treatment

These new treatments are based upon recently revealed pathologic facts in cerebral anoxic–ischemia.

These data are explained in the chapter "Biochemical Mechanisms of Hypoxic–Ischemic Brain Damage" (Saliba and Laugier, this volume), but we repeat here the essential facts:

- Anaerobic catabolism of glucose, with the disappearance of energy reserve and failure of ionic pumps with cellular edema

- Postanoxic increase in the synthesis of oxygen free radicals with the deterioration of cell-membrane unsaturated fatty acids
- Increase in the level of excitatory amino acids such as glutamate, allowing the intracellular flow of calcium responsible for a pattern of enzymatic reactions with delayed cellular death

With this in mind, the most hopeful treatment is to break this pattern of delayed cellular death and particularly to stop the intracellular calcium flow.

Oxygen Free Radical Inhibitors [15]

Two types of action in newborn animals are being studied: (1) the destruction of oxygen free radicals by enzymes such as superoxide dismutase, endoperoxidase, and catalase (it is not yet known if they cross the cerebromeningeal barrier) and (2) the inhibition of the production of prostaglandins and xanthine which are responsible for an increase in levels of oxygen free radicals. Two kinds of drugs have been tried: an inhibitor of cyclooxygenase and phospholipase – indomethacin – [9] and an inhibitor of xanthine oxidase – allopurinol – [19].

The controlled studies show that infusion of these drugs before hypoxia-ischemia (especially allopurinol) reduces cerebral injuries.

Excitatory Amino Acid Antagonists [2, 12, 17]

The most promising method of treatment seems to be the action upon excitatory amino acids either by preventing their release, or by blocking the postsynaptic receptors. These blocking molecules (phencyclidine, dextromethorphan, ketamine, and MK 801) decrease the extent of cerebral injuries in adult and newborn rats, even when the drug (especially MK 801) is infused 1 h after hypoxia-ischemia. However, only the cerebral areas rich in *N*-methyl-D-aspartate (NMDA) receptors are protected and the results vary with the animal species studied. If MK 801 seems to be effective in the newborn rat, some side-effects connected with a change in the density and distribution of receptors have been described; these side-effects must be evaluated before human clinical trials.

Calcium Channel Blockers [13, 25]

Many calcium channel blockers have been tested and the more interesting drugs seem to be nimodipine and flunarizine. Prescribed preventively, they reduce neuropathologic injuries in animals. If given 5 min after anoxia, cerebral activity and evoked potentials are more improved in treated lambs than in the control group.

Even if these drugs improve the prognosis of vasospasm in adult meningeal hemorrhage, they have the side effect of hypotension (responsible for cerebral hypoperfusion) which prevents their prescription in neonates.

If all these treatments are theoretically interesting, they must be studied before a clinical trial with neonates. The best treatment of encephalopathy is the prevention of anoxia, i.e., very careful surveillance of each delivery and rapid and effective resuscitation of the neonate.

References

1. Altman DI, Young RSK, Yagel SK (1984) Effects of dexamethasone in hypoxic-ischemic brain injury in the neonatal rat. Biol Neonate 46:149–156
2. Ben Ari Y (1991) Les canaux ioniques dans l'anoxie cérébrale: cibles possibles d'agents protecteurs. Médecine/Sciences 7:43–51
3. Bingham WF (1986) The limits of cerebral dehydration in the treatment of head injury. Surg Neurol 25:340–345
4. Brain Resuscitation Clinical Trial I Study Group (1986) Randomized clinical study of thiopental loading in comatose survivors of cardiac arrest. N Engl J Med 314:397–402
5. Callahan DJ, Engle MJ, Volpe JJ (1990) Hypoxic injury to developing glial cells: protective effect of high glucose. Pediatr Res 27:186–190
6. Campbell AGM, Milligan JB, Talner NS (1968) The effect of pretreatment with phenobarbital, meperidine or hyperbaric oxygen on the response to anoxia and resuscitation in newborn rabbits. J Pediatr 72:518–527
7. Dearden NM, Gibson JS, McDowall DG, Gibson RM, Cameron MM (1986) Effect of high-dose dexamethasone on outcome from severe head injury. J Neurosurg 64:81–88
8. Donn SM, Goldstein GW, Schork A (1986) Asphyxia neonatorum: a national survey of management practices (Abstr). Pediatr Res 20:461
9. Edwards AD, Wyatt JS, Richardson C, Potter A, Cope M, Delpy DT, Reynolds EOR (1990) Effects of indomethacin on cerebral haemodynamics in very preterm infants. Lancet 2:1491–1495
10. Eyre JA, Wilkinson AR (1986) Thiopentone-induced coma after severe birth asphyxia. Arch Dis Child 61:1084–1089
11. Goldberg RN, Moscoso P, Bauer CR, Bloom FL, Curless RG, Burke B, Bancalari E (1986) Use of barbiturates therapy in severe perinatal asphyxia: a randomized controlled trial. J Pediatr 109:851–856
12. Greisen P, Cilio MR, Schlogel X, Evrard P (1991) Les mécanismes de la souffrance cérébrale du foetus et du prématuré. Prog Neonatol 11:203–226
13. Kucharczyk J, Mintorovitch J, Moseley ME, Asgari MS, Sevick RJ, Derugin N, Norman D (1991) Ischemic brain damage: reduction by sodium-calcium ion channels modulator RS 87476. Radiology 179:221–227
14. Laugier J, Bloc D (1989) Pression intracrânienne normale et pathologique. In: Relier JP, Laugier J, Salles BL (eds) Médecine périnatale. Flammarion, Paris, pp 246–249
15. Leffler C-W, Busjia DW (1987) Arachidonic acid metabolites and perinatal cerebral hemodynamics. Semin Perinatol 11:31–42
16. Levene MI, Evans DH (1985) Medical management of raised intracranial pressure after severe birth asphyxia. Arch Dis Child 60:12–16
17. McDonald JN, Silverstein FS, Johnson MW (1989) Neuroprotective effects of MK 801. TCP, PCP, and CPP against *N*-methyl-D-aspartate-induced neurotoxicity in an in vivo perinatal rat model. Brain Res 490:33–40
18. Mujsce DJ, Stern DR, Vannucci RC et al. (1988) Mannitol therapy in perinatal hypoxic-ischemic brain damage. Ann Neurol 24:338A
19. Palmer C, Vannucci RC, Towfighi J (1990) Reduction of perinatal hypoxic-ischemic brain damage with allopurinol. Pediatr Res 27:332–336

20. Pryds O, Greisen G, Lou H, Friis-Hansen B (1990) Vasoparalysis associated with brain damage in asphyxiated term infants. J Pediatr 117:119–125
21. Rosner MJ, Coley IB (1986) Cerebral perfusion pressure, intracranial pressure and head elevation. J Neurosurg 65:636–641
22. Ruth V, Virkola K, Paetan R, Raivio KO (1988) Early high-dose phenobarbital treatment for prevention of hypoxic-ischemic brain damage in very low birth-weight infants. J Pediatr 112:81–86
23. Vannucci RC (1990) Current and potentially new management strategies for perinatal hypoxic-ischemic encephalopathy. Pediatrics 85:961–968
24. Volpe JJ (1990) Brain injury in the premature infant: it is preventable. Pediatr Res 27:528–533
25. Werner C, Hoffman WE, Kochs E, Rabito SF, Miletich DJ (1991) Captopril improves neurologic outcome from incomplete cerebral ischemia in rats. Stroke 22:910–914

V. Prognosis

Neurodevelopmental Outcome

C.L. Fawer and A. Calame

Introduction

Predicting the outcome of perinatal asphyxia is an important issue for neonatologists as well as for parents. After the initial resuscitation and the acute phase of intensive care, most parents will ask "How will my baby develop?, will he be normal?".

While the outcome of preterm infants is nowadays to a great extent clear, assessing the prognosis of full-term neonates is far more difficult and controversial. Over the years, the clinical concept of asphyxia has varied in its definition, classification, and incidence, as well as in its inferences about outcome. Most of our knowledge has been based on experimental data. Furthermore, the evaluation of follow-up studies encounters numerous problems because they have been performed with reference to a variety of points:

- Some are epidemiological analyses which provided useful figures at regional or national levels for health organizations (Stanley 1979; Hagberg et al. 1984)
- Studies do not always distinguish between preterm and full-term infants. It has been reported that low-Apgar scores are directly related to gestational age and therefore have different implications for full-term and preterm infants (Nelson and Ellenberg 1981; Catlin et al. 1986)
- Prenatal and perinatal variables are often confounding factors (Naeye et al. 1989)
- Several techniques and assessments have been developed in an attempt to establish a reliable criteria for the prediction of neurological sequelae in the neonatal period scoring systems, clinical signs, imaging techniques, EEG, near-infrared spectroscopy (NIRS), spectroscopy, and cerebral blood flow (CBF) measurements

  These studies have precise approaches and provide useful information on specific fields (Lipp-Zwahlen et al. 1985; Lipper et al. 1986; Wyatt et al. 1989; Azzopardi et al. 1989; Portman et al. 1990).

In practice, the clinicians will base their judgement on both clinical examination and various investigation procedures that are available to them. Prognosis of perinatal asphyxia will be discussed in this chapter with regard to brain maturation.

Preterm Infants

Numerous studies have been performed to identify perinatal factors that might predict the outcome in preterm infants. The most powerful predictors include perinatal asphyxia together with a lower gestational age, the need for ventilation, and perinatal complications such as hypoxemia, acidosis, hypercapnia, and seizures (Levene et al. 1982; Skouteli et al. 1985; Haas et al. 1986; Watkins et al. 1988). All these factors are known to have potential adverse effects on the developing brain. In most neonatal units, the early prognosis of preterm infants is based nowadays on the ultrasonographic images, whatever the clinical course. The predictive value of this imaging procedure has been extensively investigated and is now well recognized (Graham et al. 1987; Cooke 1987; Nwaesei et al. 1988; Bozinski et al. 1988; Stewart et al. 1988; Fawer and Calame 1991).

In our experience with preterm infants who have suffered from asphyxia, three different situations can be distinguished. Firstly, the infant may develop extensive lesions within a few hours, usually of a hemorrhagic nature. Below a 28-week gestation, intraventricular hemorrhage and intracerebral hemorrhage are frequent and most often associated with early death. In most units, large hemorrhages have also been associated with a high incidence of later sequelae in survivors. In the literature, several reports raised ethical questions about diagnostic procedures and intervention in very immature infants (Bissenden 1986; Campbell 1982; The Lancet 1988; Vaux 1986).

Secondly, the sequential ultrasound scans may remain constantly normal despite severe asphyxia and perinatal complications. This is also a challenge for the neonatologist to maintain intensive care, especially in the tiniest and most vulnerable babies. The clinician will have to prove that the child will develop normally later.

Thirdly, the infant may present an isolated hemorrhage and/or early increased echogenicities of varying size and extent in the periventricular white matter. The evolution of the echogenicities is variable. Some are transient and of good prognosis, others may persist for several weeks. The most extensive will evolve into cystic cavitations. However, in the early neonatal period, the prediction of the future in the presence of increased echogenicities is uncertain and the neurodevelopmental outcome cannot be established with total confidence.

In a recent study, asphyxiated preterm infants of 34 weeks' gestation (n = 87) were compared to nonasphyxiated preterm infants (n = 110) of similar birth weight and gestational age (Table 1). Asphyxiated preterm infants presented a respiratory distress syndrome more frequently (80%) and required mechanical ventilation more often (44%). At discharge, auditory defect was more frequent. The incidence of ultrasound abnormalities was also higher in infants with asphyxia (54%). While the incidence of isolated hemorrhage (grade I and grade II according to Levene's classification) was similar in both groups, lesions of periventricular leukomalacia (PVL) were more frequent (41%) in the asphyxiated preterm infants.

Table 1. Comparison between asphyxiated and nonasphyxiated preterm infants

	Asphyxia (n = 87)	No asphyxia (n = 110)	p value
Birth weight in g ± 1 SD	1538 ± 437	1564 ± 373	NS
Gestational age in weeks ± 1 SD	31.3 ± 1.96	31.8 ± 1.57	NS
M/F	42/45	50/60	NS
Bronchopulmonary dysplasia	5 (6%)	1 (0.9%)	NS
Retrolental fibroplasia	1 (1.1%)	2 (1.8%)	NS
Auditory defect	6 (7%)	2 (1.8%)	NS
Socio-economic class ± SD	2.77 ± 1.3	2.98 ± 1.3	NS
Abnormal ultrasound findings	47 (54%)	40 (36%)	0.002
Isolated hemorrhage	10 (11.5%)	18 (16%)	NS
Periventricular leukomalacia	36 (41%)	22 (20%)	0.002
Major handicaps	14 (16%)	5 (5.5%)	0.01

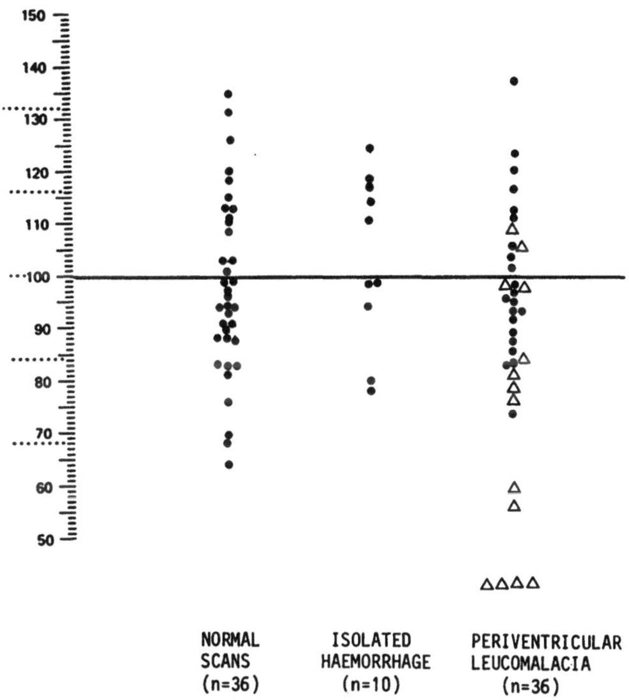

Fig. 1. Ultrasound changes, general intellectual index (McCarthy), and major handicaps (△) in asphyxiated preterm infants

Figures 1 and 2 show the relationship between ultrasound findings and handicaps in asphyxiated and nonasphyxiated preterm infants. Altogether we were able to fully assess 180 children at 5 years of age (McCarthy children abilities). The incidence of major handicap is higher in the asphyxiated

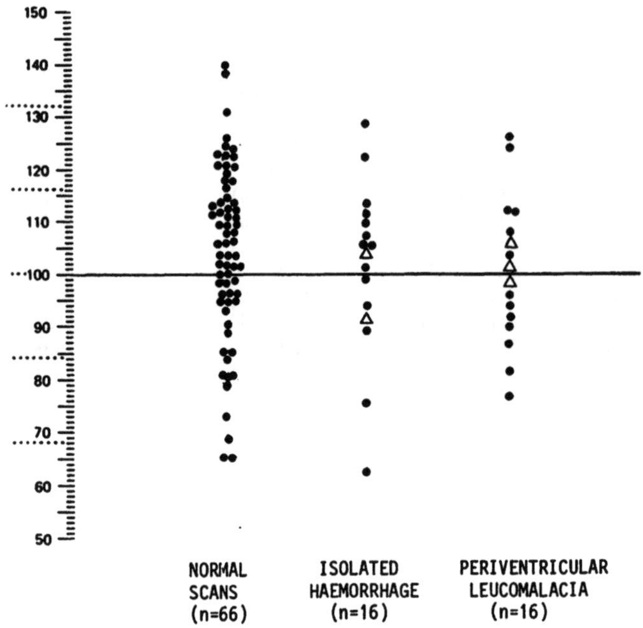

Fig. 2. Ultrasound changes, general intellectual index (McCarthy), and major handicaps (△) in non-asphyxiated preterm infants

group (16%). The most powerful ultrasound predictors for handicaps were found to be PVL lesions. On the other hand, a favorable scan is found to be a good predictor of normal neurodevelopmental progress, even though severe asphyxia has occurred.

As previously reported (Fawer et al. 1987), this study confirms the close relationship between the type and severity of handicaps and the localization and extent of PVL lesions. Neurological sequelae were found in children with frontal-parietal and frontal-parietal-occipital leukomalacia.

In conclusion, the outcome of asphyxia among preterm infants depends mainly on ultrasound findings. Consistently normal scans and isolated hemorrhage carry a good prognosis. Parents can be informed of the strong likelihood that their baby will have no neurodevelopmental impairment. Extensive cystic and persistent echogenic PVL lesions are associated with a poor prognosis. At discharge from hospital, a comprehensive assessment and a complete review of all data collected up to 40th postmenstrual week are important for establishing the final prognosis. The occurrence of subsequent neurodevelopmental sequelae should be anticipated. Discussion with and the counseling of parents should than take place, leading to a therapeutic program. The neonatologist should also give parents an opportunity to discuss the clinicial course and prognosis of their infant.

Term Infants

Birth asphyxia is the most important cause of avoidable mortality and morbidity in full-term neonates. The mortality rate and incidence of handicap depend on the severity of the insult and effectiveness of resuscitation. Prognosis of encephalopathy following birth asphyxia is poorly predicted in the literature either by umbilical pH, Apgar scores, depression of spontaneous respiration, or other biological or clinical parameters (Finer et al. 1981, 1983; Nelson and Ellenberg 1981; Dijxhoorn et al. 1985; Aylward et al. 1989). Because of intensive care, the early neurological examination has limitations.

The assessment of postasphyxial encephalopathy among term neonates should be based on a combined approach including a clinical evaluation, brain imaging, and EEGs. Measurements of circulation may also offer useful additional information.

We reviewed our recent cases of birth asphyxia from which thirty-five full-term infants were included in the study. Their mean gestation age was 39.8 weeks, mean birth weight 3168 g, and 37% were small for date. All presented with abnormal neurological signs according to Sarnat's classification. All were mechanically ventilated.

The following methods available in our unit were applied:

1. Ultrasonography: Ultrasound scans were classified as (a) normal scans, (b) transient and focal increased echogenicities and/or variations in cerebral spinal fluid spaces – these abnormalities usually disappeared by the end of the first week of life, (c) diffuse echogenicities involving both hemispheres, and (d) a prenatal anomaly was diagnosed at birth in one infant. A computed tomography was performed at 15 days of life in survivors with ultrasound abnormalities and confirmed in all cases the ultrasound findings. A postmortem examination was performed in most infants who died.
2. Doppler technique: Measurements of CBF velocities were made in the anterior cerebral arteries. The resistance index (RI) was calculated according to the Pourcelot formula $\frac{S - D}{S}$ (S, systolic amplitude; D, diastolic amplitude). The RI was considered abnormal when it fell below two standard deviations, which means below 0.55 (Levene et al. 1989). A low RI is always associated with increased "systolic" velocities. Several works have suggested that it is due to vasomotor paralysis and impaired autoregulation (Levene et al. 1989; Pryds et al. 1990).
3. Electroencephalography: The first EEG recording, usually obtained within the first 24–48 h of life, was classified according to Monod and Dreyfuss-Brisac: (a) normal EEG, (b) a moderately abnormal EEG is defined as a slow EEG or as the presence of electrical discharge, and (c) a severely abnormal EEG is defined as low-voltage inactivity, burst suppression, or paroxysmal EEGs.

Table 2. Neonatal data (ultrasound, Doppler, and EEG) and outcome

Neonatal data	*n*	Follow-up	*n*
Normal scan	8	Survivors	17
Transient abnormalities	5	Normal	11
Diffuse echogenicities[a]	21	Minor handicaps	2
Prenatal anomaly	1	Major handicaps[b]	4
Resistance index <0.55[a]	25		
Abnormal EEG[a]	24	Deaths[b]	18

[a] Abnormal parameters.
[b] Adverse outcome.

Data are shown in Table 2. In the neonatal period, eight neonates had normal scans. Five presented transient abnormalities, 21 had diffuse-increased echogenicities, and one a prenatal anomaly. Twenty-five neonates had at least one RI below 0.55. Twenty-four had abnormal EEGs.

Among the 17 survivors, 11 developed normally, two presented with minor handicaps, and four with major handicaps. Eighteen died in the neonatal period. To calculate specificity, sensitivity, and positive predictive value, diffuse-increased echogenicity and low-index abnormal EEGs were considered as abnormal parameters. Major handicaps and death were considered as adverse outcome.

Table 3 shows the relationship between ultrasound findings and outcome. Six of the eight neonates who had normal scans developed normally and two died. Infants with transient abnormalities did well. Most neonates with diffuse-increased echogenicities had a bad outcome. Four developed a major handicap (two cerebral palsy, one multihandicap, and one epilepsia) and 15 died. Two did well. The neonate with the prenatal anomaly died. Table 4 illustrates the relationship between abnormal RI and outcome. Results of RI measurements are variable. Most abnormal low indices are associated with a bad outcome. There are three false negatives and six false positives. Table 5 shows the relationship between EEG and outcome. Almost all neonates with normal EEGs or moderately abnormal EEGs did

Table 3. Ultrasound changes and outcome of asphyxiated full-term neonates

Ultrasound	Outcome			
	Normal ($n = 11$)	Minor handicap ($n = 2$)	Major handicap ($n = 4$)	Death ($n = 18$)
Normal scans ($n = 8$)	4	2	–	2
Transient focal abnormalities ($n = 5$)	5	–	–	–
Diffuse-increased echogenicities ($n = 21$)	2	–	4	15
Prenatal anomaly ($n = 1$)	–	–	–	1

Table 4. Normal and abnormal[a] resistance indexes, and outcome in asphyxiated full-term neonates

Doppler RI	Outcome			
	Normal ($n = 11$)	Minor handicap ($n = 2$)	Major handicap ($n = 4$)	Death ($n = 18$)
Normal RI ($n = 10$)	5	–	1	2
	2	–	–	–
Abnormal RI ($n = 25$)	4	2	3	16

[a] RI < 0.55.

Table 5. EEG and outcome in asphyxiated full-term neonates

EEG	Outcome			
	Normal ($n = 11$)	Minor handicap ($n = 2$)	Major handicap ($n = 4$)	Death ($n = 16$)
Normal ($n = 9$)	8	–	–	1
Moderately abnormal EEG ($n = 2$)	1	1	–	–
Severely abnormal EEG ($n = 22$)	2	1	4	15

Table 6. The sensitivity, specificity, and positive predictive value of neonatal findings in determining an adverse outcome

	Diffuse-increased echogenicities %	Abnormal RI %	Abnormal EEG %
Sensitivity	91	86	95
Specificity	85	54	54
Positive predictive value	91	76	76

Positive predictive value of the association of diffuse echogenicities, abnormal RI and abnormal EEG = 87%.

well, except one. A severely abnormal EEG was associated either with major handicap or death, except in three cases.

The sensitivity, specificity, and predictive value of neonatal findings in determining adverse outcome can be calculated from these data (Table 6). The sensitivity of diffuse-increased echogenicities, abnormal RI, and abnormal EEGs were found to be high. The clinician can therefore be very confident in predicting a bad outcome from abnormal values. The fairly high specificity of ultrasound suggested that children destined to be normal are unlikely to have abnormal ultrasound.

The specificity of low RI and abnormal EEG is lower. This indicates that there are a number of false positives. Further analysis of our data showed that if more than one low RI is considered, the specificity rises to 77%. A similar analysis was performed for abnormal EEGs. If the evolution of the EEG over the first few days of life and the disappearance of severe EEG abnormalities are considered, the specificity rises to 77%.

The positive-predictive value represents the probability of a neonate having a bad outcome if the parameter is abnormal. The positive-predictive value for increased echogenicities is high, whereas it is lower for abnormal RI and EEG. If, in the same infants, there is an association of diffuse echogenicities, an abnormal RI and abnormal EEG, the positive-predictive value will be 87%.

This study confirms other data and seems to indicate that if one wishes to define an early prognosis of postasphyxial encephalopathy in a population of term neonates, it can be based with confidence on a combined evaluation of ultrasound, EEG, and Doppler findings. The sensitivities are high. If the specificity of ultrasound is high, specificities of Doppler and EEG examination are lower, but might be improved if sequential RI measurements and EEGs are considered.

In conclusion, the prediction of normal outcome after perinatal asphyxia can be based on normal brain imaging within the first 15 days of life, normal EEG within the first 3 days of life, and neurological recovery during the first week of life. Extensive lesions are, in our experience, associated in most cases with severe encephalopathy and severe EEG abnormalities, thereby predicting a poor outcome. The prediction from moderate encephalopathy is, however, much less accurate. Apparently normal brain imaging in term infants does not guarantee an intact survival, but neurological examination as well as early EEGs will aid in the prognosis. As the cerebral lesions may appear several days after the insult, brain imaging should be performed sequentially over the first 2 weeks of life to ensure accuracy in the diagnosis.

In each individual situation, nevertheless, establishing an early prognosis of asphyxia will remain a difficult task and requires the greatest care. With improved techniques for the assessment of a prognosis, there is a need in each neonatal unit to define ethical guidelines and to develop criteria for engaging, withholding, or withdrawing intensive care. Such decisions are complex because they concern not only a medical diagnosis but also moral judgement, emotional distress, and legal aspects.

Acknowledgements. We thank Prof. P.A. Desplands for interpreting the EEGs, Mrs. M. Forcada and L. Jaunin for their collaboration, and M. Marion for preparing the manuscript. This work was supported by the Swiss National Science Foundation no 32-30215.90.

References

1. Aylward GP, Verhulst SJ, Bell S (1989) Correlation of asphyxia and other risk factors with outcome: a contemporary view. Dev Med Child Neurol 31:329–340

2. Azzopardi D, Wyatt JS, Cady EB, Delpy DT, Baudin J, Stewart AL, Hope PL, Hamilton PA, Reynolds EOR (1989) Prognosis of newborn infants with hypoxic-ischemic brain injury assessed by phosphorus magnetic resonance spectroscopy. Pediatr Res 25:445–451
3. Bissenden JG (1986) Ethical aspects of neonatal care. Arch Dis Child 61:639–641
4. Bozynski MEA, Nelson MN, Genaze D et al. (1988) Cranial ultrasonography and the prediction of cerebral palsy in infants weighing <1200 grams at birth. Dev Med Child Neurol 30:342–348
5. Campbell AGM (1982) Which infants should not receive intensive care? Arch Dis Child 57:569–571
6. Catlin EA, Carpenter MW, Braun BS, Magfield SR, Shaul OW, Goldstein M, Or W (1986) The Apgar score revisited: influence of gestational age. J Pediatr 109:865–868
7. Cooke RWI (1987) Early and late cranial ultrasonographic appearances and outcome in very low birthweight infants. Arch Dis Child 62:931–937
8. Dijxhoorn MJ, Visser GHA, Huisjes HJ, Filder V, Touwen BCL (1985) The relation between umbilical pH values and neonatal neurological morbidity in full-term appropriate-for-dates infants. Early Hum Dev 11:33–42
9. Editorial (1988) Limitations of care for very low birthweight infants. Lancet 4:1257–1258
10. Fawer CL, Calame A (1991) Significance of ultrasound appearances in the neurological development and cognitive abilities of preterm infants at five years. Eur J Pediatr 150: 512–520
11. Fawer CL, Diebold P, Calame A (1987) Periventricular leucomalacia and neurodevelopmental outcome in preterm infants. Arch Dis Child 62:30–36
12. Finer NN, Robertson CM, Richards RT, Pinnell LE, Peters KL (1981) Hypoxic-ischemic encephalopathy in term neonates: perinatal factors and outcome. J Pediatr 98:112–117
13. Finer NN, Robertson CM, Peters KL, Coward JH (1983) Factors affecting outcome in hypoxic-ischemic encephalopathy in term infants. Am J Child 137:21–25
14. Graham M, Levene MI, Trounce JQ, Rutter N (1987) Prediction of cerebral palsy in very low birthweight infants: prospective ultrasound study. Lancet 2:593–596
15. Haas G, Aspiron B, Leidig E, Buchwald-Saal M, Mentzel H (1986) Obstetrical and neonatal risk factors in very low birthweight infants related to their neurological development. Eur J Pediatr 145:341–346
16. Hagberg B, Hagberg G, Olow I (1984) The changing panorama of cerebral palsy in Sweden. IV. Epidemiological trends 1959–78. Acta Paediatr Scand 73:433–440
17. Levene MI, Fawer CL, Lamont R (1982) Risk factors in the development of intraventricular haemorrhage in the preterm neonate. Arch Dis Child 57:410–417
18. Levene MI, Fenton AC, Evans DH, Archer LNJ, Shortland DB, Gibson NA (1989) Severe birth asphyxia and abnormal cerebral blood-flow velocity. Dev Med Child Neurol 31: 427–434
19. Lipp-Zwahlen AE, Deonna T, Micheli JL, Calame A (1985) Prognostic value of neonatal CT scans in asphyxiated term babies: low-density score compared with neonatal neurological signs. Neuropediatrics 16:209–217
20. Lipper EG, Voorhies TM, Ross G, Vannucci RC, Auld PAM (1986) Early predictors of one-year outcome for infants asphyxiated at birth. Dev Med Child Neurol 28:303–309
21. Naeye RL, Peters EC, Bartholomew M, Landis JR (1989) Origins of cerebral palsy. Am J Dis Child 143:1154–1161
22. Nelson KB, Ellenberg JH (1981) Apgar scores as predictors of chronic neurologic disability. Pediatrics 68:36–44
23. Nwaesei CG, Allen AC, Vineer MJ, Brown SJ, Stinson DA, Evans JR, Byrne JM (1988) Effect of timing of cerebral ultrasound on the prediction of later neurodevelopmental outcome in high-risk preterm infants. J Pediatr 112:970–975
24. Portman RJ, Carter BS, Gaylord MS, Murphy MG, Thieme RE, Merenstein GB (1990) Predicting neonatal morbidity after perinatal asphyxia: a scoring system. Am J Obstet Gynecol 162:174–182
25. Pryds O, Greisen G, Lou H, Friis-Hansen B (1990) Vasoparalysis associated with brain damage in asphyxiated term infants. J Pediatr 117:119–125

26. Skouteli HN, Dubowitz LMS, Levene MI, Miller G (1985) Predictors for survival and normal neurodevelopmental outcome of infants weighing less than 1001 grams at birth. Dev Med Child Neurol 27:588–595
27. Stanley FJ (1979) An epidemiological study of cerebral palsy in Western Australia, 1956–1975. I. Changes in total incidence of cerebral palsy and associated factors. Dev Med Child Neurol 21:701–713
28. Stewart A, Hope PL, Hamilton P et al. (1988) Prediction in very preterm infants of satisfactory neurodevelopmental progress at 12 months. Dev Med Child Neurol 30:53–63
29. Vaux KL (1986) Ethical issues in caring for tiny infants. Clin Perinatol 13:477–484
30. Watkins A, Szymonowicz W, Yu VYH (1988) Significance of seizures in very low birthweight infants. Dev Med Child Neurol 30:162–169
31. Wyatt JS, Edwards AD, Azzopardi D, Reynolds EOR (1989) Magnetic resonance and near infrared spectroscopy of perinatal hypoxic-ischemic brain injury. Arch Dis Child 64: 953–963

Subject Index

MIX
Papier aus verantwortungsvollen Quellen
Paper from responsible sources
FSC® C105338

If you have any concerns about our products,
you can contact us on
ProductSafety@springernature.com

In case Publisher is established outside the EU,
the EU authorized representative is:
Springer Nature Customer Service Center GmbH
Europaplatz 3, 69115 Heidelberg, Germany

Printed by Libri Plureos GmbH
in Hamburg, Germany